Textbook of Pharmaceutical Biotechnology

Textbook of Pharmaceutical Biotechnology

Chandrakant Kokate
MPharm, PhD, FGAES (Germany)
Vice-Chancellor
KLE University, Belgaum, Karnataka
Former Vice-Chancellor, Kakatiya University, Warangal, AP
Former President, Pharmacy Council of India

Sunil S Jalalpure
MPharm, PhD
Professor
Department of Pharmacognosy
College of Pharmacy, KLE University
Belgaum, Karnataka

Pramod J Hurakadle
MPharm, PhD
Associate Professor and Head
Department of Pharmaceutical Biotechnology
College of Pharmacy, KLE University
Belgaum, Karnataka

CBS Publishers & Distributors Pvt Ltd

New Delhi • Bengaluru • Chennai • Kochi • Kolkata • Mumbai
Hyderabad • Jharkhand • Nagpur • Patna • Pune • Uttarakhand

Disclaimer

Science and technology are constantly changing fields. New research and experience broaden the scope of information and knowledge. The authors have tried their best in giving information available to them while preparing the material for this book. Although all efforts have been made to ensure optimum accuracy of the material, yet it is quite possible some errors might have been left uncorrected. The publisher, the printer and the authors will not be held responsible for any inadvertent errors, omissions or inaccuracies.

Textbook of Pharmaceutical Biotechnology

ISBN: 978-93-54660-34-4

CBS Edition: 2021

Copyright © Authors and Publisher

All rights reserved. No part of this book may be reproduced or transmitted in any form or by any means, electronic or mechanical, including photocopying, recording, or any information storage and retrieval system without the permission, in writing, from the authors and the publisher.

Published by **Satish Kumar Jain** and produced by **Varun Jain** for

CBS Publishers & Distributors Pvt Ltd

4819/XI Prahlad Street, 24 Ansari Road, Daryaganj, New Delhi 110 002, India.
Ph: 011-23289259, 23266861, 23266867   Website: www.cbspd.com
Fax: 011-23243014   e-mail: delhi@cbspd.com;
                                              cbspubs@airtelmail.in

Corporate Office: 204 FIE, Industrial Area, Patparganj, Delhi 110 092
Ph: 011-4934 4934   Fax: 011-4934 4935   e-mail: publishing@cbspd.com;
                                                   publicity@cbspd.com

Branches

- **Bengaluru:** Seema House 2975, 17th Cross, K.R. Road, Banasankari 2nd Stage, Bengaluru 560 070, Karnataka, India
  Ph: +91-80-26771678/79   Fax: +91-80-26771680   e-mail: bangalore@cbspd.com
- **Chennai:** 7, Subbaraya Street, Shenoy Nagar, Chennai 600 030, Tamil Nadu, India
  Ph: +91-44-26680620, 26681266   Fax: +91-44-42032115   e-mail: chennai@cbspd.com
- **Kochi:** 42/1325, 1326, Power House Road, Opp KSEB, Ernakulum, Kochi 682 018, Kerala, India
  Ph: +91-484-4059061-65, 67   Fax: +91-484-4059065   e-mail: kochi@cbspd.com
- **Kolkata:** 6/B, Ground Floor, Rameswar Shaw Road, Kolkata-700014 (West Bengal), India
  Ph: +91-33-2289-1126, 2289-1127, 2289-1128   e-mail: kolkata@cbspd.com
- **Mumbai:** PWD Shed, Gala no 25/26, Ramchandra Bhatt Marg, Next to JJ Hospital Gate no. 2, Opp. Union Bank of India, Noorbaug
  Mumbai-400009, Maharashtra, India
  Ph: +91-22-66661880/89   e-mail: mumbai@cbspd.com

Representatives

- Hyderabad   0-9885175004   • Jharkhand   0-9811541605   • Nagpur   0-9421945513
- Patna   0-9334159340   • Pune   0-9623451994   • Uttarakhand   0-9716462459

Printed at Chaman Enterprises, Daryaganj, Delhi, India

Preface

Last two decades have witnessed rapid change in our understanding of biotechnology. The substantial scientific work that has been accomplished in recent years in this branch of biosciences has revolutionized the entire concept of biotechnology related to pharmacy. The basic and applied biotechnology has facilitated the emergence of novel drug moieties of biological origin that are capable of combating a wide spectrum of human ailments. The ultimate biomedical objective of biotechnology is to contribute a rational relationship between biochemical moieties and spectrum of therapeutic effects they generate.

Keeping in view the past, present and future projections of pharmaceutical biotechnology, it is our sincere attempt to provide comprehensive text of the subject covering different facets of biotechnology for the benefit of the students of different disciplines of biosciences. The chapters on medicinal plant biotechnology, genetics and molecular biotechnology, recombinant DNA technology, hybridoma technology and monoclonal antibodies, antibiotics, applications of pharmaceutical biotechnology, immobilization of enzymes, nanobiotechnology, recent advances in biopharmaceuticals, and list of pharma–biotech industries in India and their products are important added features of this book.

We hope that the students of degree and postgraduate courses of pharmacy, biotechnology and other branches of biosciences shall be benefited by our efforts. We have enriched the text by using diagrams, figures and tables, with greater emphasis on simple and understandable language.

– *Kokate, C.K.*
– *Jalalpure, S.S.*
– *Hurakadle, P.J.*

About the Authors

Prof. Chandrakant Kokate

Prof. Chandrakant Kokate is currently the Vice-Chancellor of KLE University, Belgaum, Karnataka. He completed his BPharm in first class from Nagpur University in 1967, obtained his MPharm degree in first class and with gold medal from University of Saugar in 1969. He was awarded his PhD in 1972 and postdoctoral research fellowship in Germany (1977–78). He continued his research as a visiting scientist in Germany (1986).

He was the President of Pharmacy Council of India (constituted under the Act of Parliament – Pharmacy Act, 1948) for a decade, simultaneously also officiating President of the Council. He was also the President of Indian Pharmaceutical Association (2000–02). He was the Chairman of All India Board for Pharmacy Education of All India Council for Technical Education (AICTE) and also served in the Executive Committee of AICTE.

Prof. Kokate thas the honour of being the first pharmacist in the country to be appointed as the Vice-Chancellor of Indian Universities—the Kakatiya University, Acharya Nagarjuna University (additional charge) and currently at KLE University. He has served as the Chairman of the Southern Regional Committee of AICTE and Chairman of the Recruitment Board of Regional Selection Committee for Defence Research and Development Organization (DRDO). He is serving as the Chairman of National Assessment and Accreditation Council (NAAC)/University Grants Commission (UGC) peer teams.

He is the recipient of several awards/recognitions at the national/international and provincial levels in the profession of pharmacy. These include Eminent Pharmacist Award of Indian Pharmaceutical Association (IPA); Lifetime Achievement Award of Association of Pharmaceutical Teachers of India (APTI); Prof. M.L. Schroff Memorial Award of Indian Hospital Pharmacists Association (IHPA); Acharya P.C. Ray Memorial Award; National Herbal Academy Award; Indira Priyadarshini National Award; Dr K.M. Parikh Memorial Award; Best Teacher Award of Govt. of Andhra Pradesh; Distinguished Alumnus and Lifetime Achievement Award of University of Saugar; Shikshak Bhushan Award of Goa; Sivananda Eminent Citizen Distinguished Service Award, Andhra Pradesh; Fellowship of German Academic Exchange service (DAAD), Indian Pharmaceutical Association, Indian Society for Technical Education (ISTE), National Herbal Academy, Andhra Pradesh Akademi of Sciences, and many others. He was the Leader of Indian Delegation of Pharmacists to 18th Federation of Asian Pharmaceutical Association (FAPA) at Sydney, Australia, and International Pharmaceutical Federation (FIP) Conference at Bangkok. He has successfully supervised 17 students for PhD and 40 students for master's programme in pharmacy. He has authored 6 books and published 115 research papers in national and international journals. He has also served as an expert on the panels of Union Public Service Commission (UPSC), UGC, AICTE, Pharmacy Council of India, Department of Biotechnology (DBT), Department of Science and Technology (DST), Council for Scientific and Industrial Research (CSIR), Indian Drug Manufacturers' Association (IDMA), Indian Council for Agriculture Research

(ICAR), Indian Society for Technical Education (ISTE), Indian Council for Medical Research (ICMR) and Public Service Commission (PSC) of states.

Prof. Sunil S. Jalalpure

Prof. Sunil S. Jalalpure is Professor, Department of Pharmacognosy, College of Pharmacy, KLE University, Belgaum. He completed his BPharm from Karnataka University, Dharwad, and obtained his MPharm and PhD degrees from Rajiv Gandhi University of Health Sciences, Bangalore, Karnataka. He has undergone research training at Rhodes University, Grahamstown, South Africa, on Sophisticated Analytical Instruments.

He received travel grant from AICTE under Young Scientist Category to present research paper in World Congress on Medicinal Plants at Cape Town, South Africa. He has received financial support from various funding agencies like UGC, AICTE, DST and ICMR for research projects. He is member of several pharmacy professional societies and has chaired several scientific sessions at national and International conferences. He has worked as organising secretary for several national and International conventions. He has been invited to visit Australia, Thailand, Malaysia and South Africa to attend and present research papers at conferences.

His areas of research interests include isolation/characterization of active principles from medicinal plants and their pharmacological screening for various biological activities, and training the research students in Pharmacognosy, Phytochemistry and Biotechnological aspects with modern tools and techniques.

Prof. Jalalpure has published over 46 scientific papers in national and international journals and presented 67 papers at national and international conferences. He has 2 intellectual property rights to his credit. He has successfully guided 34 MPharm and 1 PhD students towards successful completion of their courses.

Dr Pramod J. Hurakadle

Dr Pramod J. Hurakadle is currently Associate Professor and Head, Department of Pharmaceutical Biotechnology, College of Pharmacy, KLE University, Belgaum. He completed his BPharm degree from Kuvempu University, Shimoga, and obtained his MPharm degree with distinction from Rajiv Gandhi University of Health Sciences, Bangalore. He was awarded his PhD in Pharmacy from Rajiv Gandhi University of Health Sciences, Bangalore, under AICTE-sponsored quality improvement programme. He also received a Diploma in Clinical Research from Catalyst Clinical Services Ltd, New Delhi.

He has been awarded Batch of the Best Award from Zydus-Indon Cadila Health Care Ltd, Ahmedabad. He is also a member for various professional bodies, e.g. APTI, ISP, KSPC, etc., and is working as General Secretary, IPA, Belgaum branch. He has also worked as joint secretary to several national and international workshops/conferences, sponsored by ICMR, AICTE, UGC, IPA, etc.

His areas of research interest include herbal formulations/excipients, biosurfactants and screening of herbal extracts for various pharmacological activities. Since recently he is working jointly with Regional Medical Research Centre, ICMR, Belgaum, on tissue culture studies of medicinal plants from Western Ghats region.

Dr Hurakadle is guiding 5 students for PhD and has supervised 12 students for master's programme in pharmacy. He has published several research papers in national and international journals and has presented at national and international conferences. He has also been invited as speaker for international conferences at Malaysia and South Africa.

Contents

CHAPTER 1: Historical Perspectives and Milestones 1
    Introduction 1
    Discovery of Cell and Chromosome Theory of Inheritance 1
    Major Branches of Biotechnology 2
    Advances in Biotechnology 4
    Significant Milestones in Biotechnological Research 4
    Historical Contributions of Some Leading Scientists 4
    Genetic Evaluation 10
    Historical Perspectives of Plant Tissue Culture 13
    Questions 14

CHAPTER 2: Immunology and Immunological Preparations 15
    Introduction 15
    How the Immune System Works 16
    Principles 19
    Types of Immunity 21
    Immune Tolerance 22
    Hypersensitivity 23
    Antigen–Antibody Reactions and Their Applications 24
    Immunomodulators 26
    Applications of Immunomodulators 29
    Vaccines 30
    Questions 34

CHAPTER 3: Environmental Biotechnology 35
    Introduction 35
    Bioremediation 36
    Management of Environmental Pollution 36
    Management of Air, Waste Gases and Water Pollution 40
    Questions 42

CHAPTER 4: Proteins and Proteomics 43
Introduction 43
Proteins 44
Proteomics 50
Questions 53

CHAPTER 5: Animal Cell and Tissue Biotechnology 55
Introduction 55
Primary and Established Cell Lines 56
Kinetics of Cell Growth 57
Animal Tissue Culture: Media Requirements 60
Questions 73

CHAPTER 6: Medicinal Plant Biotechnology 75
Genetics as Applied to Medicinal Herbs 75
Mutation 77
Polyploidy 79
Chemodemes (Chemical Races) 81
Hybridization 81
Genetic Engineering and Recombinant DNA Technology in Plants 82
Plant Tissue Cultures as Sources of Biomedicinal Compounds 86
Historical Development 88
Types of Cultures 89
Culture Medium 90
Surface Sterilization of Explants 91
Establishment of Cultures 92
Phytopharmaceuticals in Plant Tissue Cultures 94
Bioproduction of Useful Metabolites in Hairy Root and Multiple Shoot Cultures ... 95
Questions 97

CHAPTER 7: Microbial Culture and Fermentation Process 99
Introduction 99
Microorganisms 100
Microbial Culture 100
Quality Control of Microbial Culture Media 107
Questions 109

CHAPTER 8: Genetics and Molecular Biotechnology 111
Introduction 111
DNA as Segment of Molecular Heredity 113
Genes 116
Mutations 123

Molecular Biotechnology 124
Questions 127

CHAPTER 9: Recombinant DNA Technology 129

Introduction 129
Outline of Recombinant DNA Technology 130
Introduction of the Recombinant DNAs into a Suitable Host Cell and Their Identification 132
Applications of Recombinant DNA Technology 134
Production of Human Follicle-Stimulating Hormone 139
Disease Diagnosis by Recombinant DNA Methods 139
Recombinant DNA Technology for Improving Milk Production 140
Applications in the Pulp and Paper Industry 140
Applications in Antibiotic-Producing Microorganisms 140
Genetic Engineering as a Part of Recombinant DNA Technology 141
Questions 142

CHAPTER 10: Microbial Transformation 143

Introduction 143
Types of Reactions Mediated by Microorganisms 145
Design for Biotransformation 149
Biotransformation Process and Its Improvement with Special Reference to Steroids 150
Questions 153

CHAPTER 11: Hybridoma Technology and Monoclonal Antibodies 155

Introduction 155
Principle for Creation of Hybridoma Cells 156
Large-Scale Production of Monoclonal Antibodies 158
Applications of Monoclonal Antibodies 160
Questions 165

CHAPTER 12: Antibiotics 167

Historical Development of Antibiotics 167
Antimicrobial Spectrum 171
Standardization of Antibiotics 172
Screening of Soil for Organisms Producing Antibiotics 174
Fermenter or Bioreactor 176
Design of Industrial Fermentation Process 178
Fermenter Design and Control 180
Isolation of Mutants 181
Factors Influencing Rate of Mutation 182
Isolation of Fermentation Products 183
Questions 187

CHAPTER 13: Genetic Recombination 189
Introduction 189
Methods of Gene Transfer in Plants 190
Horizontal Gene Transfer Technology 200
Protoplast Fusion 205
Gene Cloning 207
Development of Hybridoma by Monoclonal Antibodies 208
Some Important Medicines Produced by Biotechnology 211
Gene Transfer in Humans 213
Applications of Gene Transfer 213
Questions 216

CHAPTER 14: Medicine and Edible Vaccines 217
Introduction 217
Edible Vaccines 219
Questions 225

CHAPTER 15: Broad Applications of Pharmaceutical Biotechnology 227
Introduction 227
Applications of Biotechnology in the Production of Biomolecules 228
Applications of Biotechnology in the Diagnosis of Diseases 231
Applications of Biotechnology in Agriculture 231
Applications of Ultrasound in Biotechnology 232
Testing of Pharmaceuticals 232
Questions 232

CHAPTER 16: Biotechnology in Drug Discovery 233
Introduction 233
Current Trends in Modern Pharmaceutical Analysis for Drug Discovery 235
Recent Pharmacokinetic Advances in Drug Discovery and Development 235
Natural Product Drug Discovery 235
Questions 237

CHAPTER 17: Immobilization of Enzymes 239
Introduction 239
History of Enzymes 240
Enzyme Structure 241
Enzyme Kinetic 242
Bioproduction of Enzymes 246
Enzyme Production 248
Enzyme Immobilization 249
Immobilization of Plant Cells 255

Enzyme Engineering 259
Applications of Enzymes 260
Biosensors 262
Questions 267

CHAPTER 18: Nanobiotechnology 269
Introduction 269
Computational Gene 270
Nanolithography 272
Nanomedicine and Nanodevices 274
Medical Applications of Molecular Nanotechnology 277
Questions 286

CHAPTER 19: Recent Advances in Biopharmaceuticals 287
Introduction 287
Advantages of DNA Technology in the Production of Biopharmaceuticals 288
Biomolecules Produced Through Recombinant DNA Technology 289
Biopharmaceuticals of Animal Origin 290
Biopharmaceuticals of Plant Origin 292
Pharmaceutical Substances of Microbial Origin 293
Recent Advances in Biopharmaceuticals 300
Questions 303

Appendix 305
Glossary 313
Bibliography 321
Index 327

CHAPTER 1
Historical Perspectives and Milestones

CHAPTER OUTLINE

- Introduction
- Discovery of Cell and Chromosome Theory of Inheritance
- Major Branches of Biotechnology
- Advances in Biotechnology
- Significant Milestones in Biotechnological Research
- Historical Contributions of Some Leading Scientists
- Genetic Evaluation
- Historical Perspectives of Plant Tissue Culture
- Questions

INTRODUCTION

The term *biotechnology* represents a fusion or an alliance between biology and technology. It is a newly discovered discipline for old-age practices. Biotechnology, as an integral component of biosciences, has derived its strength from different disciplines of science and technology. It is an interdisciplinary pursuit with multidisciplinary applications.

DISCOVERY OF CELL AND CHROMOSOME THEORY OF INHERITANCE

Robert Hooke is credited with the discovery of cell. An important discovery made by Robert Brown (1831) was the presence of a small sphere within the cells of orchid roots. In 1839, Hugo von Mohl and J. Purkinje named the jelly-like substance as *protoplast*. In 1885, Virchow explained that cells are derived from pre-existing cells.

In 1902, cytologists Walter Sutton and Theodor Boveri independently came to the conclusion that the behaviour of chromosomes at meiosis can serve as the cellular basis of both segregation and independent assortment. The theory of heredity of chromosomes was further expanded by Thomas H. Morgan on fruit fly drosophila. The term *factor* as the basic unit of inheritance was replaced with the term *gene* by Johannsen in 1909. The term *mutation* was coined in 1901 by Hugo de Vries to explain the variation he observed in the plant *Oenothera lamarckiana* (primrose).

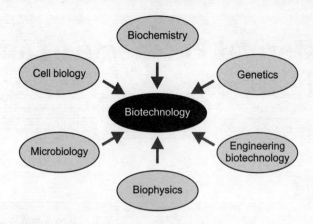

Figure 1.1 Basic strength for advancement of biotechnology.

The origin of biotechnology dates back to 6000 BC, the same year when the yeast was first used to produce beer and wine. The history of biotechnology begins with zymotechnology, which commenced with a focus on brewing techniques for beer. By World War I, zymotechnology expanded to cover larger industrial issues, and the potential of industrial fermentation gave rise to biotechnology. Yoghurt was produced from bacteria. Louis Pasteur, also known as the *father of biotechnology,* identified the role of microorganisms in fermentation. The biotechnological revolution began in the late 1970s and early 1980s when scientists understood the genetic constitution of living organisms. Biotechnology, as an applied bioscience, has been effectively utilized for better industrial therapeutic production. As a matter of fact, every discipline of science has contributed either directly or indirectly to the growth of biotechnology (Fig. 1.1).

MAJOR BRANCHES OF BIOTECHNOLOGY

Biotechnology is the application of scientific and engineering principles to the processing of materials by biological agents—microorganisms, plants and animal cells. Biotechnology can be represented as a mixture of various biological sciences (biosciences) for better service in the field of pharmaceuticals (Fig. 1.2).

Agricultural Biotechnology

It deals with a group of scientific techniques that are used to create, improve or modify plants and animals. It encompasses the knowledge of biosciences concerned with plants and agriculture, such as plant tissue culture, production of haploid plants, somaclonal variation, micropropagation cryopreservation methodology in the genetic engineering of plants, application of transgenic plants, etc.

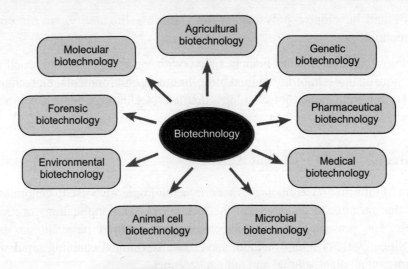

Figure 1.2 Branches of biotechnology.

Forensic Biotechnology

It is the method for detecting unique DNA pattern of organism, and it includes DNA fingerprinting.

Medical Biotechnology

It deals with the utilization diagnostic kits for the detection of different diseases. It has wide applications in human health care, and it includes information related to gene therapy, DNA in disease diagnosis, fingerprinting, specialized products of DNA biotechnology, etc.

Pharmaceutical Biotechnology

It is a major branch of biotechnology, which includes production of therapeutic proteins and hormones, fermentation products such as antibiotics, vaccines and drugs, etc.

Environmental Biotechnology

It encompasses the information on biotechnology with references to environment such as environment pollution, treatment modalities, etc.

Microbial Biotechnology

The information pertaining to fermentation technology, downstream processing, enzyme technology, microbial production of organic solvents, organic acids, antibiotics, amino acids, vitamins, foods and beverages, polysaccharides, biomass–bioenergy, etc., is covered under microbial biotechnology.

Biotechnology has several other branches also, with varied applications, such as molecular biotechnology, aquatic biotechnology, animal biotechnology, environmental biotechnology, leather biotechnology, mining and metal biotechnology, textile biotechnology, genetic engineering, etc.

ADVANCES IN BIOTECHNOLOGY

With the advent of information technology, modern technologies give rise to genomics, proteomics and cellomics that promise to develop into the technology with applications in medicine, food, agriculture, etc. The present focus of biotechnology is centred basically on genomics and postgenomic. Since 2001, different methods have been developed enabling rapid sequencing of more than 50 microbial, plant, animal and human genomes.

Newer frontiers in biotechnology include development of nucleic acid probes, production of monoclonal antibodies and molecular markers. Development of microbial inoculants covering biopesticides, biofertilizers and genetic modifications of microbes through DNA recombination are other important applications. Novel approaches such as proteomics and structure biology are contributing to the understanding of chemistry of life and disease.

Different techniques of population genetics and biometric analysis are complemented with modern methods in reproductive biology and gene diagnosis. Artificial insemination, in vitro fertilization and embryo transfer are important applications of biotechnology in animal breeding.

SIGNIFICANT MILESTONES IN BIOTECHNOLOGICAL RESEARCH

During the 20th century, the pharmaceutical industry witnessed a series of developments in science and technology, which had generated new opportunities for biotechnological discoveries. The important historical developments in biotechnology, recent milestones in biotechnology and chronological sequences of important biotechnological products (drugs, vaccines and other therapeutics) are listed in Tables 1.1, 1.2 and 1.3, respectively.

HISTORICAL CONTRIBUTIONS OF SOME LEADING SCIENTISTS

1. Antony van Leeuwenhoek (1632–1723) is the greatest of all microscopists. He was the first to observe and accurately describe the shape of human red blood cells. He observed and

Table 1.1 Important historical developments in biotechnological research

Year	Scientist	Discovery
1928	Alexander Fleming	Discovery of penicillin from common moulds
1944	Avery, MacLeod and McCarty	Identification of DNA as the genetic material
1953	Watson and Crick	Determination of DNA structure
1958	Meselson and Stahl	Semiconservative application of DNA
1961	Jacob and Monod	Lac operon model for gene regulation
1972	Khurana, et al.	Synthesis of *tRNA* gene
1975	Kohler and Milstein	Production of monoclonal antibodies
1976	Sanger and Gilbert	Techniques to develop DNA sequence
1978	Joshua Lederberg	Production of insulin in *E. coli*
1983	Kary Mullis	First artificial chromosome synthesized
1989	UC Davis and researchers	Recombinant vaccine against deadly rinderpest virus
1998	Thomson and Gearhart	Technique for culturing embryonic stem cells
2010	Craig Venter	Creation of first self-replicating synthetic bacterial cell

Table 1.2 Recent milestones in biotechnology

Year	Discovery
1977	First genome sequence
1983	Use of Ti plasmids to genetically transform plants
1987	Gene transfer by biolistic transformation
1988	Development of polymerase chain reaction (PCR)
1990	Official launching of human genome project
1994–95	Genetic and physical maps of human chromosome elucidated
1996	First eukaryotic organism sequence
1997	First mammalian, Dolly (a sheep), developed by nuclear cloning
2000	First plant genome sequence (*Arabidopsis thaliana*)
2001	Human genome, the first mammalian genome sequence
2002	First crop plant genome sequenced
2003	Mouse genome, the experimental model to men, sequenced
2005	Artificial cell for use in nanotechnology, nanobiotechnology, blood substitutes, regenerative medicine and gene therapy
2006	Artificial cell encapsulated bone marrow stem cells regenerated liver resulting in recovery

Table 1.3 Chronological sequence of important biotechnological products (drugs, vaccines and other therapeutics)

Year of discovery	Product	Company name
1982	Humulin	Eli Lilly and Company
1986	Digibind	Burroughs Wellcome
	Roferon-A	Hoffmann-La-Roche
	Intron-A	Schering-Plough
	Recombivax HB	Merck
1989	Engerix-B	SmithKline Beecham
1990	Actimmune	Genetech
1991	Leukin	Immune
1995	Follitropin alpha	Ares Serono
	Betaferon	Schering AG
1996	Humalog	Eli Lilly
1997	Com hep A and B vaccines	GlaxoSmithkline Biologicals
1998	Simulect	Novartis Pharmaceuticals
1999	Procomvax	Sanofi Pasteur MSD
	Intron A	SP Europe
2000	ViraferonPeg	SP Europe
	Insulin glargine	Aventis Pharma Deutschland GmbH
	Hexavac	Pasteur Merieux MSD
	Luveris	Merck Serono Limited
	Fasturtec	Sanofi-Synthelabo
	Metalyse	Boehringer Ingelheim International GmbH
2001	Tenecteplase	Boehringer Ingelheim International GmbH
	Nespo	Dompe Biotec S.p.A.
	Aranesp	Amgen Europe B.V.
	Nonfact	Sanquin
	Replag	TKT Europe-5S AB
2002	Anakinra	Amgen Europe B.V.
	Neulasta	Amgen Europe B.V.
	Pegfilgrastim	Novatech Biopharmaceutical Co Ltd
	Neupopeg	Dompe Biotec S.p.A.
	Ambirix	GlaxoSmithKline
	Ultratard	Novo Nordisk
	Actraphane	Novo Nordisk
	Forsteo	Eli Lilly and Company
	Teriparatide	Eli Lilly and Company
	Somavert	Pfizer Limited

(Continued)

Year of discovery	Product	Company name
2003	Zevalin	Schering AG
	Advate	Baxter AG
	Humira	Abbott Laboratories
2004	Trudexa	Abbott Laboratories
2005	Fendrix	GlaxoSmithKline
	Comb vaccine	GlaxoSmithKline Biologicals
2010	Coagulation factor VIIa	EMD Biosciences

measured a large number of minute living organisms, including bacteria and protozoa and communicated them to the Royal Society of London (1684).

2. John Needham (1713–81) was the greatest supporter of the theory of spontaneous generation. Spontaneous generation is the hypothesis that some vital force contained in or given to organic matter can create living organisms from inanimate objects.
3. Louis Pasteur (1822–95) first demonstrated that air contains microscopically observable organized structures. He passed large quantities of air through a tube that contained a plug of guncotton to serve as filter. The guncotton was then removed and dissolved in a mixture of alcohol and ether, and the sediment was examined under a microscope. He found that this sediment contains not only organic matter, but also large number of small microorganisms.

Louis Pasteur opened the field of sterilization by stating that boiling rendered fluid sterile. He introduced the method of sterilizing glassware by dry heat at 170°C. Louis Pasteur in 1880 isolated the bacterium responsible for chicken cholera and grew it in pure culture. He invented the vaccines for anthrax and rabies. He knew that the causative agent of rabies attacks the brain and spinal cord.

Antony van Leeuwenhoek

John Needham

Louis Pasteur

4. Augustino Bassi (1773–1856) presented convincing evidence that living organism was the cause of disease. He demonstrated that a fungus that caused a disease in silkworm could be transmitted from one silkworm to another.

5. Lord Joseph (1827–1912), a famous English surgeon, first introduced antiseptic for the prevention and cure of wound healing.
6. Robert Koch (1843–1910) was a German physician who isolated *Bacillus anthracis*, the causative agent of anthrax. He was one of the first scientists to demonstrate the role of bacteria in causing diseases.

      Augustino Bassi   Lord Joseph   Robert Koch

7. Richard Petri (1852–1921) designed a special plate to hold solid culture media. This plate has great significance in microbiology and is referred to as *Petri plate*.
8. Paul Ehrlich (1904) found that the dye trypan red was active against trypanosomes that cause African sleeping sickness. This dye with antimicrobial activity was referred to as a *magic bullet*.
9. Sahachiro Hata (1909) introduced the drug salvarsan for the treatment of syphilis caused by *Treponema pallidum*.

      Richard Petri   Paul Ehrlich   Sahachiro Hata

The period from 1857 to 1914 is considered as the *Golden Age of Microbiology,* because significant advances made during this period led to the establishment of microbiology as an important discipline of science.

The discovery of microorganisms responsible for wide-spectrum ailments in human beings had been a significant contribution of microbiologists across the globe. It paved way for pinpointed

Table 1.4 Milestones in discovery of causative agents

Year	Causative agent	Discoverer	Disease
1874	*Mycobacterium leprae*	Hansen	Leprosy
1877	*Actinomyces bovis*	Bollinger	Actinomycosis
1879	*Neisseria gonorrhoeae*	Albert Neisser	Gonorrhoea
1880	*Salmonella typhi*	Eberth	Typhoid fever
1880	*Plasmodium* spp.	Laveran	Malaria
1882	*Mycobacterium tuberculosis*	Robert Koch	Tuberculosis
1883	*Vibrio cholerae*	Robert Koch	Cholera
1885	*Clostridium tetani*	Arthur Nicolaier	Tetanus
1894	*Yersinia pestis*	Alexander Yersin	Plague
1900	Avian influenza virus	WHO	Bird flu
1906	*Bordetella pertussis*	Border and Gangou	Whooping cough
1983	Human immunodeficiency virus	Luc Montagnier	AIDS
1983	*Helicobacter pylori*	R. Warren and B. Marshall	Gastritis
1984	Human T-cell virus	Robert, et al.	Leukaemia
2003	SARS coronavirus	Carlo Urbani	Respiratory disease

Abbreviations: SARS, severe acute respiratory syndrome; WHO, World Health Organization.

Table 1.5 Discovery of antibiotics

Year	Source	Discoverer	Antibiotics
1929	*Penicillium notatum*	Alexander Fleming	Penicillin
1944	*Streptomyces griseus*	Walksman, et al.	Streptomycin
1947	*Streptomyces venezuelae*	P.R. Burkholder	Chloramphenicol
1949	*Streptomyces fradiae*	Waksman and Lechevalier	Neomycin
1950	*Streptomyces noursei*	Hazen and Brown	Nystatin
1956	*Streptomyces nodosus*	Gold, et al.	Amphotericin A
2000	*Micromonospora* spp.	Fernandez-Chimeno, et al.	Cytotoxic macrolide, IB-96212

research, which ultimately resulted in the discovery of wide range of antibiotics. The chronologically ordered milestones in discovery of causative agents and discovery of antibiotics are given in Tables 1.4 and 1.5, respectively.

Gerhardt Domagk found that prontosil—an azo dye from *para*-aminobenzene-sulphonamide— was active against specific bacteria. The first success in treating streptococcal infection was reported in 1935. Sir Alexander Fleming accidentally discovered a substance produced by *Penicillium notatum*. He extracted a compound from the fungus and named it *penicillin,* which could destroy several pathogenic bacteria. The commercial production of penicillin was taken up by US firms in 1941 during World War II. S.A. Waksman (1944) reported production of streptomycin from two different strains of actinomycetes.

GENETIC EVALUATION

Historical Perspective

Gregor Johann Mendel (1900) is known as the *father of genetics*. W. Bateson and R.C. Punnet (1906) reported the first case of linkage in sweet pea and proposed the presence or absence theory. The British physician, Sir Archibald Garrod (1908) first proposed one gene–one product hypothesis. W. Johnson (1909) coined the term *gene* that acts as hereditary unit. T.H. Morgan (1926) discarded all the previous adjusting theories and put forward the particulate gene theory. In 1933, Morgan was awarded Nobel Prize for his research in explaining gene theory.

In 1940, Beadle and Tatum proposed one gene–one protein hypothesis, which explained that one gene encodes one protein and is known as *overlapping gene* (genes within genes). In 1955, Benzer found that cultures of T4 bacteriophage formed plaques on other plates of *E. coli*. G.W. Beadle and E.L. Tatum (1958) with Lederberg received a Nobel Prize for their contribution to physiological genetics.

Discovery of genetic code was possible with the significant contributions made by Francis Crick, Severo Ochoa, M.W. Nirenberg and Har Gobind Khorana early in 1960. For this work, Khorana shared Nobel Prize with Nirenberg and Holley in 1968. Jacob and Monod (1961), explained regulation of gene activity. In 1962, Benzer coined the term *mutant* to denote the smallest unit of chromosome that undergoes mutational changes. Shapiro and coworkers (1969) published the first picture of isolated genes. Thomas Cech in 1986 discovered that pre-rRNA isolated from ciliated protozoa *Tetrahymena thermophila* is self-splicing.

The International Human Genome Project began in 1990 with the following objectives:
1. Developing ways of mapping the human genome at increasing fine level of precision
2. Storing the information in databases and developing tools for data analysis
3. Addressing the ethical, legal and social issues that may arise from execution of this project

Robert and Sharp in 1993 independently hybridized the mRNA of adenovirus with their progeny or DNA segments of virus.

Gene Cloning

Gene cloning is a technological tool used for identifying, isolating and copying a gene coding for a valuable polypeptide with an objective of making available gene for analysis or for production of protein. In 1972, the first recombinant molecule was reported to generate DNA fragments by an enzyme *lipase*. Recombinant technology has resulted into large-scale production of vaccines, hormones and blood clotting factors. A new era was started with the discovery of hybridoma technology, a method for producing pure, identical antibodies against specific antigens. George Milstein and Cesar Kohler (1975) created hybridomas by the fusion of cancer cell with antibodies producing lymphocytes from immunized animal, which resulted in the production of monoclonal antibodies.

Gene Therapy

Discovery of DNA structure is the most important historical achievement in the field of biotechnology. Watson and Crick in 1953 proposed the structure for DNA, for which they were awarded Nobel Prize. According to these two scientists, the DNA molecule consists of deoxyribose sugar, nitrogen bases and phosphoric acid. Erwin Chargaff (1948) used the technique of paper chromatography for revealing the basic composition of DNA. He discovered that in the DNA of different types of organisms, the total amount of purines is equal to the total amount of pyrimidines.

Many attempts in human gene therapy made as early as in 1979 were not successful. The gene therapy involving genetic manipulations may provide new approaches for treating a disorder caused by pathogen that is resistant to the conventional drug.

Protein Biosynthesis

It is the process through which cells build proteins. The term is also used to refer only to protein translation. But more often it refers to a multistep process, beginning with amino acid synthesis and transcription of nuclear DNA into messenger RNA, which is then used as input to translation. The proteins can be synthesized directly from genes by translating mRNA. When a protein needs to be available on short notice or in large quantities, a protein precursor is produced. A *pro-protein* is an inactive protein containing one or more inhibitory peptides that may be activated when the inhibitory sequence is removed by proteolysis during posttranslational modification. It contains a signal sequence (*N*-terminal signal peptide) that specifies its insertion into or through membranes (Fig. 1.3).

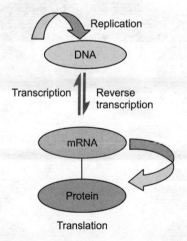

Figure 1.3 Protein biosynthesis.

DNA Replication

It is the basis for biological inheritance. It reflects a fundamental process occurring in all living organisms to copy their DNA. The process is known as *replication* as each strand of the original double-stranded DNA molecule serves as template for reproduction of the complementary strand. As a result of this, two identical DNA molecules are produced from a single double-stranded DNA molecule. Cellular proofreading and error toe-checking mechanisms ensure near-perfect fidelity for DNA replication. The DNA replication in a cell is initiated at specific locations in the genome, known as *origin*. Unwinding of DNA at the origin and synthesis of new strands form a replication

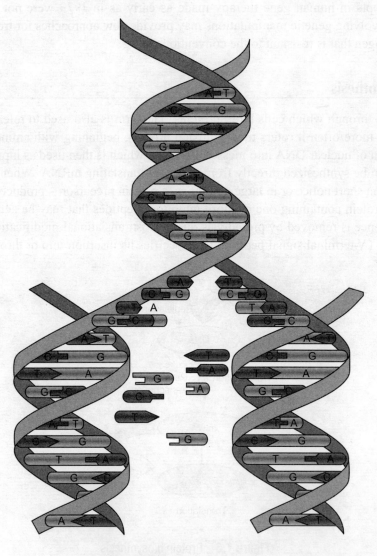

Figure 1.4 DNA replication.

fork. *DNA polymerase* is the enzyme that synthesizes the new DNA by adding nucleotides matching to the template strand. In addition to this, a number of other proteins are associated with the fork and assist in the initiation and continuation of DNA synthesis. DNA replication can also be achieved in vitro (outside a cell). DNA polymerases isolated from cells and artificial DNA primers may be used to initiate the process of DNA synthesis at known sequences in a template molecule. The PCR is a common laboratory technique that employs such artificial synthetic process in a cyclic manner to amplify a specific target DNA fragment from a pool of DNA (Fig. 1.4).

The DNA double helix is stabilized by hydrogen bonds between the bases attached to the strands. The four bases found in DNA are adenine (A), cytosine (C), guanine (G) and thymine (T). These four bases are attached to the sugar/phosphate groups to form the complete nucleotide.

Cytosine and thymine are six-membered rings and are known as *pyrimidines*, whereas adenine and guanine are fused five- and six-membered heterocyclic compounds (*purines*). A fifth pyrimidine base known as *uracil* (U) normally takes the place of thymine in RNA and differs from thymine by lacking a $-CH_3$ group on its ring.

HISTORICAL PERSPECTIVES OF PLANT TISSUE CULTURE

The cell therapy advanced by Schleiden and Schwann (1838) revealed the principle of tissue culture. According to biologist Gautheret (1985), the discovery of tissue culture could be traced to the pioneering experiment conducted by Hennery–Louis on wound healing in plant showing spontaneous callus formation on the decorticated region of the Elm plant.

According to Haberlandt's hypothesis (1902), a cell is capable of autonomy and has potential of totipotency. The term *totipotency* was first coined by Morgan. Hannig started his research work by taking embryogenic tissue instead of single cells. He excised nearly matured embryos of some crucifers such as *Raphanus sativus*, *Raphanus landra* and successfully cultivated (in vitro) them on artificial medium consisting of mineral salts and sugar.

Symun (1908) established the basis for callus culture and to some extent also for micropropagation Kotte and Robbins (1922) simultaneously put forward a new approach to tissue culture and reported that a true in vitro culture could be developed using meristematic cells. White (1934–39) carried out in vitro techniques for tissue culture by changing the nature of nutrient medium. Michael (1939) demonstrated the role of sodium nitrate in protoplast fusion. In the same year, Steward et al. successfully raised a large number of plantlets from carrot root suspension cultures. Overbeek et al. (1941) used coconut milk for embryo development and callus formation in datura, and it proved to be a very important turning point in the development of embryo culture. Later on, this was helpful in the development of several hybrids. Muir and associates (1954) reported that pieces of callus of *Tagetes erecta* and *Nicotiana tabacum* could be cultured in the form of cell suspensions. Skoog and Miller (1957) proposed the concept of hormone-controlled organ formation. In 1960, Cocking introduced protoplasmic plant tissue cultures.

QUESTIONS

1. Define biotechnology. Enlist different branches of biotechnology.
2. Mention in brief the contributions of some leading scientists in biotechnology.
3. Discuss the significant milestones in biotechnological research.
4. Give a brief account of discovery of cell and chromosome theory of inheritance.

CHAPTER 2: Immunology and Immunological Preparations

CHAPTER OUTLINE

- Introduction
- How the Immune System Works
- Principles
- Types of Immune System
- Immune Tolerance
- Hypersensitivity
- Antigen–Antibody Reactions and Their Applications
- Immunomodulators
- Applications of Immunomodulators
- Vaccines
- Questions

INTRODUCTION

Immune System

The word *immune* in Latin refers to 'protection'. The immune system of body is a built-in protection against attack by foreign substances known as *antigens*. An antigen is a substance that causes the body to attack and induce an immune response. Examples of antigens include bacteria or viruses. Unfortunately, the immune system does not discriminate between harmful foreign substances (such as bacteria and viruses) and transplanted organs.

Immunity is defined as the resistance to disease, especially infectious disease. The collection of cells and molecules that mediate resistance to infection is referred to as *immune system*. The coordinated reaction of these cells and molecules to infectious microbe is known as *immune response*. The physiological function of the immune system is to prevent infection and to eradicate established infections.

Edward Jenner (1778) observed that milkmaids, who developed cowpox, were immune to smallpox. The classic work of Russian scientist Elie Metchnikoff (1845–1916) is credited with the biological hypothesis that marked a new stage in the history of immunobiology. Louis Pasteur (1822–95) developed vaccines for anthrax and rabies. Paul Ehrlich (1854–1915) reported that certain substances in the blood serum secreted by special cells under the influence of microbes and their toxins contribute to the defence mechanism of human body.

A transplanted organ is seen by the immune system of the body as a foreign substance that needs to be eliminated. To protect the transplanted organ from attack, medications are used to suppress or hold back the body's immune response. These medications are known as *immunosuppressants*, e.g. Neoral, Prograf, OKT3, Imuran, prednisone, Cellcept and rapamycin.

HOW THE IMMUNE SYSTEM WORKS

The immune system is able to differentiate between substances that belong to our body and those that are foreign. Once a substance is recognized as foreign, the immune system tries to eliminate it from the body. It is therefore a practice that immunosuppressant medications must be administered after an organ transplant. A basic understanding of the immune system helps us understand why taking these medications as instructed is critical to the success of organ transplant.

Lymphocytes are white blood cells primarily responsible to defend our immune system. These are the only cells in the body that have the ability to recognize specific antigens (foreign substances). There are different kinds of lymphocytes in the body. B lymphocytes react to antigens in solution such as blood. T lymphocytes react against infected or foreign cells such as a transplanted organ. The different cells of the immune system working together to resist or overcome an infection in the body are given in Figure 2.1.

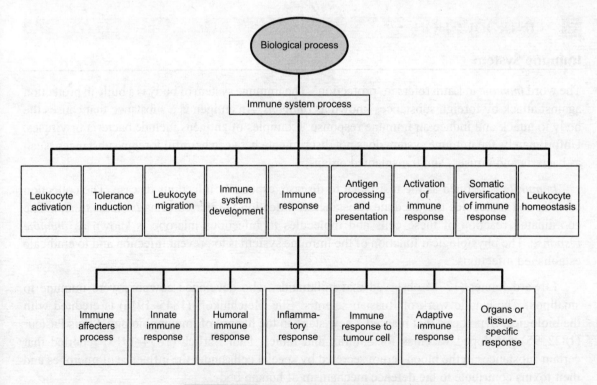

Figure 2.1 Structure of immune system.

T and B lymphocytes are present in all the tissues in human body. They remain inactive until they recognize a specific antigen. The receptors on the surface of lymphocytes bind with the antigen. When this coupling takes place, the immune response is initiated and the lymphocyte destroys the antigen. Immunosuppressant agents prevent the lymphocyte and the antigen from binding, therefore preventing the immune system from destroying the transplanted organ.

Organs of Immune System

A number of organs are involved in the development of an immune response, and they impart different functions (Fig. 2.2).

Different organs of the body play their roles in the development of an immune system.

Lymphoid System

These organs provide appropriate microenvironment for the development and maturation of antigen-sensitive lymphocytes. Maturation of T lymphocytes occurs in the thymus, whereas that of B lymphocytes takes place in the bone marrow. Secondary lymphocytes are the sites for the initiation of immune response, e.g. spleen, tonsils, lymph nodes, appendix, etc.

Cells of the Immune System (Fig. 2.3)

Two types of lymphocytes occur, namely B cells and T cells.
1. **B lymphocytes (B cells)**: B cells are produced in the bone marrow. They constitute a major class of lymphocytes producing immunoglobulin, and they are primarily involved in humoral

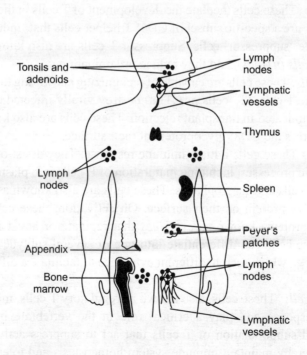

Figure 2.2 Organs of immune system in human body.

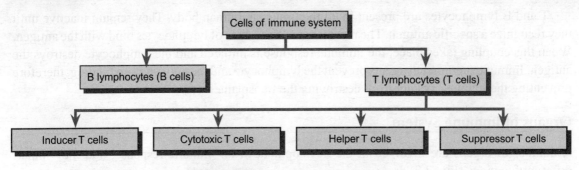

Figure 2.3 Cells of immune system.

immunity. The B cells possess the capability to specifically recognize each antigen and produce antibodies against it.

2. **T lymphocytes (T cells)**: T cells are also produced in the bone marrow but leave the bone marrow and mature in the thymus. T cells belong to a group of white blood cells known as *lymphocytes* that play a crucial role in the cell-mediated immunity. They can be differentiated from other lymphocyte types, such as B cells and natural killer cells (NK cells), by the presence of a special receptor on their cell surface *T-cell receptor* (TCR). The abbreviation T in 'T cell' represents *thymus*, which is the principal organ responsible for the maturation of T cells.

T cells are broadly of the following four types:

a. *Inducer T cells:* These cells mediate the development of T cells in the thymus. Suppressor–inducer T cells are a specific subset of $CD4^+$ T helper cells that 'induce' $CD8^+$ cytotoxic T cells to become 'suppressor' cells. Suppressor T cells are also known as $CD25^+$–$Foxp3^+$ *regulatory T cells (nTregs)*, and they reduce inflammation.

b. *Cytotoxic T cells:* These cells are capable of recognizing and killing the infected or abnormal cells. Cytotoxic T cells (T_C cells or CTLs) destroy virally infected cells and tumour cells and are also implicated in transplant rejection. These cells are also known as $CD8^+$ *T cells*, since they express the $CD8^+$ glycoprotein at their surface.

c. *Helper T cells:* These cells initiate immune response. They assist other white blood cells in immunologic processes, including maturation of B cells into plasma cells and activation of cytotoxic T cells and macrophages. These cells are also known as $CD4^+$ *T cells* as they express the CD4 protein on their surface. On activation, these cells divide rapidly and secrete small proteins known as *cytokines* that regulate or assist in the active immune response. These cells can differentiate into one of several subtypes, including T_H1, T_H2, T_H3, T_H17 or T_{FH}, which secrete different cytokines to facilitate a different type of immune response.

d. *Suppressor T cells:* These cells, also known as regulatory T cells, mediate the suppression of immune response and play a critical role in the vertebrate immune system. They are specialized subpopulation of T cells that act to suppress activation of the immune system and thereby maintain immune system homeostasis and tolerance to self-antigens. The immunosuppressive potential of these cells can be harnessed therapeutically to treat autoimmune diseases and also to facilitate transplantation tolerance and in cancer immunotherapy.

Table 2.1 Human immunoglobulins: Types, molecular weight and functions

Type	Molecular weight (Da)	Major function
IgA	1,60,000	Protect the body surface
IgD	1,80,000	B-cell receptor
IgE	1,90,000	Humoral sensitivity and histamine release
IgG	1,50,000	Mostly responsible for humoral immunity
IgM	9,00,000	Humoral immunity, serves as first line of defence

Human Immunoglobulins

Immunoglobulins are group of proteins capable of mediating humoral immunity. Immunoglobulins are of five different types (Table 2.1).

1. **Immunoglobulin A (IgA)**: It occurs as a single or double unit held together by J chain. It is mostly found in the body secretion, such as saliva, tears, sweat, milk, etc. It is capable of preventing the foreign substances from entering into the body cells.
2. **Immunoglobulin D (IgD)**: It is composed of a single Y-shaped unit. It is present in a low concentration in the circulation, and it functions as B-cell receptor.
3. **Immunoglobulin E (IgE)**: The molecules of IgE tightly bind with mast cells that release histamines and cause allergy. It is normally present in very low concentration in blood.
4. **Immunoglobulin G (IgG)**: It is the most abundant (75–80%) class of immunoglobulin. It is the only immunoglobulin capable of crossing the placenta and transferring the immunity of mother to the developing fetus.
5. **Immunoglobulin M (IgM)**: It is the first antibody to be produced in response to an antigen and is the most effective against attacking microorganism. It is the largest immunoglobulin composed of five Y-shaped units held together by a polypeptide chain.

PRINCIPLES

Innate Immunity

Innate immunity is a type of host defence that is always present in healthy individuals. It is prepared to block the entry of microbes and rapidly eliminate them if they succeed in entering host tissue. The main components of innate immunity are physical barriers, soluble mediators capable of directly inhibiting foreign microorganisms and specialized cells endowed with recognition receptors and the ability to phagocytise and kill microorganisms.

Physical barriers are sometimes underestimated as part of the immune system. An intact skin offers very effective protection from infection. The superficial wounds can be easily colonized by foreign invaders. In addition to skin, the other physical barriers are (1) the blood coagulation system that stops bleeding and forms a protective clot over wounds and (2) internal mucosae that is protected by mucus and continuously cleansed by the action of ciliated cells that sweep away bacteria and other materials. The mildly acidic pH of skin inhibits bacterial proliferation, and very acidic pH of stomach juices sterilize ingested material.

The varieties of soluble molecules present on the skin, in secretions, in blood and throughout the body kill bacteria and inhibit viral infections. The important examples are (1) defences (natural antibiotics), (2) the lysozyme (an enzyme that lyses the wall of some bacteria), (3) complement (a cascade of proteins activated by bacterial components that make holes in bacterial membranes), (4) opsonins (complement fragments and other proteins that coat bacteria and facilitate phagocytosis) and (5) some ancient cytokines such as type I interferons that are released by virus infected cells and induce an antiviral state in neighbouring cells.

The basic cellular mechanisms of innate immunity are cellular migration towards invaders and phagocytosis followed by intracellular destruction of the ingested microorganism. If phagocytes release lytic substances in the environment, extracellular killing is also possible. In mammals, the two main cell types are granulocytes that contain lytic substances packed in intracellular granules (also known as *polymorphonuclear cells* because nucleus is of irregular shape) and macrophages. Neutrophil granulocytes have an important role in defence against bacterial infections, and basophils have an antiparasitic role. Macrophages derive from circulating monocytes that emigrate to practically all tissues and organs, sometimes under different names (e.g. Kupffer cells in the liver). In addition to immune functions, macrophages play important roles in the turnover of aging cell components, like red blood cells that are continuously phagocytosed by splenic macrophages. Natural killer (NK) cells are nonphagocytic elements that kill virus-infected cells. It is worth noting that the evolution of NK cells is relatively recent and goes in parallel with the evolution of lymphocytes, to which NK cells closely resemble morphologically.

Recognition in innate immunity is mediated by interactions with pathogen-associated molecular patterns as expressed by microorganisms. The cellular receptors are referred to as *pattern recognition receptors* and include toll-like receptors (TLRs), mannose receptors (MRs) and seven-transmembrane spanning receptors (TM7s). TLR recognizes bacterial and viral nucleic acids, flagellin, bacterial peptides, lipopolysaccharide (LPS) and other bacterial components. MRs are capable of binding carbohydrate moieties on several pathogens, such as bacteria, fungi, parasites and viruses. TM7 receptors are usually activated by bacterial peptides or by endogenous chemokines.

Adaptive Immunity

Adaptive immunity is a system of host defence that is stimulated by microbes that invade tissues. It adapts to the presence of microbial invaders.

The evolution of adaptive immunity is related to the evolution of vertebrates, starting from sharks. It is, therefore, deduced that part of the evolutionary success and long lifespans of vertebrates could be attributed to improved defences against exogenous pathogens. The functions of adaptive immunity are performed by a new type of cell (the lymphocyte) that recirculates amongst blood, tissues and a specialized circulatory system (the lymphatic vessels) and resides in specialized organs (lymphoid organs, which include thymus, spleen and lymph nodes).

The hallmarks of adaptive immune responses are specificity, immune memory and immune tolerance. A lymphocyte population is made up of millions of clones, each specific for a different antigen. Specificity is encoded in clonotypic antigen receptors generated by a process of DNA rearrangement that includes random events to produce billions of variant molecules from a relatively small pool of DNA sequences.

Specificity permits a considerable economy in adaptive immune responses to pathogens. It is because only the clones expressing receptors for a specific microorganism are activated upon infection, whereas all other clones remain inactive. In addition to the proliferation of specific lymphocyte clones, the first encounter with a given antigen (primary immune response) leaves behind a population of memory cells that respond more promptly and more efficiently to subsequent encounters (secondary immune response). The presence of random events in the generation of specific antigen receptors implies the risk of producing autoreactive receptors.

Cellular–Humoral Immunity

The humoral immune response (HIR) is an aspect of immunity that is mediated by secreted antibodies produced in the cells of the B lymphocyte lineage (B cell). The B cells (with costimulation) transform into plasma cells that secrete antibodies. The costimulation of the B cell can be from another antigen-presenting cell, like a dendritic cell. This entire process is aided by $CD4^+$ T helper 2 cells, which provide costimulation. The secreted antibodies are capable of binding to antigens on the surfaces of invading microbes (such as viruses or bacteria), which flag them for destruction.

Humoral immunity is so named because it involves substances found in the humors, or body fluids. Humoral immunity refers to antibody production and the accessory processes that accompany it, including T_h2 activation and cytokine production, germinal centre formation and isotype switching, affinity maturation and memory cell generation. It also refers to the effector functions of antibody that include pathogen and toxin neutralization, classical complement activation and opsonin promotion of phagocytosis and pathogen elimination.

TYPES OF IMMUNITY

Immunity can be broadly classified into natural and acquired. They can be further classified into active and passive.

Natural Immunity

Nature has given certain individuals, species and races some advantage by providing them with immunity against certain disease. Natural immunity is resistance to a disease possessed by an individual.

Acquired Immunity

Only with the protection offered by natural immunity it is difficult to survive. Hence, immunity is provided by stimulating an individual's antibody production. It is also further classified into active and passive immunity.

1. **Active immunity**: Active immunization entails the introduction of a foreign molecule into the body that causes the body itself to generate immunity against the target. This immunity comes from the T cells and the B cells with their antibodies.

   Active immunization occurs naturally when a person comes in contact with a microbe. If the person has not come into contact with the microbe earlier and has no premade antibodies for defence (like in passive immunization), the immune system shall eventually create antibodies and other defences against the microbe. The next time, the immune response against this microbe can be very efficient. This is the case in many of the childhood infections that a person only contracts once, but then is immune to that infection.

   Artificial active immunization is where the microbe or parts of it is injected into the person. If whole microbes are used, they are known as *pretreated, attenuated vaccine*.

2. **Passive immunity**: In passive immunization, presynthesized elements of the immune system are transferred to a person. Hence, the body does not need to produce these elements itself. The antibodies can be used for passive immunization. This method of immunization begins to work very quickly, but it is short lasting, because the antibodies are naturally broken down. Moreover, if there are no B cells to produce more antibodies, they subsequently disappear. When antibodies are transferred from mother to fetus during pregnancy, passive immunization occurs physiologically to protect the fetus before and shortly after birth.

   Artificial passive immunization is normally administered by injection. It is used if there has been a recent outbreak of a particular disease or as an emergency treatment for toxicity (e.g. for tetanus). Although there is a high chance of anaphylactic shock because of immunity against animal serum itself, the antibodies can be produced in animals (serum therapy). If available, humanized antibodies produced in vitro by cell culture are preferred.

IMMUNE TOLERANCE

It is the process by which the immune system does not attack an antigen. It can be either 'natural' or 'self-tolerance', where the body does not mount an immune response to self-antigens or induced tolerance, tolerance to external antigens may be created by manipulating the immune system. It is of three types: central tolerance, peripheral tolerance and acquired tolerance.

1. **Central tolerance**: Central tolerance occurs during lymphocyte development. It operates in the thymus and bone marrow. The T and B lymphocytes that recognize self-antigens are deleted before they develop into fully immunocompetent cells, preventing autoimmunity. This process is most active in fetal life, but continues throughout life as immature lymphocytes are generated.

   In mammals, the process occurs in the thymus (T cells) and bone marrow (B cells), when maturing lymphocytes are exposed to self-antigens. The self-antigens are present in both organs due to endogenous expression within the organ and importation of antigen due to circulation from peripheral sites. In case of T cell central tolerance, additional sources of antigen are made available in the thymus by the action of the transcription factor AIRE.

Positive selection occurs initially when naive T cells are exposed to antigens in thymus. For this, T cells that have receptors with sufficient affinity for self-MHC molecules are selected. Other cells that do not show sufficient affinity to self-antigens shall undergo a deletion process, known as death by neglect that involves apoptosis of the cells.

Negative selection of T cells with a very high affinity of self-MHC molecules is induced to energy or lineage divergence to form T regulatory cells.

2. **Peripheral tolerance**: Peripheral tolerance is a system of immunological tolerance developed after T and B cells mature and enter the periphery.
3. **Acquired tolerance**: Acquired or induced tolerance refers to the immune system's adaptation to external antigens characterized by a specific nonreactivity of the lymphoid tissues to a given antigen. One of the most important natural kinds of acquired tolerance is immune tolerance in pregnancy, where the fetus and the placenta must be tolerated by the maternal immune system. In adults, tolerance is usually induced by repeated administration of large doses of antigen. It is most readily induced by soluble antigens administered either intravenously or sublingually. Immunosuppression also facilitates the induction of tolerance.

In clinical practice, acquired immunity is important in organ transplantation, when the body must be forced to accept an organ from another individual. The failure of the body to accept a foreign organ is known as *transplant rejection*. To prevent rejection, a variety of medicines are used to produce induced tolerance.

Oral tolerance is one of the most important forms of acquired tolerance. The specific suppression of cellular and/or humoral immune reactivity to an antigen with prior administration of the antigen by oral route is probably evolved to prevent hypersensitivity reactions to food proteins and bacterial antigens present in the mucosal flora. It is of immense immunological importance, since it is a continuous natural immunologic event driven by exogenous antigen. Oral tolerance is evolved to treat external agents that gain access to the body via natural route. The limitation of oral tolerance is attributed to the development and pathogenesis of several immunologically based diseases, including inflammatory bowel disease (Crohn's disease and ulcerative colitis).

HYPERSENSITIVITY

Not all immune responses against an antigen produce a desirable resistance. When the sensitivity is beyond limits, it is termed as *hypersensitivity*. When an adaptive immune response occurs in an exaggerated or inappropriate form causing tissue damage, the condition is termed as hypersensitivity. Hypersensitivity (also known as *hypersensitivity reaction*) refers to undesirable (damaging, discomfort-producing and sometimes fatal) reactions produced by the normal immune system.

Hypersensitivity reactions require a presensitized (immune) state of the host. The comparative account of hypersensitivity types is given in Table 2.2.

Table 2.2 Comparisons of hypersensitivity types

Type	Alternative names	Often-mentioned disorders	Mediators
1	Allergy (immediate)	1. Atopy 2. Anaphylaxis Asthma	IgE
2	Autoimmune disease, receptor mediated	1. Grave's disease 2. Myasthenia gravis	IgM or IgG (complement)
3	Immune complex disease	1. Serum sickness 2. Arthus reaction 3. Systemic lupus erythematosus (SLE)	IgG (complement)
4	Delayed-type hypersensitivity (DTH)	1. Contact dermatitis 2. Mantoux test 3. Chronic transplant rejection	T cells
5	Autoimmune disease, receptor mediated	1. Grave's disease 2. Myasthenia gravis	IgM or IgG (complement)

Abbreviations: IgE, immunoglobulin E; IgG, immunoglobulin G; IgM, immunoglobulin M; SLE, systemic lupus erythematosus.

 ANTIGEN–ANTIBODY REACTIONS AND THEIR APPLICATIONS

Antigen

An *antigen* is a foreign particle that enters the body. This could be a disease-causing agent such as part of a bacterium or virus or could be a particle such as pollen or dust. When an antigen enters the body, the immune system produces antibodies against that antigen.

Antibodies

An *antibody* is a protein made by the body's immune system. Antibodies react with specific antigens to enable the antigens to be removed from the body. Antibody binds to specific antigen.

The multicellular organisms have evolved defence mechanisms against the onslaught of viruses, bacteria and other parasites. The immune system of vertebrates combines evolutionarily innate (or natural) immunity with newer adaptive responses against specific antigens. The two systems are tightly integrated, and several molecules and cells, in addition to their fundamental role in one system, act as interfaces between the two.

Antigen–Antibody Reaction

The antigen–antibody reaction is a bimolecular association similar to an enzyme–substrate interaction. It is with the distinction that it does not lead irreversible chemical alteration. The reaction involves various noncovalent interactions between the antigenic determinant or epitope of the antigen and variable region of the antibody (Fig. 2.4).

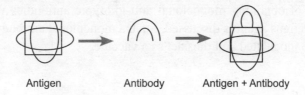

Figure 2.4 Antigen and antibody reaction.

Improved methods are available for the detection of antigens and haptens by the use of the passive haemolysis inhibition test. The test is capable of detecting nanogram quantities of proteins (e.g. ferritin, IgE) and haptens (e.g. folinic acid, methotrexate). The method is also useful for studying quantitative and qualitative aspects of antibody–antigen interaction.

Applications of Antigen–Antibody Reactions

1. The detection of antibody to hepatitis C virus in prospectively followed transfusion recipients with acute and chronic non-A and non-B hepatitis is possible. The antibody (anti-HCV) to hepatitis C virus that causes non-A and non-B hepatitis by radioimmunoassay is measured in prospectively followed transfusion recipients and their donors.
2. A method is described for quantitative analysis of proteins with a charge differing from that of the bulk of the immunoglobulins. It is rapid, suitable for serial analysis and requires minute concentration of protein antigen. The amount of antiserum needed is slightly less than that in the radial immunodiffusion techniques. The method utilizes the difference between the rate of electrophoretic migration of proteins and their antibody complexes in agarose gel.
3. A sensitive and specific radioimmunoassay technique is developed for the precursor-specific peptide segment located at the amino end of bovine type III procollagen. Human material showed high cross-reactivity in this assay. Two forms of human procollagen peptides were detected in body fluids. The larger peptide (45K) was found in serum and ascites, and it resembled the whole precursor-specific segment, which is presumably released from human type III procollagen by a single enzymatic cleavage. The smaller peptide (10K) was found mainly in urine, indicating that further degradation of circulating procollagen peptides is required prior to their passage through the kidney.
4. An immunocytochemical technique for demonstration of intracellular antigens (secretory proteins) on thin sections is reported. By using this technique, different secretory proteins of the exocrine and endocrine pancreas are localized. The protein A-gold technique is proposed as a general method for visualization of antigenic sites on thin sections.
5. An in vitro technique is developed for assessing the chemotactic activity of soluble substances on motile cells. The antibody–antigen mixtures when incubated at 37°C in medium containing fresh (i.e. noninactivated) normal rabbit serum exert a strong chemotactic effect on rabbit polymorphonuclear leukocytes. The results indicate that when antibody–antigen complexes are incubated (37°C) in fresh serum, a heat-stable (56°C) substance(s) is produced, which acts directly as a chemotactic stimulus on the polymorphs.
6. It is evidenced that some anti-idiotypic antibodies may possess one or more antigenic determinants (epitopes) that are closely related to epitope(s) on the original antigen (antigen X).

The possibility of obtaining monoclonal anti-idiotypic antibodies with epitopes similar to those on an infectious agent is discussed. Such a monoclonal antibody might elicit protective antibodies when inoculated and thus act as a vaccine.

Haptens

Hapten is a small molecule that can elicit an immune response only when attached to a large carrier such as a protein. The carrier may be one that also does not elicit an immune response by itself. Once the body has generated antibodies to a hapten-carrier adduct, the small-molecule hapten may also be able to bind to the antibody. It will not initiate an immune response. Sometimes, the small-molecule hapten may block immune response to the hapten-carrier adduct by preventing the adduct from binding to the antibody. This process is known as *hapten inhibition*.

The concept of haptens emerged from the work of Karl Landsteiner.

The first haptens used were aniline and its carboxyl derivatives (*o*-, *m*- and *p*-aminobenzoic acid). Other haptens used in molecular biology applications include fluorescein, biotin, dinitrophenol and digoxigenin.

Urushiol, a toxin from poison ivy when absorbed through the skin, undergoes oxidation to generate the actual hapten, a reactive molecule known as *quinone* that reacts with skin proteins to form hapten adducts. Usually, the first exposure only causes sensitization with proliferation of effector T cells. After a second exposure later, the proliferated T cells can become activated, generating an immune reaction and producing the typical blisters.

Hydralazine, a blood pressure–lowering drug can induce autoimmune disease in certain individuals. The anaesthetic gas halothane can cause a life-threatening hepatitis. The penicillin-class drugs can cause autoimmune haemolytic anaemia.

Hapten Inhibition

It is the inhibition of a type III hypersensitivity response towards a molecule by introducing a fraction of it that works as a hapten. The rest of the molecule is regarded as the carrier, and the entire molecule as the adduct. The hapten binds up antibodies towards that molecule without causing the immune response, leaving less antibodies to bind to the molecule itself. An example of a hapten inhibitor is dextran 1, which is a small fraction (1 kDa) of the entire dextran complex. It is enough to bind antidextran antibodies, but insufficient to result in the formation of immune complexes and resultant immune responses.

 IMMUNOMODULATORS

Immunomodulators are drugs that either suppress or stimulate the *immune system* (*immunosuppressants*). *Immunomodulators* may be defined as substances either of biological or synthetic origin that can stimulate, suppress or modulate any components of the immune system including, both innate and adaptive arms of immune response. One of the starting points for conceptualizing immunomodulation has been the search for agents that could be used for the treatment of cancer.

Classification of Immunomodulators

Immunomodulators are of three types (Fig. 2.5):
1. Immunoadjuvants
2. Immunosuppressants
3. Immunostimulants

1. **Immunoadjuvants**: These agents are used for enhancing efficacy of vaccines, and therefore they could be considered as specific immune stimulants. Immunoadjuvants hold the promise of being the true modulators of the immune response. Some examples of immunomodulators are as follows:
   a. *Glucocorticoids:* e.g. Prednisolone.
      It exhibits lympholytic activity and possesses anti-inflammatory property. It is used as first-line immunosuppressive therapy in solid and haematopoietic stem cell transplant.
   b. *Calcineurin inhibitors:* e.g. Cyclosporine and tacrolimus.
      It is a fat-soluble peptide antibiotic. It binds to cyclophilin and inhibits a cytoplasmic phosphatase calcineurin necessary for activation of a T cell-specific transcription (NFAT), which causes synthesis of IL-2 by activated T cells. It shows very low incidence of bone marrow toxicity. It is used in renal, pancreatic and liver transplantation and also in psoriasis and asthma.

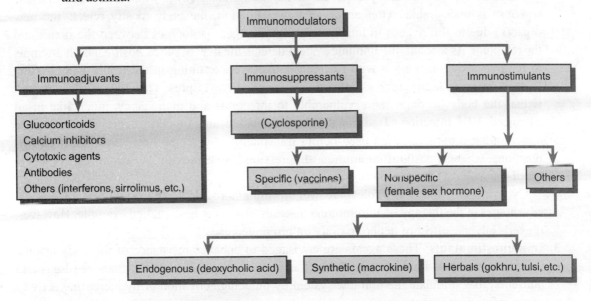

Figure 2.5 Types of immunomodulators.

c. *Cytotoxic agents:* e.g. Azathioprine, cyclophosphamide, dactinomycin, leflunomide, methotrexate, mercaptopurine and vincristine.
   Mercaptopurine interferes with purine nucleic acid metabolism and incorporates false nucleotide. It is used in renal allograft, rheumatoid arthritis (RA), systemic lupus erythematosus (SLE), idiopathic thrombocytopaenic purpura (ITP), Crohn's disease and glomerulonephritis. Leflunomide inhibits pyrimidine synthesis.
d. *Antibodies:* e.g. antilymphocyteglobulin (ALG), Antithyroglobulin (ATG) like muromonab-CD3, immune globulin intravenous (IGIV), monoclonal antibodies like daclizumab and basiliximab.
e. *Others:* e.g. Interferons, mycophenolate mofetil, sirolimus and thalidomide.
   Interferon (IFN) exhibits IFN a-immune enhancing action. It is used in melanoma, RCC, IFN b-multiple sclerosis and IFN g-chronic granulomatous disease. *Sirolimus* inhibits protein kinase and T cell response to interleukin-2. *Mycophenolate mofetil* inhibits inosine monophosphate dehydrogenase that is a key enzyme in guanine nucleotide synthesis. It is used in steroid refractory graft. *Thalidomide* inhibits angiogenesis, reduces phagocytosis and enhances cell-mediated immunity. It increases levels of IL-10 and is used in multiple myeloma, graft-versus-host disease, myelodysplastic syndrome, prostate cancer and colon cancer.

2. **Immunosuppressants**: An immunosuppressant is a substance that causes suppression of the immune system. It may be either exogenous, as immunosuppressive drugs, or endogenous (testosterone). There is an increased susceptibility to infectious diseases and cancers, where the immune system function is suppressed. The undesirable immunosuppressants (polychlorinated biphenyls [PCB] and other pollutants, herbicide, DDT, etc.) are usually referred to as *immunotoxins*.

   Immunosuppressants are prescribed in autoimmune diseases when a normal immune response is undesirable. After an organ transplantation, the body mostly rejects the new organ(s) due to differences in human leukocyte antigen haplotypes between the donor and the recipient. As a result, the immune system detects the new tissue as *hostile*, and it attempts to remove it by attacking it with recipient leukocytes, resulting in the death of the tissue. In such cases, immunosuppressants are used as countermeasures. The side effect, however, is that the body becomes more vulnerable to infections and malignancy, much like in an advanced HIV infection. Traditionally, myeloablative and immunosuppressive high-dose conditioning programmes are used before transplant to overcome host-versus-graft (HVG) reactions, whereas postgrafting immunosuppression has dealt with the second barrier, GVH (graft-versus-host) reactions.

   Cyclosporine, the low molecular weight immunoregulant, offers several theoretical advantages in the therapy of autoimmune diseases. RA is the best-studied example. However, it exerts adverse effect of nephrotoxicity on prolonged use.

3. **Immunostimulants**: These agents are envisaged to enhance resistance of the body against infection. Immunostimulants, also known as immunostimulators, are substances (drugs and nutrients) that stimulate the immune system by inducing activation or by increasing activity of any of its components.

Immunostimulants may be broadly categorized into the following three groups:
a. *Specific immunostimulants:* They provide antigenic specificity in immune response, e.g. as vaccines or any antigen.
b. *Nonspecific immunostimulants:* They augment immune response of other antigen or stimulates components of the immune system without antigenic specificity, e.g. prolactin, growth hormone and vitamin D.
c. *Other immunostimulants:*
   a. Endogenous immunostimulants: Deoxycholic acid, a stimulator of macrophage
   b. Synthetic immunostimulants: Macrokine, a stimulator of macrophages
   c. Herbal immunostimulants: *Aloe vera, Asparagus racemosus, Azadirachta indica,* beta-glucan, *Curcuma longa,* medicinal mushrooms, *Echinacea, Ocimum sanctum, Panax ginseng, Picrorhiza kurroa, Tinospora cordifolia, Withania somnifera,* etc.

APPLICATIONS OF IMMUNOMODULATORS

Immunomodulators as a group of therapeutics are systematically used in the treatment of different human ailments with varied mechanism of action. This class of drugs offers wide spectrum of therapeutic utility.

1. **In the treatment of different human ailments**: Immunomodulators are very useful in the systematic treatment of inflammation, infection and cancer. Topical immunotherapy with immunostimulatory agents are effective in the treatment of inflammatory, infectious and cancerous skin diseases. Immune enhancers such as imidazoquinolines and CpG-sequences (cytosine–phosphate–guanine) possess adjuvant property that is capable of improving conventional (protein) and DNA vaccination against cancer, atopy and allergies.

    Besides topical contact sensitizers (e.g. diphencyprone or dinitrochlorobenzene), newer agents of the imidazoquinoline family, such as imiquimod and resiquimod, act by inducing cytokine secretion from monocytes or macrophages (interferon-12, interleukin-12, tumour necrosis factor-α). The locally generated immune milieu leads to a T_h1-dominance and cell-mediated immunity that have been used clinically to treat viral infections such as human papillomavirus (HPV), herpes simplex virus (HSV), mollusca and cancerous lesions, including initial squamous cell and basal cell carcinoma in immunocompetent and immunosuppressed patients.

2. **In treatment of cancer**: Systemic activation of macrophages can be achieved by the administration of liposomes containing immunomodulators. These activated macrophages to the tumouricidal state can recognize and destroy neoplastic cells and leave normal cells unharmed. Similar to any particle, liposomes are cleared from the circulation by phagocytic cells. The multiple administrations of such liposomes have caused eradication of cancer metastases in several rodent-tumour systems.

    The recent development of new immunomodulatory agents has opened promising therapeutic options. PF-3512676, lenalidomide and NGR-TNF drugs belonging to three different classes of immunomodulatory agents are capable to affect tumour blood vessels with different mechanisms, and they play potential roles in the NSCLC treatment strategy.

3. **In treatment of cutaneous lymphoma**: Immunomodulators are effective in treatment of cutaneous lymphoma in early stages of disease. Immunomodulatory therapies in cutaneous lymphoma include interferons, interleukin-2, cyclosporine A, monoclonal antibodies, autologous bone marrow transplantation, fusion toxins, thymopentin and extracorporeal photopheresis.
4. **In treatment of AIDS**: Development of promising new agents and the discovery of synergy with cytokines or cell products continue to accelerate the pace of research related to utility of these therapeutic agents in treatment of cancer and AIDS. Although, clinical trials have shown the ability of chemical modulators to modulate the human immune system, so far these have generally not fulfilled the therapeutic promise generated in animal models for the treatment of human diseases. While the variation in spectrum of results between animal models and human trials is obvious, the basis is not apparent. The differences of species in elimination kinetics, presence of active drug at the site of action and the development of tachyphylaxis have been postulated as reasons for the minimal activity of these agents in humans.
5. **In treatment of respiratory disorders**: Many immunomodulators are undergoing human clinical trials for the therapy of asthma and allergic ailments. Novel therapeutic approaches include immunostimulatory oligodeoxynucleotides, IL-TLR-4 and -9 agonists, oral and parenterally administered cytokine blockers and specific cytokine receptor antagonists. Transcription factor modulators targeting spleen tyrosine kinase (Syk), peroxisome proliferator-activated receptor γ and nuclear factor κB are also being evaluated for the treatment of asthma. The anti-IgE monoclonal antibody (mAb) omalizumab is in use for the treatment of allergic asthma. But, its potential role for other allergic diseases has yet to be clearly defined.

VACCINES

Formulation of Vaccines

The vial is vigorously shaken to obtain a uniform suspension prior to withdrawing each dose. Whenever solution and container permit, the vaccine is inspected visually for particulate matter and/or discolouration prior to administration. If problems are noted (e.g. vaccine cannot be resuspended), the vaccine should not be administered.

The diluent supplied with the vaccine is only used. The volume of the diluent shown on the diluent label on the vial of lyophilized vaccine is injected and gently agitated to mix thoroughly. The entire contents are withdrawn and administered immediately after reconstitution. The single dose MPSV, varicella and zoster vaccines if not used within 30 min after reconstitution are discarded.

Unused reconstituted MMR vaccine and multidose MPSV vaccine may be stored at 35–46°F (2–8°C) for a limited time. The reconstituted MPSV vaccine must be used within 35 days. The reconstituted MMR vaccine must be utilized within 8 h.

Standardization and Evaluation of Vaccines: WHO Approach

Keeping in view the global concept of standards, there is a need for regular review of the uses of vaccines. The updating of the information should be related to following parameters in accordance with WHO guidelines:
1. Product-specific standards and assays
2. Assignment of unitage: International units
3. Collaborative studies: Suitability of the standards for their intended use
4. Continuous improvements of the assays and comparability studies

From Standards to Vaccines of Assured Quality

Setting norms and standards and promoting their implementation are core activities of World Health Organization (WHO). WHO norms and standards are the basis for the regulation of biologicals worldwide. WHO is committed to support countries to ensure that 100% of vaccines used in all national immunization programmes are of assured quality (Fig. 2.6).

Standards for Different Stages of Vaccine Development

Vaccine Characterization: Setting standards for monitoring consistency of vaccines production are described in the following points:
1. **Preclinical evaluation**: It distinguishes most promising candidates on the basis of quantifiable parameters (e.g. immunogenicity).

Figure 2.6 WHO standards for evaluation of vaccines and biotherapeutics.

2. **Clinical evaluation**: The comparison of different vaccine formulations is carried out in terms of protective efficacy and/or safety.
3. **Lot release**: The specifications on a lot-to-lot basis are defined.

Principles for Biological Standardization

The reference standard should be assigned a value in arbitrary rather than absolute units and the exceptions need to be justified.
1. The unit is defined by a reference standard with a physical existence.
2. In the establishment of the standard, a variety of methods are used. The value is assigned to the standard. Therefore, the definition of the unit is not necessarily dependent on a specific method of determination.
3. The behaviour of the reference standard should resemble as closely as possible the behaviour of test sample in the assay systems to test them.

The prerequisite for development of vaccines and approach for their standardization are given in Figure 2.7.

Vaccine Storage

The safe handling and storage of vaccine are important for safety purposes. In many cases, an effective vaccine is a matter of life or death. While most of the precautions are common sense, a checklist helps to organize the actions in the most efficient and effective manner.

Vaccine Storage Checklist

The Immunization Action Coalition has developed a checklist of safety precautions for the safe handling of vaccines. The checklist can serve as a base document to be modified to fit the specific needs of many medical and even veterinarian facilities.

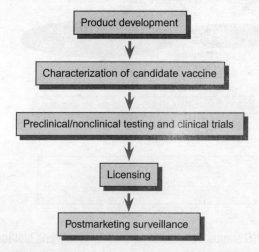

Figure 2.7 Vaccine development and evaluation.

Some important steps to be taken for proper storage of vaccines and handling practices of vaccines are as follows:
1. Vaccines should be stored between 2 and 8°C.
2. They should be kept in a proper container. Never leave them out on the counter or the floor.
3. Check vaccine expiry dates regularly.
4. Vaccines should be protected from light as exposure to ultraviolet light shall cause loss of potency.
5. Freezing may cause loss of potency and can cause hairline cracks in the container, leading to contamination of the contents.
6. Vaccine stocks should be placed within the refrigerator, so that those with shorter expiry dates are used first.
7. Return unused vaccine to the refrigerator immediately after the required dose has been drawn up. Mark the date on all multi dose vials when the first dose is withdrawn. Once opened, multi dose vials must be used within 30 days (unless otherwise indicated on the product monograph). Aseptic technique for the withdrawal of vaccines must be adopted at all times.
8. Take vaccines out of the refrigerator only when ready to administer.
9. Any expiry stock should be labelled clearly, removed from the refrigerator and destroyed as soon as possible.
10. Always move vaccines with shorter expiry dates to the front of the refrigerator so that they can be used first.
11. Protect all vaccines from direct sunlight and fluorescent light.
12. Always check expiry dates before its use. Remove expired vaccines and return them to vaccine supply source.

Vaccine Cold Storage

For vaccines to be effective, it is necessary that they are stored within the temperature range recommended by their manufacturers (2–8°C) to ensure that they remain potent. The vaccination cold chain refers to all the materials, equipments and procedures involved in maintaining vaccines under the required storage conditions from manufacturer to administration.

The maintenance of the cold chain in the delivery of vaccines in certain developed countries is subject to a very vigorous and robust procedure. In the unlikely event of a break in this cold chain during the delivery process, the vaccines can be traced and recalled. Individual patients who have received them can be identified within a matter of hours using tested recall procedures.

All vaccines supplied clearly display the recommended storage conditions on the label and on the summary of product characteristics (SPC). It is a good practice to physically check the labelled storage conditions for each delivery before it is put away. Once vaccines are delivered to providers, the maintenance of the cold chain and the storage of vaccines may, on occasion, fall short of manufacturers' recommendations.

QUESTIONS

1. What are antibodies?
2. Define immunity. Give the difference between active and passive immunity.
3. Define cellular and humoral immunity.
3. What are antigens? What properties are essential for antigenicity?
4. How do primary and secondary immune responses differ?
5. What are immunoglobins? List the different classes of immunoglobulins.
6. Define the terms antigen and antibody.
7. Explain the various antigen–antibody reactions.
8. Define hypersensitivity. What are the various types of hypersensitivity reactions?
9. What are the T and B cells? What is their role in immunity?
10. What are antigens? Discuss the principle of antigen–antibody reactions. Add a note on their applications in modern therapeutics.
11. Discuss the various immunological and diagnostic preparations.
12. What do you understand by hypersensitivity?
13. Give the applications of antigen–antibody reactions.
14. Discuss immunity and give its classification. Write a note on immunological tolerance.
15. Differentiate cell-mediated and humoral immunity.
16. Write short notes on immunodeficiency, immunological tolerance, humoral immunity, hypersensitivity, passive immunity, immune response and cell-mediated immunity.
17. Discuss cellular immunity in detail.
18. Write brief note on haptens.
19. What are vaccines? Explain their preparation and standardization.
20. Explain the role of immunoglobulins in the body.
21. Compare antigen–antibody containing preparation.
22. Define haptens and explain their role as carrier systems.
23. Discuss humoral immunity in brief.
24. Discuss various methods for standardization of antimicrobials.
25. What are the important properties of immunoglobulins?
26. What is the role of IgE in allergy?

CHAPTER 3: Environmental Biotechnology

CHAPTER OUTLINE

- Introduction
- Bioremediation
- Management of Environmental Pollution
- Management of Air, Waste Gases and Water Pollution
- Questions

INTRODUCTION

Environmental biotechnology refers to the development, use and regulation of biological systems for remediation of contaminated environments (land, air, water) and for environment-friendly processes (green manufacturing technologies and sustainable development). Environmental biotechnology employs a diverse set of methodology approaches to explore and exploit the natural biodiversity of microorganisms and their enormous metabolic capacities.

It is a rapidly expanding area of biotechnology that involves introduction of selected microorganisms into the environment for a diversity of purposes, such as rhizobium and mycorrhizal inoculants, enhanced silage production, pollutant dispersal, mineral leaching, degradation of crop residue in situ, improved nitrogen status of soil, etc. The environmental biotechnology, as compared to other areas of biotechnology, is focused on achieving microbial function in complex environment that is not subject to the precise experimental control as visible in bioreactors. The scope of environmental biotechnology is enlarged to address issues such as environmental monitoring, restoration of environmental quality, waste management, substitution of the nonrenewable resource base with renewable resources, etc. Environmental biotechnology has been successfully applied to major oil spills and numerous contaminated sites, thereby reducing the environmental damage caused by pollutants and decreasing the risk they pose to human health. Different methods of aerobic and anaerobic biological treatments are being successfully employed in the treatment of domestic and industrial waste. It has played significant role in improving human environment by converting complex organic matter from the waste into simpler one that can be recycled (e.g. single-dwelling septic tanks, municipal trickling filters, activated sludge system, lagoons, etc.).

Recombinant DNA methodology can also be effectively used to genetically engineer microorganisms with the capacity to degrade many of these compounds.

BIOREMEDIATION

Bioremediation refers to the use of biological systems to reduce pollution from air, aquatic or terrestrial systems. Plants and microorganisms are the biological systems generally used for bioremediation. It is an emerging technology that offers significant potential for cost-effective and eco-friendly treatments of contaminated water and soils. Bioremediation is one of the few processes that actually destroys pollutants, converting organic matter into CO_2, water, chloride and other materials.

The techniques employed for bioremediation can be broadly classified as discussed below.

Ex-Situ Bioremediation

It involves design, construction and operation of an engineered reactor above the ground. It involves the removal of waste materials and their collection at a place to facilitate microbial degradation. The process has many limitations and disadvantages.

In-Situ Bioremediation

Natural Bioremediation
This action is initiated other than monitoring the sites for the degradation by microorganisms.

Engineered In-Situ Bioremediation
The materials are added to stimulate growth of organisms. The compounds used successfully for bioremediation are petroleum hydrocarbons, benzene, toluene, ethyl benzene, xylene, alcohols, kerosene, esters, polynuclear aromatic hydrocarbons, creosote, chlorinated aromatic hydrocarbons, chlorofluorocarbons, chlorinated benzene, polychlorinated benzene, polychlorinated biphenyls, phenols, chlorinated phenols, nitroaromatics, pesticides, alachlor, atrazine, etc.

MANAGEMENT OF ENVIRONMENTAL POLLUTION

The strategic approaches for management of environmental pollution are based on following five considerations:
1. CO_2 Reduction of atmospheric
2. Sewage treatment by bacteria and algae
3. Eutrophication and phosphorous pollution
4. Management of metal pollution
5. Immobilized cells in the management of pollution

Reduction of Atmospheric CO_2

There are mainly two types of approaches adopted for reduction of carbon dioxide in the atmosphere:
1. Photosynthesis
2. Biological classification

Photosynthesis

The utilization of CO_2 for photosynthesis by plants is the most significant process to reduce CO_2 content in the atmosphere.

$$6CO_2 + 6H_2O \xrightarrow[\text{Chlorophyll}]{\text{Sunlight}} C_6H_{12}O_6 + 6O_2 \uparrow$$

Biological Classification

Certain organisms present in the deep sea such as corals, green and red algae, etc., are capable of storing CO_2 through a process of biological classification.

The overall bioprocess may be represented as follows:

$$H_2O + CO_2 \longrightarrow H_2CO_3$$
$$Ca^{2+} + 2HCO_3^- + OH^{-1} \longrightarrow CaCO_3 + CO_3^{2-} + H^+ + H_2O$$
$$\text{(insoluble)} \quad \text{(soluble)}$$

Sewage Treatment by Bacteria and Algae

Most algae are quite flattened, which maximize the surface area for absorbing water, minerals and sunlight. The vegetative form of algae is represented by thallus. For single-celled algae, the thallus is just the single cell. In multicelled algae, the thallus represents the entire continuous organism. In most of the wastewater treatment plants of recent times, degradation of organic matter is effected through bacterial activity. When bacterial activity is promoted by aerobic bacteria (bacteria that survive in oxygenated environment)—involving arrangements supplying oxygen to water—the process is known as *aerobic treatment*. The process of supplying air may be different, such as through fixed or floating aerators or through diffused aeration. The purpose is simply to aerate water so that bacteria survive and eat away the organic matter and clean up water (Fig. 3.1).

Anaerobic wastewater treatment is the biological treatment of wastewater without the use of air or elemental oxygen. Many applications are directed towards the removal of organic pollution in wastewater, slurries and sludges. The anaerobic microorganisms convert organic pollutants into a gas containing methane and carbon dioxide, i.e. *biogas*.

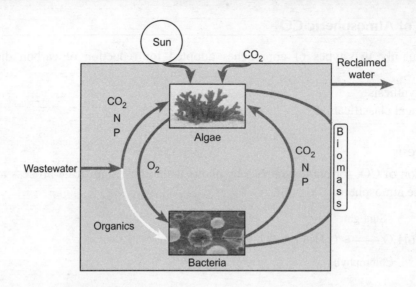

Figure 3.1 Sewage treatment by bacteria and algae.

Eutrophication and Phosphorus Pollution

Eutrophication is a scientific term that describes the overfertilization of lakes with nutrients and the changes associated therewith. In other terms, it is the 'bloom' or great increase of phytoplankton in a water body. Negative environmental effects include anoxia or loss of oxygen in water with severe reduction in fish and other animal populations. The plants have high nitrogen requirement, so addition of nitrogen compounds stimulates plant growth. Ecosystems having more nitrogen than the plants would require are known as *nitrogen-saturated ecosystems*. Such kind of saturated terrestrial ecosystems contribute both inorganic and organic nitrogen to freshwater and marine eutrophication, where nitrogen is also a limiting nutrient. Phosphorus is much more important as a limiting nutrient in aquatic systems. When present in very high levels, phosphates are toxic to people or animals. In freshwater lakes and rivers, phosphorus is often found to be the growth-limiting nutrient, as it is present in low concentration as compared to the needs of plants. If excessive amounts of phosphorus and nitrogen are added to the water, they facilitate algae and aquatic plants to grow in large quantities. When these algae die, bacteria decompose them and use up the oxygen liberated. This process is referred to as *eutrophication*. The loss of oxygen in the bottom waters can free phosphorus previously trapped in the sediments, further increasing the available phosphorus (Fig. 3.2).

Management of Metal Pollution

Environment pollution with heavy metals, such as lead, cadmium, mercury, etc., causes several toxic manifestations, including cancer, in living organisms. Some characteristic features of metal pollution are disussed ahead.

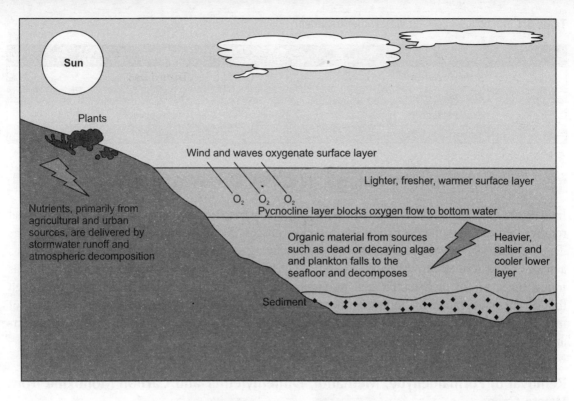

Figure 3.2 Eutrophication process.

1. Bioaccumulation
2. Biomagnifications
3. Biomethylation

Plants and microorganisms are also used for control of metal pollution. As small but significant amounts of mercury still remain in water, existing techniques of mercury removal by precipitation or ion exchange are expensive and not sufficiently efficient. Researchers discovered that many bacteria develop high tolerance to heavy metals, which relate to the binding of these metals to proteins, e.g. metallothionein that binds mercury. As naturally thriving mercury-tolerant bacteria are rare and cannot be grown easily in culture, researchers at Cornell University, New York, inscrted the metallothionein gene into *E. coli*. A sufficiently large number of genetically engineered bacteria can thus treat mercury polluted water inside a bioreactor. By using this process, mercury can be removed from polluted water down to a few nanograms per litre. Once the bacteria die, they are incinerated to recuperate the accumulated pure mercury.

Immobilized Cells in Management of Pollution

The technique that uses immobilized cells—particularly microbial whole cells, for the control of environment pollution—is of recent origin. The immobilized cell lines can be effectively used for the treatment of wastewater and for recovery of metal (Table 3.1).

Table 3.1 Immobilized cells for wastewater treatment

Immobilized cells	Pollutants
Arthrobacter sp.	Triethyl lead
Thiosphaera sp.	Ammonia
Escherichia coli	Chlorophenol
Rhizopus arrhizus	Copper

MANAGEMENT OF AIR, WASTE GASES AND WATER POLLUTION

Industrial waste gas treatment systems were originally based on cheap compost-filled filters that removed odours. Despite their short life and slow processing rates, such filters are still in use. The development of *biorubbers*, wherein the pollutants are washed out using a cell suspension was a welcome feature. *Biotrickling filters*—in which the pollutant is degraded by microorganisms immobilized on an inert matrix and provided with an aqueous nutrient film trickling through the device—was another important development of commercial success. The selection of microorganisms that are more efficient against metabolizing pollutants has also led to better air- and gas-purifying biofilters.

Removal of Formaldehyde, Methanol, Dimethylether and Carbon Monoxide from Waste Gases

Bioreactors

A series of biofilters and biotrickling filters are commonly used in bioreactors. The conventional biofilters have a packing volume between 1.0 and 3.4 L, whereas the biotrickling filters contain packing material between 1.0 and 2.0 L. Lava rock is commonly employed as filter bed. The bioreactor systems are operated with a downward flow at room temperature, and fed with a nutrient solution.

Biocatalysts

Two different types of inocula are used for the batch studies. The sludge from the wastewater treatment plant of a formaldehyde resin-producing facility is usually employed. The sludge is also employed as inoculum in almost all bioreactor studies. Additionally, a pure *Oligotropha carboxidovorans* OM5 culture is employed for a series of batch studies. *O. carboxidovorans* OM5 is a carboxidotrophic Gram-negative bacillus able to catalyse the biodegradation of carbon monoxide.

Batch Experiments

The batch assays are carried out in 635 mL glass vials, containing 100 mL inoculum and the nutrient solution used for the bioreactor studies. The pH is adjusted to 7.5, and distilled water is added to

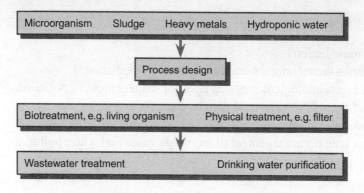

Figure 3.3 Management of water pollution.

make up to 150 mL. After adding the substrate (formaldehyde, methanol, dimethylether or carbon monoxide), the vials are sealed and maintained at 30°C with constant shaking at 200 rpm. The gas samples of the headspace are periodically taken in order to measure the substrate concentration. Pollutant concentrations in the liquid could be calculated by means of Henry's law. All the vials are prepared in duplicate, including for 'blank' and 'control' assays. *Blank vials* are identical to the experimental vials with the highest pollutant concentration but have been previously submitted to sterilization at 120°C for 30 min. *Control vials* also have the same composition but contain distilled water instead of sludge.

Management of Water Pollution

The conventional wastewater treatment methods are being re-examined in view of major changes that have taken over the past few years. Among these are growing problems of worldwide energy and food shortages and nutrients not being removed by conventional secondary processes causing algal blooms and other problems in the receiving waters. Biotreatment systems utilizing living organisms are being preferred since they are ecologically sound, cheap and applicable in areas without land constraints. Filter feeders are promising in this area since they can remove suspended organic matter and bacteria, even in the size range of microns (Fig. 3.3).

Aerobic Biotreatment of Wastewater Pollution

Aerobic bioprocesses have been viewed as effective methods for wastewater treatment to remove biodegradable organics and, to a lesser extent, nutrients and phosphorus. On the basis of microbial development, these processes can be broadly classified into the following two categories:

1. Suspended growth processes
2. Attached growth processes

The fixed film processes offer good protection against toxic substances and provide higher treatment efficiencies. They also cause long retention of biomass solids.

QUESTIONS

1. What is bioremediation?
2. Discuss in brief the techniques employed for bioremediation.
3. Discuss the biotechnological methods used for management of pollution.
4. Write a short note on sewage treatment by bacteria and algae.
5. Write a short note on immobilized cells in management of pollution.
6. Write a short note on management of air, waste gases and water pollution.

CHAPTER 4: Proteins and Proteomics

CHAPTER OUTLINE

- Introduction
- Proteins
- Proteomics
- Questions

INTRODUCTION

Proteomics refers to the identification and quantification of all proteins present in cells and organisms. It is the study of the proteome, the protein complement of the genome. The terms *proteomics* and *proteome* were first coined by Marc Wilkins and colleagues in the early 1990s. *Genomics* and *genome* describe the entire collection of genes in an organism.

After genomics, proteomics is considered to be important in the study of biological systems. It is much more complicated than genomics. While genome of an organism is more or less constant, the proteome differs from cell to cell and varies from time to time. This is because of the fact that distinct genes are expressed in distinct cell types. It means that even the basic set of proteins that are produced in a cell needs to be determined. In the past, this was done by mRNA analysis. It is now established that mRNA is not always translated into protein, and the amount of protein produced for a given amount of mRNA depends on the gene it is transcribed from and on the current physiological state of the cell (Fig. 4.1).

Proteomics includes the characterization and functional analysis of all proteins that are expressed by the genome at a certain moment, under certain conditions. Since expression levels of many proteins strongly depend on complex but well-balanced regulatory systems, the proteome, unlike the genome, is highly dynamic. This variation depends not only on the biological function of a cell, but also on signals from its environment.

Protein chemistry involves the study of structure and functions of proteins and most commonly manifests in the field of physical biochemistry. Proteomics is the study of multiprotein systems, in which the focus is on the interplay of multiple, distinct proteins and their roles as part of longer systems. The difference between protein biochemistry and proteomics is reflected in Table 4.1.

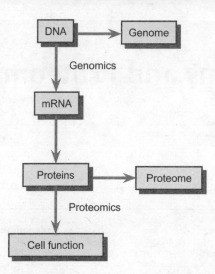

Figure 4.1 Biochemical context of genomics and proteomics.

Table 4.1 Differences between protein biochemistry and proteomics

Protein biochemistry	Proteomics
Individual chemistry in biological context	Complex mixture of genome. Study of protein compliment.
Complete sequence analysis	Partial sequence analysis
Emphasis on structure and function	Emphasis on identification by database matching
Structural biology	Systems biology

PROTEINS

Proteins are organic compounds made up of amino acids arranged in a linear chain and folded into a globular form. The amino acids are joined together by the peptide bonds between the carboxyl and amino groups of adjacent amino acid residues. The sequence of amino acids in a protein is defined by the sequence of a gene, which is encoded in the genetic code. Proteins were first described by the Dutch chemist Gerardus Johannes Mulder and were named by the Swedish chemist Jons Jakob Berzelius in 1838.

Proteins are involved in almost all cellular processes and fulfil many functions. They are involved in enzyme catalysis, transport, mechanical support, organelle constituents, storage reserves, metabolic control, protection mechanisms and osmotic pressure.

Structure of Proteins

Proteins are macromolecules consisting of one or more polypeptides, each polypeptide consisting of a chain of amino acids linked together by peptide bonds. A polypeptide chain folds up, assuming

Table 4.2 Structure, function and uses of important proteins

Proteins	Total no. of amino acids	Molecular mass	No. of polypeptide chains	Function/use
Insulin (human)	51	5800	2	Complex nature and biological regulation of blood glucose level
Lysozyme (egg)	129	13,900	1	Used for degrading peptidoglycan in bacterial cell walls
Human erythropoietin	165	36,000	3	Stimulates red blood cell production
Chymotrypsin	241	21,600	3	Digestive proteolytic enzyme
Haemoglobin	574	64,500	4	Used for gas transport
Tumour necrosis	471	52,000	3	Used as mediator of inflammation and immunity
Hexokinase	800	1,02,000	2	Used for phosphorylating selected monosaccharides

a specific three-dimensional shape that is unique to it. The conformation adopted is dependent on the polypeptide's amino acid sequence, and this conformation is largely stabilized by multiple and weak noncovalent interactions. A structure protein cannot be totally predicted from its amino acid sequence. Its conformation can, however, be determined by techniques such as X-ray diffraction and nuclear magnetic resonance spectroscopy.

The shape into which a protein naturally folds is known as its native conformation. Although many proteins can fold unassisted through the chemical properties of their amino acids, others require the aid of molecular chaperones (proteins) to fold into their native states.
The proteins are classified broadly into the following:
1. Simple proteins
2. Conjugated proteins

Simple Proteins
These consist exclusively of polypeptide chain(s) with no additional chemical components.

Conjugated Proteins
In addition to their polypeptide moieties, these proteins contain one or more nonpolypeptide constituents known as *prosthetic group*. Commonly occurring prosthetic groups in association with proteins include carbohydrates, vitamin derivatives, phosphate groups and metal ions.

The total number of amino acids, molecular mass, number of polypeptides present and functions of some protein are given in Table 4.2.

Primary Structure of Proteins

Polypeptides are linear and unbranched polymers potentially containing up to 20 different monomer types linked together in a precise predefined sequence (Fig. 4.2). The primary structure of a polypeptide is related to its amino acid sequence, along with the exact positioning of any disulphide bonds present.

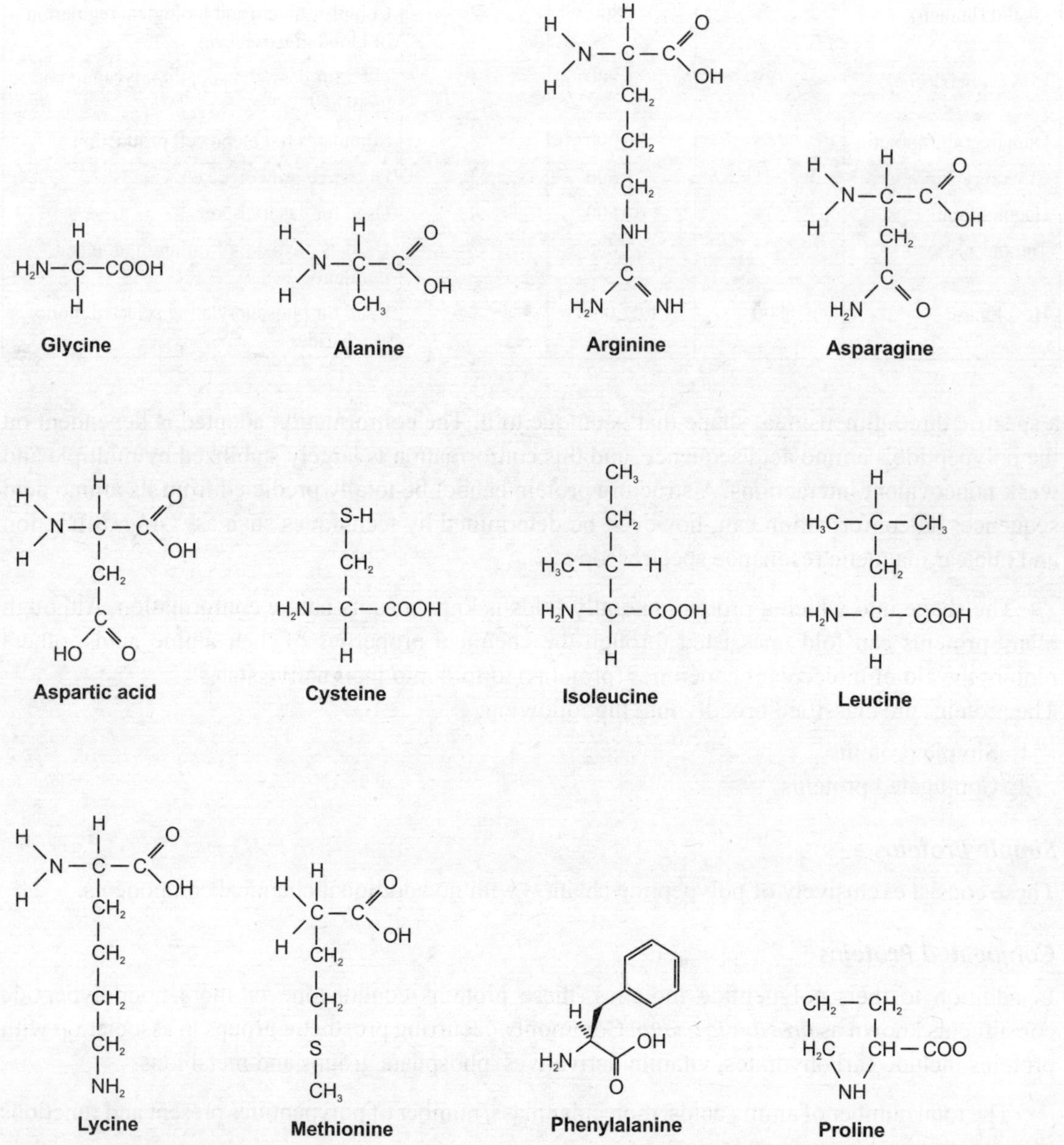

Figure 4.2 Chemical structures of amino acids found in proteins. *(Continued...)*

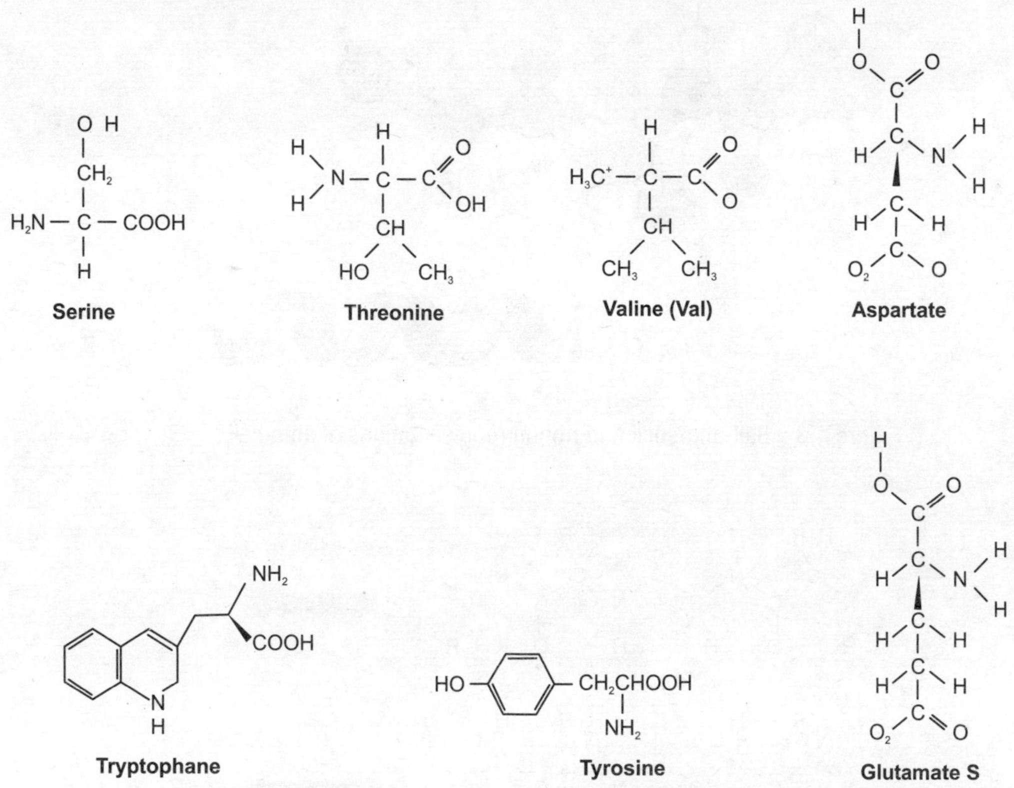

Figure 4.2 Chemical structures of amino acids found in proteins.

Secondary Structure of Proteins

These are regular repeating local structures stabilized by hydrogen bonds. The most common examples are the a-helix, b-sheet and turns. Secondary structures are local. Therefore, many regions of different secondary structures can be present in the same protein molecule. The a-helix contains 3,6-amino acid residues in a full turn, which is approximately 0.56 nm in length along the long axis of the helix. The participating amino acid side chains protrude outwards from the helical backbone. Amino acids that are most conducive with a-helix formation are alanine, leucine, methionine and glutamate. The helical structure is stabilized by hydrogen bonding, with every backbone C=O group forming a hydrogen bond with N–H group, four residues ahead of it in the helix. The stretches of a-helix found in globular (i.e. tightly folded and approximately spherical) polypeptides can vary in length from a single helical turn to greater than 10 consecutive helical turns (Figs. 4.3 and 4.4).

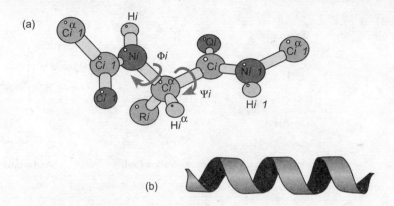

Figure 4.3 Ball-and-stick and ribbon representations of an *a*-helix.

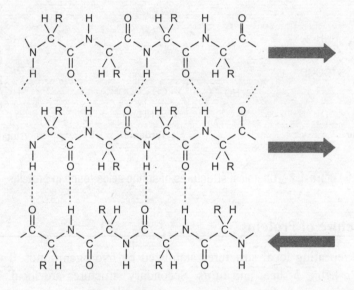

Figure 4.4 Structure of *b*-sheet.

Tertiary Structure of Proteins

The term *tertiary structure* is often used synonymously with the term *fold*. It is generally stabilized by nonlocal interactions, most commonly the formation of a hydrophobic core, but also through salt bridges, hydrogen bonds, disulphide bonds and even posttranslational modifications. The tertiary structure controls the basic function of the protein. A polypeptide's tertiary structure refers to its exact three-dimensional structure, relating the relative positioning in space of all constituent atoms of polypeptide to each other. The tertiary structure of small polypeptides (approximately 200 amino acid residues or less) usually forms a single discrete structural unit. However, when the three-dimensional structure of many larger polypeptides is examined, the presence of two or more structural subunits within the polypeptide becomes apparent.

Quaternary Structure of Proteins

This structure is formed by several protein molecules (polypeptide chains), usually called *protein subunits* that functions as a single protein complex.

Methods of Studying Proteins

A particular protein can be studied by developing an antibody that is specific to that modification. For example, there are antibodies (pan antibodies) that recognize only certain proteins when they are tyrosine phosphorylated. These can be used to determine the set of proteins that have undergone the modification of interest. The posttranslational modification of interest can be determined by subjecting a complex mixture of proteins to two-dimensional gel electrophoresis (2-DE). In this technique, proteins are electrophoresed first in one direction and then in another. This allows small differences in a protein to be visualized by separating a modified protein from its unmodified form.

Recently, another approach called *PROTOMAP* has been developed that combines SDS-PAGE (sodium dodecyl sulphate-polyacrylamide gel electrophoresis) with shotgun proteomics. This enables the detection of changes in gel migration, such as those caused by proteolysis or posttranslational modification. To an important degree, antibodies to particular proteins or to their modified forms have been used as common analytical tools in biochemistry and cell biology studies. The technique such as *ELISA* can be used for quantitative determination of protein. For proteomic study, more recent techniques such as *matrix-assisted laser desorption/ionization (MALDI)* are employed for rapid determination of proteins in particular mixtures.

Most proteins function in collaboration with other proteins. One goal of proteomics is to identify which proteins interact. Several methods are available to probe protein–protein interactions. The traditional method is yeast two-hybrid analysis. New methods are protein microarrays, immunoaffinity chromatography, followed by mass spectrometry (MS), dual polarization interferometry and experimental methods, such as phage display and computational methods.

The laboratory should have access for undertaking the following assays:

1. A rapid, reliable assay for the target protein
2. Purity determination
3. Total protein determination
4. Assay for impurities that must be removed

The importance of a reliable assay for the target protein cannot be overemphasized. The purity of the target protein is most often estimated by SDS-PAGE, capillary electrophoresis, reversed phase chromatography or mass spectrometry. For large-scale protein purification, assay of target protein and critical impurities is often essential.

PROTEOMICS

Proteomics is a systematic research approach aiming to provide the global characterization of protein expression and function under given conditions. Proteomics techniques have been widely adopted in the areas of biology and medicine. Like any new technology, when it is first introduced, proteomics has also been promoted with much hope and promise. Proteomic technology is used to elucidate protein changes between healthy and diseased states. The application of proteomic technique to the field of medicine is slowly transforming the way biomarker discovery is conducted.

Clinical Applications of Proteomics

In medical research, it has become increasingly apparent that cellular processes, particularly in diseases, are determined by multiple proteins. Hence, it is important not to focus on one single gene product (one protein) but to study the complete set of gene products (the proteome). In this way, the multifactorial relations underlying certain diseases may be unravelled potentially identifying therapeutical targets. For many diseases, characterization of the functional proteome is crucial for elucidating alterations in protein expression and modifications.

Proteomics in Lab Diagnosis

The rapid spread of proteomic technology—which principally consists of 2-DE with in-gel protein digestion of protein spots and identification by MS—has provided an explosive amount of diagnostic results. Proteomics plays an important role in the translation of genomics to clinically useful applications, especially in the areas of diagnostics and prognostics. The emergences of new techniques such as protein and peptide microarrays are useful in laboratory diagnosis. The role of proteomics in the development of new laboratory diagnostics and the implications for routine diagnosis and monitoring of diseases are much evident today. The application of high-resolution mass spectrometers and other quantitative proteomic methodologies to laboratory diagnosis is a reality today.

In the coming decades, proteomic technologies shall broaden our understanding of the underlying mechanisms and further our ability to diagnose, prognosticate and treat many diseases.

Proteomics in the Discovery of Protein Biomarkers

Proteomics, with its high-throughput and unbiased approach to the analysis of variations in protein expression patterns (actual phenotypic expression of genetic variation), promises to be the most suitable platform for biomarker discovery. Biomarkers of drug efficacy and toxicity are key factors in the drug development process. The success of biomarker discovery effort depends upon the quality of samples analysed, the ability to generate quantitative information on relative protein levels and the ability to readily interpret the data generated. Proteomic technology is widely used in biomarker discovery and pathogenic studies. Mass spectral-based proteomic technologies are ideally suited for the discovery of protein biomarkers in the absence of any prior knowledge of quantitative changes in protein levels.

Proteomics in Neurology

In neurology and neuroscience field, proteomics has applications in neurotoxicology and neurometabolism, as well as in the determination of specific proteomic aspects of individual brain areas and body fluids in neurodegeneration. Investigations of brain protein groups in neurodegeneration—such as enzymes, cytoskeleton proteins, chaperones, synaptosomal proteins and antioxidant proteins—are in progress as phenotype-related proteomics.

Proteomics in Diagnosis

In the diagnosis and treatment of kidney disease, a priority is the identification of disease-associated biomarkers. Combining important analytical techniques of 2-DE with more sophisticated techniques such as MS has enabled considerable progress in cataloguing and quantifying proteins. This type of measurement has been made possible in urine and various kidney tissue compartments in both normal and diseased physiological states.

Given the dynamic range and complexity of proteomics, completely defining the proteome in various biological compartments (e.g. tissues, serum and urine) in both health and diseases remains a major challenge.

Proteomics in the Study of Tumour Metastasis

The molecular and cellular mechanisms underlying tumour metastasis, which is the dominant cause of death in cancer patients, are still difficult to describe. The identification of protein molecules with their expressions correlated to metastatic process is helpful in understanding the metastatic mechanisms and thus, facilitating the development of strategies for the therapeutic interventions and clinical management of cancer. The combination of proteomics with other experimental approaches in cell biology, biochemistry, chemistry and molecular genetics together with the development of new technologies in existing methodologies shall have focused applications in studying cancer metastasis.

Proteomics in Neurotrauma

Proteomics provides a valuable approach to evaluate the posttraumatic central nervous system (CNS). Although a comprehensive assessment of all methods for protein analysis cannot be provided, some of the newer proteomic technologies that have propelled this field into the limelight and that are available to most researchers in neurotrauma are described.

Neurotrauma results in complex alterations to the biological systems within the nervous system, and these changes evolve over time. Exploration of the 'new nervous system', which follows injury, requires methods that can both fully assess and simplify this complexity. Proteomics is likely to be very useful for developing diagnostic predictors after CNS injury and for mapping changes in proteins after injury in order to identify new therapeutic targets.

Proteomics in Urological Cancer Research

The comparison of the proteins present in a diseased tissue sample with the proteins present in a normal tissue sample is possible with proteomics analysis. Proteins that are altered either qualitatively or quantitatively between the normal and diseased sample are likely to be associated with the disease process. These proteins after identification can serve the purpose of diagnostic markers for early detection of the disease. Alternatively, they can be prognostic markers to predict the outcome of the disease, or they may be used as drug targets for the development of new therapeutic agents.

Proteomics in the Study and Treatment of Autoimmune System

Proteomics provides a powerful tool to characterize autoreactive B cell responses in diseases, including rheumatoid arthritis, autoimmune diabetes, multiple sclerosis and systemic lupus erythematosus. Autoantibody profiling may be useful in classification of individual patients and subsets of patients based on their 'autoantibody fingerprint', discovery and characterization of candidate autoantigens, examination of epitope spreading and antibody isotype usage in addition to tailoring antigen-specific therapy.

Proteomics in Diabetes Research

The techniques involved in proteomics allow the global screening of complex samples of proteins and provide qualitative and quantitative evidences of altered protein expression. This lends to the investigations of the molecular mechanisms underpinning disease processes and the effects of treatment. When proteins undergo nongenetically determined alterations, such as alternative splicing or posttranslational modifications (phosphorylation or glycosylation), it may affect their function. Although abnormalities in splicing or posttranslational modifications may cause a disease process, they can also be a consequence. Diabetes has high blood glucose, which glycosylates hundreds or thousands of proteins, including HbA_{1c}, which is used to monitor diabetes.

Proteomics in Cardiovascular Research

The analysis undertaken by proteomic technologies at the organ, subcellular and molecular levels have revealed dynamic, complex and subtle intracellular processes associated with heart and vascular diseases. The power and flexibility of proteomic analyses, which facilitate protein separation, identification and characterization, should hasten our understanding of these processes at the protein level. Evolution of proteomic techniques has permitted detailed investigations into molecular mechanisms underlying cardiovascular diseases and facilitating identification not only of modified proteins but also of the nature of their modifications.

Proteomics in Investigating Plants

Proteomics offers novel techniques to study all facets of protein structure and functions in plants. The application of proteomics in plant pathology is more evident with techniques such as 2-DE and MS being used to characterize cellular and extracellular virulence and pathogenicity factors. It is useful to identify changes in protein levels in host plant upon infection by pathogenic organism. The whole genome sequences are now available for plants, such as for *Arabidopsis thaliana, Oryza sativa* and others.

QUESTIONS

1. What is protein engineering?
2. Write in brief about the denaturation of proteins.
3. Write a note on differences between protein biochemistry and proteomics.
4. Differentiate primary and secondary structures of proteins.
5. Define proteomics and give its applications.

CHAPTER 5: Animal Cell and Tissue Biotechnology

CHAPTER OUTLINE

- Introduction
- Primary and Established Cell Lines
- Kinetics of Cell Growth
- Animal Tissue Culture: Media Requirements

INTRODUCTION

The animal cell and tissue culture is a technique wherein the tissues/cells taken out from the body of animals are grown in a suitable medium and maintained in laboratory. It is an amalgamation of art and science, covering wide range of activities that make a rough distinction between fundamental and applied research. At present, tissue culture is an indispensable tool, particularly in the fields of medical and agricultural sciences.

The foundation of animal cell and tissue culture was laid at the beginning of the 20th century, when cell was shown to divide and survive in vitro. The initiation of animal cell and tissue culture was, however, reported by Ross Harrison (1907) using frog as a source of tissue. Alexis Carrel (1912) used tissue and embryo extract as culture medium. For around 50 years thereafter, tissue explants, rather than cells, were used for culture techniques. After 1950, mainly dispersed cell cultures were utilized. If the explant maintains its structure and function in culture, it is described as an *organotypic culture*. Irrespective of the animal tissue from which it is derived, it is described as a *histotypic culture*.

Advantages and Limitations of Tissue Culture Methods

There are several advantages and limitations of animal tissue culture. Due to certain limitations, this technique should not be used indiscriminately.

Advantages

1. The homogeneity of cell types is retained (achieved through serial passages).
2. The behaviour of cells in culture is easily interpreted and regulated properly.

3. The legal, moral and ethical questions of animal experimentation can be avoided.
4. Smaller quantities of reagents are required.

Limitations

1. It needs controlled physiochemical environment (pH, temperature, osmotic pressure, O_2, CO_2, etc).
2. For same quantity of animal tissue, the technique is expensive. Therefore, reasons for its use should be compelling.
3. It indicates unstable aneuploid chromosome constitution.

PRIMARY AND ESTABLISHED CELL LINES

Primary Cell Lines

Besides the specialized cells of the tissue, a piece of tissue from the organism is usually quite complex and contains connective tissue cells, a variety of blood cells and reticuloendothelial cells. When a culture is established, all these survive for at least a day or two. In usual culture conditions, many cells begin to migrate almost immediately. The first cells to do so are often macrophages followed by fibroblasts that migrate in a radial manner from the explant. Many of the specialized cells, however, remain immobile in the connective tissue stroma. Nerve cells, for example, usually remain in the explant, although axons may migrate from it. The subsequent fate of these cells varies enormously. Many of them are quite short-lived, e.g. most of the blood cells die and disappear from the culture within 2 or 3 days. Other cells, such as neurons and muscle cells, frequently persist in culture, sometimes for months without division and then eventually die. Still other cells begin to divide rapidly and continue to do so for some time. However, many of these also die after certain period that may vary from few weeks to months.

If the cells multiply repeatedly for a long time, they can often be 'passaged'. It is usually achieved by first obtaining the cells in suspension and then inoculating them into a new culture vessel along with fresh medium. As cells are 'passaged' in this way, the culture is known as *primary cell lines*.

Advantages

The advantages of primary cell cultures are as follows:

1. The expense and inconvenience of maintaining an established cell stock are obviated.
2. The culture can be sustained well in medium of relatively simple composition.
3. Massive quantities of tissue are obtained conveniently for short-term studies.
4. They are particularly suitable for vaccine production, since the probability of in vitro transformation of cells to malignancy is minimized.

Limitations

The use of primary cell cultures has following inherent limitations:

1. Long-term experiments concerning the biological properties of cells cannot be carried out.

2. The cultures may be contaminated with latent viruses (e.g. foamy virus in monkey kidney tissues).
3. The degree and range of sensitivity to viral infection may exceed as compared to common established cell lines.
4. Mixed populations of cells may confuse interpretation of experimental findings.

Established Cell Lines

Frequently, primary cell lines go on dividing at a high rate for a long time and can be passaged repeatedly. Sometimes, cell lines can be cultured for such a long time that they apparently develop the potential for subculturing in vitro. Such cell lines are called *established cell lines*. A cell line is not designated as an 'established cell line' unless, it has been subcultured at least 70 times at an interval of 3 days between two subcultures.

The transition from primary cell line to established cell line is smooth and gradual in some instances. But in others, the latter are established from a primary cell line by a dramatic event called *cell alteration* or *transformation*. During this process of transformation, a few rapidly growing colonies of transformed cells appear that quickly outgrow the rest of the culture and become the predominant cell type. The established cells formed this way differ in many respects from the primary cell lines from which they are derived. The primary cell lines have the normal number of chromosomes, whereas, such established cell lines almost invariably have an unusual number. There are some interesting species differences in the stability of cells. The human fibroblast is a classical example of particularly stable primary cell lines, and it never undergoes spontaneous transformation to an established cell line. On the other hand, mouse fibroblasts almost invariably undergo a smooth transition to form established cell lines.

Established cell lines behave in a remarkably similar manner irrespective of their origin. Their distinct characteristics are given below.

1. The established cell lines are invariably aneuploid.
2. They grow into much higher densities than primary cell lines.
3. They have similar nutritional requirements, irrespective of their origin.
4. They have short-doubling timings, of the order of 12–20 h.
5. The cell lines often grow from dilute inocula.
6. The established cell lines usually show much evidence of spatial orientation.
7. It is unusual for established cell lines to show obvious specialized functions.
8. Most of the established cell lines can be established in suspension cultures, whereas, it is exceptional for primary cell lines to grow in suspension.

KINETICS OF CELL GROWTH

Established Cell Lines

The cells in culture display very much the same type of growth pattern as microorganisms. In particular, they show the classical growth demonstrated by cultures of bacteria, yeast and protozoa. The culture exhibits the following phases:

Lag Phase

This is the phase when practically no growth occurs, lasting for some hours to a few days before growth commences.

Log Phase

During this phase, growth proceeds steadily with the population doubling every 15–20 h in case of fast-growing cells.

Stationary Phase

At the end of log phase, maximum population is reached, and the growth becomes stationary. During log phase, the cell population increases according to the formula

$$N = N_0 2kt$$

$$\log N = \log N_0 + kt \log 2$$

(where, N_0 = initial inoculum, N = cell number at t hours, k = regression constant).

The factor kt is equal to the generation number (n), i.e. the number of generations involved in the increase by doubling at each generation from N_0 to N. It follows that the mean generation time (T), the time for the population to double, is the inverse of k, i.e. $T = 1/k$.

The equation can be written as

$$\log N = \log N_0 + n \log 2$$

$$n = (\log N - \log N_0)/\log 2$$

Since $1/\log 2 = 3.32$, this can be written in the following form:

$$n = 3.32 (\log N - \log N_0)$$

This calculation assumes that cells in a population divide, and they are at approximately the same rates.

The growth rate of cultured cells may vary to some extent from one cell strain to another. Considerable variation is observed under the influence of environmental conditions such as pH, temperature and osmotic pressure.

Primary Cell Lines

Primary cell lines do not behave in quite the same way as established cell lines. When established cell lines remain stationary, the medium in which they have been grown is usually inadequate for the maintenance of a fresh inoculum of cells. It is observed that cultures of primary cell lines stop growing before the medium is exhausted. It is mainly because of the phenomenon called as *contact inhibition*. This is a very common property of primary cell lines. The cell movement is affected by contact inhibition. In addition to this, cell division, DNA synthesis, RNA and protein synthesis are also much reduced or eliminated due to contact inhibition. The effect of contact inhibition on the kinetics of growth of primary cell lines is that they exhibit ideal kinetics only when they are in a very dilute inoculum. When transformed, the primary cell lines show absence of great diminution of contact inhibition, and growth of transformed cells reaches higher densities.

Interaction of Cells

Besides contact inhibition, another interesting interaction is the ability of similar cells to recognize and adhere to each other. When amphibian embryonic cells were disaggregated in calcium- and magnesium-free salt solution, separated cells—when allowed to mix together—first adhered in a random manner and then sorted themselves out into groups of like-cells. Eventually, the entire embryo was reconstructed and the same phenomenon was demonstrated with mammalian and avian cells. In this context, two different theories were put forward: (a) a kind of intercellular cement sticks like-cells to each other and (b) adhesion phenomenon is an intrinsic property of the cell membrane.

Metabolism of Cultured Cells

As in the tissues of intact organism, the metabolism of glucose in cultured cells proceeds by way of glycolytic and Krebs pathways. Many cells exhibit a marked tendency to accumulate lactic acid and keto acids in the medium. The cultured cells with the aid of necessary enzymes can build nucleic acids from simple compounds in a medium (e.g. formate, glycine and bicarbonate). The cultured cells have a lot of variable needs for oxygen. To a great extent, the energy requirements can be met by the breakdown of 6-carbon carbohydrate (glucose) into 3-carbon compounds (lactic acid and pyruvic acid) that require no oxygen. The cells may be raised in the complete absence of oxygen for rather short periods of time. Tissue culture cells are capable of synthesizing lipids from simple chemical substances such as acetate. Some cells are also capable of synthesizing cholesterol.

Genetics of Cultured Cells

While primary cell lines usually retain their diploid karyotype, transformed cell lines exhibit variation in karyotype. Shortly after transformation, the incidence of tetraploid cells increases. Aneuploid cells then make their appearance. The stable established cell lines commonly have an aneuploid karyotype with a wide spread of chromosome numbers. The emergence of aneuploidy is mainly because of nondisjunctive cell division. The morphology of the individual chromosome changes in established cell lines. It is because of chromosome breaks, fusion and translocations.

Applications of Cell Cultures

Animal cell cultures have wide spectrum of applications. They serve as model systems for biochemical, physiological and pharmacological studies and the production of growth factors, enzymes, blood factors, interferons, monoclonal antibodies, vaccines and hormones. Some important processes investigated through animal cell and tissue cultures are summarized in Table 5.1.

Table 5.1 Processes investigated in cell and tissue cultures

Type of cell	Process investigated
Monocytes and macrophages	Pinocytosis and phagocytosis
Blood lymphocytes	Karyotype analysis for detection of genetic defects in humans
Myeloma cells and B lymphocytes	Purification and characterization of specific membrane proteins (e.g. adrenergic and dopamine receptors)
Primary monkey kidney	Production of polio vaccines, hormone secretion
Kidney tubule epithelial cells	Differentiation of monolayers, monoclonal antibody production, electrical and vectorial transport of solutes
Kidney epithelial cells	Investigation of relationship between membrane polarity and budding properties of enveloping RNA viruses
Normal and transformed fibroblasts	Surface adherence properties of normal and malignant cell membranes
Mouse fibroblasts	Acute and chronic toxicity testing and metabolism of xenobiotics, vaccine production
Fibroblasts, mammalian brain cells	Identification of chemicals capable of including in chromosome aneuploidy
Transformed leucocytes: fibroblast either lymphocytes	Infection of cells with Sendai virus to produce α and β interferons
Transformed HeLa cells, mammalian cells	Radiation therapy and the design of radiosensitizers and radioprotectors

The major problem associated with the isolation of free cells and cell aggregates from organs is related to releasing the cells from their supporting matrix without affecting the integrity of the cell membrane. Following methods are used to overcome this hurdle.

Mechanical

It involves forcing of the tissue through cheese or silk cloth or shaking the tissue with glass beads in an appropriate buffer solution. This technique causes considerable cell damage and results in a low cell yield.

Biochemical

This technique is useful in overcoming the problems caused by mechanical dissociation. The enzymes collagenase and hyaluronidase in a calcium-free medium are used for hepatocytic isolation. The liver is sliced and incubated with these enzymes.

ANIMAL TISSUE CULTURE: MEDIA REQUIREMENTS

Culture Medium

Despite vast use of chemically defined media in tissue culture, it is still necessary in most cases to depend more on naturally occurring substances derived from the organisms. The naturally occurring ingredients commonly used in media preparation are: (1) blood plasma, (2) blood serum, (3) tissue extract and (4) complex natural media.

Blood Plasma

The first tissue culture was prepared in clotted frog lymph substituted with a coagulum prepared from chicken plasma. It was observed that plasma provided a complete nutrients in which cells could survive and multiply for extended period under conditions that resembled in many respects to those found in the body.

The plasma is obtained by centrifugation of whole blood before coagulation takes place. The tissue is then placed in a small quantity of the plasma. The coagulation is encouraged by the addition of a small amount of tissue extract or thrombin, because the cells in culture require a solid support for continued growth and activity. For the preparation of culture, plasma from the adult chicken is preferred to mammalian plasma as it forms a clear, solid coagulum even when diluted several times. Mammalian plasma is either too opaque for good optical work or else it fails to produce solid clots. The blood is obtained from the wing, heart or carotid artery of fowl, whereas in case of mammals it is procured from carotid artery and heart.

The use of plasma in culture preparation has the following distinct advantages:
1. It provides nutritive substrate and a supporting structure for many types of cultures.
2. It offers protection to cells and tissues from excessive traumatic damage during subculture.
3. It provides a matrix for new cells during the repair of injury in the body.
4. It also protects cells and tissues from sudden changes in the environment (at times of fluid change).
5. It provides localized pockets of conditioned medium around cells.
6. It provides a means of conditioning the surface glass for better attachment of cells.

Blood Serum

Blood serum (*plasma minus fibrinogen*) with or without other nutritive substances may be used either as the entire culture medium or as the fluid phase of a medium consisting partly of a plasma coagulum. For many years, it was assumed that whole serum was toxic and it was only plasma useful as a supportive structure and that the nutritive requirements of the cells were met with the embryo extract that was usually added to the medium. Eventually, it was established that blood serum may be effectively used in preparation of medium.

In 1928, des Ligneris reported the successful cultivation of many mammalian tissues in diluted serum. Later, Parker (1933, 1936) undertook cultivation of chick tissues in serum. Simms (1936) and Sanders (1942) introduced an ultrafiltrate of serum used as a basal medium for many purposes, including for the propagation of viruses. Fischer and coworkers (1948) proved the importance of the low molecular weight growth factors provided by serum. Harris (1959) concluded that medium 199 and NCTC 109, as well as, the simpler basal medium of Eagle (1955) are all deficient in one or more factors that occur in serum dialysate and are essential for the growth and maintenance of chick skeletal muscle fibroblasts. It was, thus, established that serum provides some of the growth factors or some of the physical conditions, or both, that are presently lacking in synthetic media.

Preparation of chicken serum

The fluid plasma from which the serum is prepared should be completely coagulated. The plasma is coagulated deliberately by adding to each tube a drop or two of embryo tissue extract or an equivalent amount of thrombin and leaving the tubes to incubate for several hours at 37°C. The coagulated plasma is broken up into fragments, and it is ground in a mortar with sterile quartz sand. After grinding, the serum is separated by centrifugation.

Preparation of mammalian serum

The mammalian blood is left at room temperature for an hour. The clot is removed by a glass rod and then centrifuged for 30 min at 3000 rpm. Lastly, the serum is separated.

Serum-free medium

The advantages and limitations of serum-free medium are briefly summarized as follows:

Advantages

1. It has the ability to make a medium selective for a particular cell type, since each cell type appears to require a different recipe.
2. It has high degree of purity of reagents and water.
3. It prefers high degree of cleanliness of all apparatus.

Limitations

1. The quality of serum varies from batch to batch and deteriorates within one year. Therefore, every batch of serum needs fresh testing.
2. If more than one cell type is used, each may require different serum batch. Therefore, many batches are to be maintained and coculturing may be difficult.
3. Serum increases the cost of the medium.
4. When cell culture is used for downstream processing to recover cell products, the presence of serum is an obstacle to purification.
5. Serum may encourage undesirable growth and may even inhibit growth in some cases.

In preparation of medium, a known recipe for a related cell type may be used. Until the medium is optimized, the individual constituents of the serum may be altered. However, the development and assessment is a time-consuming process and in some cases, it takes about 3 years for the development of a new medium.

It is possible to use an existing medium, such as RPMI 1640 or Ham's F-12 or DMEM (Dulbecco's modified minimal essential medium). The constituents, such as selenium, albumin, insulin, androgen, hydrocortisone, oestrogen, etc. may be employed for manipulation.

Tissue Extracts

The extract from embryo tissue has remarkable action in promoting cell growth and multiplication in cultures of connective tissues. The active fraction of the extract is obtained by precipitation with carbon dioxide. The activity is concentrated largely in the protein portion comprising of nucleoproteins and glycoproteins. Active nucleoprotein fractions from adult chicken heart, liver, brain and spleen have been used, but no indications of organ specificity were observed. The activity of the fraction also does not depend on its total nucleic acid content or on the age of the individual from which it is prepared. The proteases and higher molecular weight protein degradation products also possess very potent growth-promoting properties. The growth-promoting activity appears to be associated with fractions containing high concentration of nucleoproteins of ribonucleic acid type. Fractions with high content of deoxyribonucleic acid are less active.

The chick embryo extract is prepared from 10-day-old embryos (before the calcifying mechanism becomes too active). The embryos are removed from the egg and homogenized in a motor-driven homogenizer. Six to eight embryos and a measured quantity (approximately 2 mL/embryo) of balanced salt solution (BSS) are processed together. After homogenization, it is centrifuged and further diluted 10–20 times. Embryo extract may be stored indefinitely after it has been dried in the frozen state.

Complex Natural Medium
Supplemented Hank–Simms' medium
Weller and coworkers (1952) in their work with polioviruses used a combination of three parts of Hank's balanced salt and one part of Simms' ox serum ultrafiltrate in the preparation. For roller tube cultures of various human and animal tissues (embryonic, infant and adult), the complete medium consists of Hank–Simms' solution (85%), beef embryo extract (10%), horse serum inactivated at 56°C for 30 min (5–20%), streptomycin (50 mg/mL) and penicillin (50 mg/mL).

Supplemented bovine amniotic fluid medium
Milovanic and coworkers (1957, 1958) used the medium consisting of bovine amniotic fluid (37.5%), horse serum inactivated at 56°C for 30 min (20%), bovine embryo extract (5%), Hank's BSS (37.5%), streptomycin (100 mg/mL), penicillin (100 mg/mL) and mycostatin (100 mg/mL).

Serum-supplemented yeast extract medium
Various human cell lines and strains from different species have been successfully grown in this medium. The medium contains yeast extract medium (76 parts), human serum (20 parts) and sodium bicarbonate solution (4 parts).

Serum-supplemented lactalbumin hydrolysate medium
This medium consists of human/ox serum (10–20%). Earle's saline containing lactalbumin hydrolysate (0.5%) and yeast extract (0.1%).

Other Tissue Culture Media
Medium no. 612
Healy and coworkers attempted to bring it near the physiological range by increasing very considerably the levels of reducing agents present in the medium namely, cysteine (2600 times), glutathione (200 times) and ascorbic acid. The medium was designated as medium no. 612. Cysteine

and glutathione increase the rate of cell multiplication and improve the appearance of the cells. The ascorbic acid has no apparent effect.

Medium no. 635

The medium no. 612 without purines, pyrimidines, as well as, adenosine triphosphate, adenylic acid, ribose and deoxyribose constitutes medium no. 635. This medium gives a better growth response as compared to medium no. 612, but the cultures rarely survive for more than 40 days.

Medium no. 858

This medium includes all the deoxyribonucleosides, sodium glucuronate and coenzymes, in addition to medium no. 635. This medium contains only L-form of amino acids and yields 10-fold increase in the population of culture cells within 7 days. Addition of 10–20% of horse serum increase yields up to 30 times.

Medium no. 866

It is identical to medium no. 858 with supplementation of three fatty acids, namely: linoleic acid, linolenic acid and arachidonic acid. This medium does not cause any significant growth of culture cells.

Substrate Surfaces and Treatments

Disposable Plastic as a Substrate

Polystyrene is often used as the most common and cheapest substrate, which can be discarded after single use. Other plastics used for tissue culture include: polyvinyl chloride (PVC), polytetrafluoro ethylene (PTFE) or teflon, polycarbonate, thermanox (TPX), etc. This plasticware is supplied sterile and is not autoclavable. However, it is hydrophobic and needs to be treated by gamma irradiation.

Palladium and Metallic Surface as Substrates

The glass or plastic is employed as a substrate for majority of animal cell and tissue cultures. Palladium deposited on agarose (using electron microscopy shadowing equipment) is used as a substrate for growth of fibroblasts. The stainless steel discs and other metallic surfaces are also used in many cases.

Treatment of Substrate Surfaces

The surfaces of substrate are treated to improve cell attachment. The treatment is commonly done either with culture medium meant for another culture or with purified fibronectin added to the medium or with collagen. The coating with collagen can be achieved, using rat tail collagen or other commercially available form, and pouring collagen solution over the surface of the dish, draining the excess solution, and permitting the residue to dry. Thereafter, the substrate may be sterilized under UV and used either immediately or stored. An inert coating of collagen (or gelatin or poly-D-lysine), a monolayer of an appropriate cell type, may also be used to provide a matrix for the maintenance of some specialized cells.

Feeder Layers on Substrate

The substrate surface is a special type of monolayer of cells that are used partly, to supplement the culture medium for growth and partly, for conditioning of the substrate by cell products. These monolayers are also called as *feeder layers*, since they also feed the growing cultures. The feeder layer may consist of mouse embryo fibroblasts, normal fetal intestine, etc. They are effectively used for selective growth of breast and coelomic epithelium, central and peripheral neurons and also for growing cells used for the production of transgenic animals through transfection.

Gas Phase for Tissue Culture

For growth of animal cell and tissue cultures, a gas phase consisting of oxygen and carbon dioxide is also required, even though cultures may vary in their oxygen requirement. These gases also influence pH and HCO_3^- ion concentration of the culture.

Processing of Culture

The apparatus and liquids meant for tissue culture should be reserved separately for the purpose and the glassware should be washed separately. The important steps involved in preparation of culture and sterilization of glassware, equipment, reagents and media are described in the following sections.

Glassware and Apparatus

1. The soiled glassware should not be allowed to dry. It has to be cleaned in water with sodium hypochlorite to prevent microbial contamination. Automatic washing machines can also be used for this purpose.
2. A nontoxic and effective detergent should be used.
3. After washing, glassware should be thoroughly rinsed with tap water followed by deionized or distilled water and then dried upside down. The dried glassware is sterilized by dry heat (in oven at 160°C for 1 h).

Reagents and Media

1. The reagents and components of media are procured in pure form.
2. They are sterilized either by moist heat, i.e. by autoclaving (if heat-stable) or by membrane filtration (if heat-labile). During autoclaving (moist heating by steam), the container is kept sealed (if borosilicate or polycarbonate) or the caps may be left slack (if soda glass) to avoid breakage.
3. The liquid layer in container may reduce due to evaporation and should be restored with sterile distilled water after autoclaving.
4. Alternatively, presterilized culture media available in market are also used.

Preparation of Animal Material

It is essential that contamination of any kind should be avoided during experimental manipulations, by collecting tissue aseptically or sterilizing it after removal. The use of laminar flow cabinet or hood facilitates the handling of material under aseptic conditions. The embryo or other contents of uterus are free from contamination, whereas other adult tissues are contaminated with bacteria. The embryo should be removed carefully under aseptic conditions, eliminating the possibility of any contamination. The adult tissue is washed with a solution containing high concentration mixture of antibiotics (e.g. a BSS containing 100 units of penicillin and 0.5 mg of streptomycin or neomycin per mL).

Disaggregation of Culture

A primary cell culture is obtained either by allowing cells to migrate from the tissue or by disaggregating the tissue mechanically or enzymatically for the production of suspension of cells. The normal untransformed cells can survive and proliferate easily to produce primary culture when attached to a substrate. These cells are obtained by disaggregation achieved by one of the following techniques:

Physical disruption

It involves cutting the tissue into pieces.

Enzymatic digestion

Several enzymes are used for disaggregation, such as trypsin pronase, collagenase, elastase, pancreatin, mucase, papain, etc. Enzymatic disaggregation is labour intensive and causes damage to cells.

Treatment with chelating agents

The chelating agents are used for production of cell suspensions from established cultures. Epithelium tissues, which need Ca^{++} and Mg^{++} ions for their integrity, may be treated with chelating agents, such as citrate and ethylenediaminetetraacetic acid (EDTA).

Isolation of Tissue

Freshney (1987) described various protocols for isolation of tissues, such as mouse embryo, hen's egg, human biopsy material, etc. The mechanical or enzymatic disaggregation gives a much higher yield of representative cells in a short duration. For tissue isolation, the site is sterilized with 70% ethanol and the tissue is removed aseptically. The isolated tissue is stored in a refrigerator before transferring it to BSS or to a culture medium. The viable cells can be recovered from chilled tissue.

Enzymatic disaggregation

The embryonic tissue disperses more readily and gives higher yield than the newly born or adult. In several tumours, the tissue being fragile is mostly destroyed during enzyme treatment, and viable cells are obtained with great difficulty. The enzymatic disaggregation of the tissue under study is possible by the following methods.

1. **Trypsinization**: Crude trypsin is the most common enzyme used for disaggregation. Trypsinization is of two types: (1) warm trypsinization and (2) cold trypsinization. The latter is effective for many tissues and is tolerated well by a number of cells. Its residual activity is neutralized by serum of the medium or by a trypsin inhibitor (e.g. soybean trypsin inhibitor) in case of a serum-free medium. The cells are exposed to warm enzyme (36.5°C) for a fixed duration. The dissociated cells are collected every half an hour. The trypsin is removed by

centrifugation after complete disaggregation, which takes approximately 4 h. Alternatively, cold trypsinization may be used that involves soaking of tissue in trypsin to allow penetration of enzyme, followed by incubation at 36.5°C for a shorter period.

2. **Disaggregation by collagenase**: Sometimes, the disaggregation by trypsin is damaging for epithelial cells or it is ineffective against fibrous tissues. The intracellular matrix contains collagen. The enzyme collagenase is effective in the disaggregation of several normal and malignant tissues, which may be sensitive to trypsin. Crude collagenase (with contamination of nonspecific proteases) is often used with a firmly chopped tissue in complete medium. When tissue is disaggregated, collagenase is separated by centrifugation, and the cells are seeded and cultured at a high concentration.

Mechanical disaggregation

By this technique, the cell suspension is obtained more quickly as compared to enzymatic disaggregation. The mechanical disaggregation is a force-based technique. The tissue is carefully sliced, and the cells which spill out are collected. In this technique, the cells are either pressed through the sieves of gradually reduced mesh or forced through a syringe and needle, or repeatedly pipetted. This is useful in giving good yield of cells in a shorter time.

Separation of cells

The dissociated cells obtained as above are usually described as *primary cells*. They grow well when seeded on culture plates at high density. These cells represent adherent primary cultures in which the nonviable cells are removed at the first change of medium. The primary cultures can also be maintained in suspension, wherein the nonviable cells are gradually diluted out on commencement of cell proliferation. The nonviable cells can be separated from primary disaggregate by centrifugation method.

Somatic Cell Fusion

The technique of somatic cell fusion is useful in the study of the control of gene expression differentiation and in understanding the problem of malignancy, viral replication, gene mapping and production of hybridomas for antibody production. Fusion of different types of cells has been successfully achieved in animals to produce hybrid cells. The cell fusion in animals has been studied within the body, e.g., myoblasts fusing to form multinucleate muscle fibres. They can also be fused in vitro to produce binucleate heterokaryons or uninucleate hybrid cells. The cell fusion may be induced by polyethylene glycol (PEG). The removal of surface carbohydrates is also, sometimes, required for successful cell fusion. Sendai virus is also often used to induce cell fusion.

The hybrid cells were obtained from mixed cultures of two different cell lines derived from mouse. Later, cell fusion using different taxa was successfully attempted.

Tissue Culture
Primary techniques

This technique is used for cultivation of pieces of fresh tissue derived from the organism. In addition to being used for embryo and organ culture, this technique is modified to become specialized techniques. The different techniques are (1) slide or coverslip cultures, (2) carrel flask cultures and (3) roller test tube cultures.

Slide or coverslip cultures

This is the oldest method of tissue culture being used, especially for morphological studies and for cinemicrographic investigations. The slides or coverslip cultures are prepared by placing a fragment of tissue onto a coverslip that is subsequently inverted over the cavity of a depression slide. The advantages and limitations of this technique are outlined below.

Advantages

1. It is simple and relatively inexpensive technique.
2. The cells grow directly on coverslip and can be fixed and stained to make permanent slides.
3. The cells in living state are spread out in a manner suitable for microscopy and photography.

Limitations

1. The supply of oxygen and nutrients is rapidly exhausted, and the medium quickly becomes acidic.
2. The sterility of cultures cannot be maintained for a long period.
3. Very small amount of tissues can be cultured at a time.

Different techniques of slide or coverslip cultures are as follows:

1. **Single coverslip with plasma clot**: This technique involves the following steps:
   a. The medium is prepared in two parts: one containing 50% plasma in BSS and the other with 50% embryo extract in serum.
   b. Under aseptic conditions and with the use of a capillary pipette, one drop of plasma containing solution is placed in the centre of each one, or more coverslips.
   c. A fragment of tissue (explant) is transferred to this drop without crushing the tissue.
   d. The embryo extract containing solution is added to it and mixed thoroughly before clotting starts, and then the explant is located.
   e. The two small spots of petroleum jelly are placed using a glass rod near the concavity of a depression slide, and the slide is inverted over the coverslip. A gentle pressure is applied, so that jelly sticks to the coverslip.
   f. The culture medium is allowed to clot.
   g. The slide is turned over and the margins of coverslip are sealed with paraffin.
   h. It is then labelled properly and incubated at 37°C.

The single coverslip cultures are useful for short-term studies. Therefore, they are difficult to handle except by process of transfer.

2. **Single coverslip with liquid medium**: It covers the following steps:
   a. The explants are prepared and placed in culture medium in a watch glass.
   b. The explants are then drawn into the tip of a capillary pipette.
   c. One explant is deposited in the centre of each coverslip.

d. The liquid medium is spread out in a very thin circular film with explant protruding above the surface.
e. A depression slide with petroleum jelly is applied immediately and preparation turned over with a quick flip to prevent the fluid from running into the crevice between the slide and coverslip.
f. The coverslip is sealed and the slide is incubated at 37°C, upright or inverted; the tissue grows on the cover glass.

3. **Double coverslip with plasma clot**: This technique resembles the single coverslip method. In this technique a large depression slide is used, and the entire preparation is attached to it by petroleum jelly and wax in such a way that the small coverslip is not in contact with the slide at any point. The technique includes the following steps:

   a. A small drop of BSS is placed on a large coverslip.
   b. A square or round coverslip is placed over BSS in the centre of large coverslip.
   c. Thereafter, the steps enlisted above for single coverslip method are followed.

Due to limitations in use of single coverslip cultures, double coverslip cultures are recommended.

It is desirable to leave the explant in its original location to obtain long-term tissue cultures. This needs washing, feeding, patching and transfer of cultures. It involves the following steps:

1. The seal is removed using razor blade.
2. The large coverslip is removed with small one still attached and flipped over it, so that culture is uppermost in orientation.
3. The small coverslip is detached from the large coverslip using needle and forceps.
4. It is then transferred to a Columbia staining dish (Petri dish or watch glass may also be used) containing BSS.
5. The small coverslip is removed and is placed on a large coverslip. It is required to be carefully controlled.
6. The culture is fed by adding a drop of feeding solution (e.g. BSS:serum:50% embryo extract = 1: 1: 1) to the small coverslip.
7. A clean depression slide is attached using petroleum jelly.

Carrel flask cultures

The Carrel flask technique is used for establishment of a strain from fresh explants of tissue, as it has excellent optical properties for microscopic examination. The polystyrene culture flasks can also be used for this purpose. Following two types of flask techniques are possible:

1. Thick clot culture: It allows rapid growth suitable for short-term cultures.
2. Thin clot culture: It can be maintained for a considerable period.

1. **Preparation of flask cultures**: It is achieved through the following steps:

   a. Around six D-3.5 Carrel flasks are placed in a rack with their necks flamed and pointing to the right.

b. A drop of plasma is placed on the floor of flask and spread out in a circle.
   c. Desired number of explants are transferred to the plasma with the use of spatula and allowed to clot.
   d. After clotting of plasma and fixing of explants in position, extra medium is added.
   e. Around 1.2 mL of dilute plasma is added for thick clots, whereas for thin clots 1.2 mL of dilute serum is added instead of plasma and left for clotting.
   f. The flasks are gassed with gas phase (5% CO_2 in air).
2. **Renewal of medium**: It is essential to remove the medium periodically, after which it is replaced in flask cultures. The old fluid is drawn off with the help of a pipette. Around 1.2 mL of fluid medium is added as replacement, and the flask is gassed with gas phase (5% CO_2 in air).
3. **Transfer of culture**: For transfer, the culture grown in a flask culture is removed and cut into pieces. These pieces are then used for replantation as usual.

   The Carrel flask culture has the following advantages:
   a. A large number of culture can be easily prepared.
   b. A large amount of tissue can be grown with large quantity of medium.
   c. The tissue can be maintained in the same flask for months or even for years.

Test tube culture

The operations of feeding, patching and transfer of culture are carried out as in other primary explantation techniques. The cultures on plasma clots in test tube are prepared just like those in flasks, but tissues can be grown on the wall of test tube without a plasma clot. The test tubes may also be used for developing suspension cultures.

It is an economical technique, and it can be used for preparing a large number of cultures that can be placed in stationary racks or roller drums. This technique, however, provides poor optical property for microscopy. There is also high risk of contamination.

Organ Culture

It is possible to culture pieces of an organ or whole organ in vitro. In the technique of organ culture, it is essential that the tissue is never disrupted or damaged. It requires careful handling. The media used for growing organ cultures are generally the same as those used for tissue culture. The main objective of this technique is to maintain the architecture of the tissue and direct it towards normal development. The techniques of organ culture can be divided into two types: (1) those employing a solid medium and (2) those employing liquid medium.

Culture of embryonic organs

Embryonic organ culture is easier to be established as compared to normal organs from adult animals. Three techniques commonly used for establishment of embryo culture are described below.

1. **Organ cultures on plasma clots**: It involves the following steps:
   a. The plasma clot is prepared by mixing 15 drops of plasma with 5 drops of embryo extract in a watch glass.

b. The watch glass is placed on a pad of cotton wool in a Petri dish; cotton wool is kept moist to prevent excessive evaporation from the dish.
c. A small, carefully dissected piece of tissue is placed on top of plasma clot in the watch glass.

The technique is now modified. A raft of lens paper or rayon net is used on which the tissue is placed. The transfer of tissue can then be achieved by moving the raft easily. The excess fluid is removed, and the net with the tissue is placed again on fresh pool of medium.

2. **Organ cultures on agar**: The medium solidified with agar is also used for establishment of organ culture. The medium with agar provides the mechanical support for organ culture and it does not liquefy. Embryonic organs generally grow well on this medium. This medium consists of seven parts of 1% agar in BSS, three parts of chick embryo extract and three parts of horse serum. Defined media with or without serum, but with agar, are also used.

The cultures of adult organs or parts from adult animal are more difficult to establish due to their greater requirement of O_2. A variety of adult organs (e.g. liver) have been cultured using special media with a special apparatus (Towell's type II culture chamber). Since the serum is found to be toxic, serum-free media is also used. The special apparatus is devised, which permitted the use of 95% O_2.

Whole-Embryo Culture

In this technique, 40-hour-old embryos are used, and the development could be followed for another 24–48 h in vitro before the embryo dies.
The following steps are involved in the preparation of chick embryo culture:

1. A suitable defined medium is prepared and added to sterile watch glasses, placed on moist absorbent cotton wool pads in Petri dishes (as for organ culture).
2. The hen's eggs are incubated at 38°C for 42 h to provide about a dozen embryos.
3. The shell is wiped with alcohol and broken into a sterile evaporating dish, containing 50 mL chick saline or BSS.
4. A circular cut is made into the vitelline membrane around the blastoderm using scissor, and the latter is transferred to a Petri dish containing BSS.
5. The adherent vitelline membrane is removed with the aid of forceps, and the embryo is examined under the microscope to determine the stage of development.
6. The blastoderm is spread on agar gel (ventral side down), and the excess BSS is removed
7. The culture is incubated at 37.5°C.

Tissue Cultures in Biomedical Research

The cell culture methods are extensively used in both theoretical and applied genetic studies. The various genetic applications are outlined below.

In karyological studies

The karyotype population of cells or an organism is the catalogue of the chromosomes of a typical or an average cell. While karyotyping is a useful technique in experimental biology especially, for identifying the origin of cell lines, its most important application in medicine is in the early diagnosis of some congenital abnormalities. Mostly, these abnormalities result from the inheritance of an unbalanced complement of chromosomes, such as Klinefelter's syndrome (XXY), Turner's syndrome (XO) or Down's syndrome (mongolism), wherein chromosome number 21 is represented by 3 instead of 2. Some other anomalies such as translocations or deletions can also be recognized using this technique of karyology.

In identification and hereditary metabolic disorders

The precise nature of many of the genetic diseases can be determined only by demonstrating the metabolic abnormality, commonly the absence of an enzyme in tissue. Disorders such as absence of catalase in acatalasia or of α-D-galactose-1-phosphate, UDP glucose uridyltransferase in galactosaemia or failure to decarboxylate branched chain keto acid in maple syrup urine disease can be demonstrated in primary fibroblast cultures. The heterozygous locus, a deficiency of the transferase enzyme per cell can also be measured. Some human genetic disorders that can be studied in this manner are glycogen storage diseases, homocystinuria, Hurler's syndrome, Hunter's syndrome, Niemann-Pick disease, I-cell disease, orotic aciduria, Refsum's disease, xeroderma pigmentosa, etc.

In somatic-cell genetics

The studies involving somatic cell are made primarily with the following objectives:

1. For understanding behaviour of heterokaryons, following fusion of parent cells with different properties
2. For knowing the behaviour of hybrid cells derived from heterokaryons with the objective of eliciting information about interaction between different genes
3. For mapping loci on different chromosomes

In biomedicine

The study of the relationships between viruses and cells has been greatly supplemented with the use of tissue culture methods. These techniques are also useful for detection and identification of viruses from the suspected cases. In the recent years, the significance has been in the development of vaccine against poliomyelitis that is made from the viruses grown in tissue culture. In addition to viruses, other intracellular parasites that are grown in tissue culture and studied include rickettsiae, mycobacteria (especially tubercle and leprosy bacilli), lepraemurium, parasitic protozoans, etc.

The ability of tissue cultures to support the growth of viruses and to reveal their presence by lesions (that are in some cases specific) has been applied in virology for the following purposes:

1. Study of host–parasite relationship
2. Detection and identification of viruses
3. Production of viruses for vaccine manufacture

The tissue culture techniques have been very useful in understanding of cancer. The culture of canine lymph sarcoma and the development of quantitative assay procedures for viruses responsible for tumours—stimulated by the exploitation of cell culture methods by virologists—are significant achievements in biomedical sciences.

QUESTIONS

1. Write a short note on maintenance of animal cell culture.
2. Give the difference between batch and continuous culture.
3. Explain the steps involved in the establishment of animal cell culture.
4. Explain the techniques used for in vitro animal cell culture.
5. Write the different phases of bacterial growth curve.
6. Discuss methods of single-cell culture techniques and their applications.
7. Write about sterilization methods used in tissue culture works.
8. Write a note on natural culture media for animal cell culture media.
9. Write the advantage and disadvantage of animal cell culture media.

CHAPTER 6
Medicinal Plant Biotechnology

CHAPTER OUTLINE

- Genetics as Applied to Medicinal Herbs
- Mutation
- Polyploidy
- Chemodemes (Chemical Races)
- Hybridization
- Genetic Engineering and Recombinant DNA Technology in Plants
- Plant Tissue Cultures as Sources of Biomedicinal Compounds
- Historical Development
- Types of Cultures
- Culture Medium
- Surface Sterilization of Explants
- Establishment of Cultures
- Phytopharmaceuticals in Plant Tissue Cultures
- Bioproduction of Useful Metabolites in Hairy Root and Multiple Shoot Cultures

GENETICS AS APPLIED TO MEDICINAL HERBS

Gregor J. Mendel (1865) showed that certain hereditary factors operate in all biological species. The Danish biologist Wilhelm Johannsen (1909) named these factors as *genes*. The genes not only transmit hereditary characters but also mastermind the entire process of life. The genes are located in chromosomes that are themselves situated in the nucleus of the cell. It is said that the genes form the riddle and the chromosomes represent the mystery. Much of the mystery surrounding the genes was cleared up with the discovery of the structure of the DNA by J.D. Watson and Francis Crick in April 1953. The structure of the DNA resembles a long-rope ladder twisted around like a corkscrew. The two sides of the ladder are long chains of sugars and phosphates in repeated sequence (Fig. 6.1). DNA contains the bases, namely, adenine (A), cytosine (C), thymine (T) and guanine (G), that make up the genetic code that is not only complex but also extensive. In 1977, Fred Sangar pointed out that the DNA code of a virus, when decoded by computer, come to a printout of 15 metres.

The genes control all functions of cell and body growth. The two main events in the life of most cells are multiplication (by division) and synthesis of proteins. Both these operations are carried out on the basis of blueprints coded in the genes.

The science of genetics has made valuable contributions to the improvement of plants with economic, medicinal and ornamental values. This improvisation is attained by selective breeding, mutation, polyploidy and hybridization. It has important roles in mitosis and meiosis of the cells.

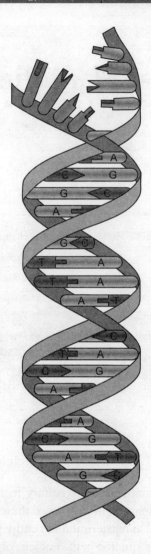

Figure 6.1 Structure of DNA.

Mitosis results in precise equal distribution of chromosomes from a parent nucleus to the daughter nuclei. Further, it maintains equilibrium in the amount of DNA contents of cells. Mitosis is usually associated with vegetative propagation of plants. The somatic number is the number of chromosomes in plant cell nucleus, which in normal conditions remains constant, but varies with different species (Fig. 6.2).

Meiosis occurs in reproductive cells of sexually reproducing species. It has a genetic significance because of formation of four haploid nuclei from a single diploid one, in two successive divisions; thus, balancing the doubling of chromosome numbers that results from fertilization. In meiosis, the crossing over provides new combinations of genetical substance. Hence, new combinations of characters in offsprings occur. The process of segregation of chromosomes also occurs in meiosis.

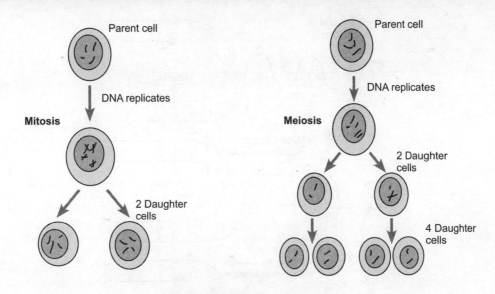

Figure 6.2 Mitosis and meiosis.

These techniques of genetic manipulations in plant cells have made much impact to improve yield of medicinal plants.

MUTATION

Mutation is represented as variation in characters of the species. It is caused either due to environmental changes or changes in hereditary constitution. Normally, as a response to environmental changes, the variations are observed, but the original traits are restored when changes in environment disappear or they are withdrawn. This type of change and restorage is not heritable and termed only as *phenotypic variations*. It is evident that between organisms of similar genotype, there are phenotypic variations. But when a change occurs in the genome of an individual that is not caused due to environment, it may make a permanent evolutionary change. This is termed as *mutation* that represents a sudden change in a genotype causing qualitative or quantitative alterations of genetic material. Mutations are again distinguished into two types, such as *chromosomal mutations* and *point mutations*. The chromosomal mutation is also called *chromosomal aberration* that leads in most cases to changes in amount or position of genetic material. On the other hand, the changes with a gene or cistron of DNA molecule cause point mutations, and they are permanent and heritable (Fig. 6.3).

Mutation that occurs due to some unknown reason from nature is known as *spontaneous mutation*. This has been observed in some plants, bacteria and viruses. Mutations can also be induced by artificial means (*induced mutation*) with certain reagents called *mutagens*. Exposure of cells to UV rays, X-rays, ionizing radiations and abnormal environment may induce mutation. The chemical mutagens commonly used are nitrogen mustard, formaldehyde, nitrous acid, ethyl ethane sulphonate, 5-bromouracil, 2-aminopurine, manganese chloride, etc.

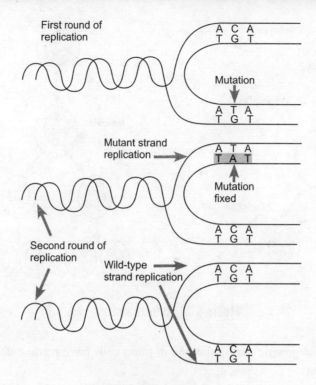

Figure 6.3 Mutation.

The mutations induce morphological and anatomical changes as well as changes in the chemical composition of medicinal and aromatic plants. In some cases, favourable changes and yields in active constituents of plants have been achieved. Mutations may give rise to building the resistance of a medicinal plant towards certain disease. But, in all these cases, the plant may become susceptible to climatic conditions, certain other diseases, retardation in growth, etc. These undesirable effects are to be eliminated by breeding and selection (Fig. 6.4).

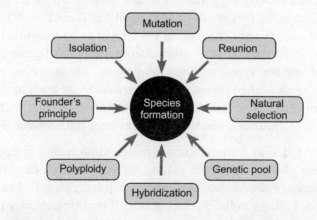

Figure 6.4 Effect of various factors on species formation.

Artificial Mutation

Artificial mutations are induced by artificial means in the living organism by exposing them to abnormal environment. These include radiations, certain physical conditions such as temperature and chemicals.

The discovery of *chemodemes* has enabled bioscientists to select high-yielding chemical strains or to eliminate chemical strains that contain toxic principles. The chemodemes are induced by breeding in some species, so as to manipulate active constituents with a view to enhance the therapeutic efficacy.

Radiation Mutations

The electromagnetic waves of short wavelength (ultraviolet light, X-rays, gamma rays, alpha and beta rays) are radiation mutagens. The X-rays and gamma rays are called *ionizing radiations*, and also include alpha particles, beta rays, thermal and fast neutrons. Due to ionizing radiations, in many cases, water molecule in a biological system releases one electron and becomes unstable and eventually splits into hydrogen ion and hydroxy radical. Hydrogen ion reacts with O_2 and produces hydroperoxyl (HO_2) radical. Both these radicals, such as hydroperoxyl and hydroxy are potent oxidizing agents. When chromosomes and their DNA are struck by such radicals, they react due to which sugar phosphate part of DNA may be impaired leading to chromosomal mutations, such as deletions, additions, breaks, inversions and translocations.

Chemical Mutations

Some chemical mutagens, like nitrogen mustard, formaldehyde, nitrous acid and ethyl ethane sulphonate alter chemical constitution of DNA bases and cause transitional substitution in DNA. But, some compounds like 2-aminopurine, urethane, 5-bromouracil, caffeine, triazine, phenol and certain other carcinogens act as a base analogue and bring out certain error mutations in DNA. In general, the chemical mutagens have profound cellular effects like production of abnormal DNA (e.g. nitrogen mustard), inhibition of deoxyribonucleotide synthesize (e.g. deoxyadenosine) and inhibition of cytochrome oxidase with resultant peroxide formation (e.g. inorganic cyanides). The artificial production of mutations is an important milestone in the development of cultivation technology of medicinal plants. The higher content of solasodine is obtained by applying radiation and chemical mutagens in *Solanum khasianum*. The chemical mutagens also increase morphine content in *Papaver somniferum*. The yield of tuber of *Dioscorea bulbifera* and its diosgenin content is increased by radiation. The medicinally important characters of *Atropa belladonna* have been enhanced by radiations and chemical mutagens. The capsicum seeds (*Capsicum annuum*) treated with sodium azide and ethyl methane sulphonate have led to plants with higher contents of capsaicin. The agronomical performance in terms of harvest index of *Mentha arvensis* var. *piperascens* (Japanese mint) has been improved by exposure to gamma radiations. The positive effects of artificial mutations have been witnessed on different species of *Mentha*.

POLYPLOIDY

The induction of polyploidy has useful effects on medicinal plants, such as digitalis, mentha, datura, hyoscyamus, belladonna, lobelia, etc. The specific number of chromosomes is a character

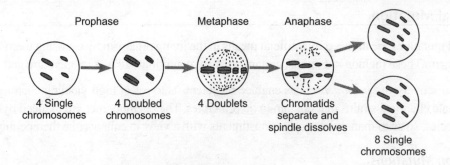

Figure 6.5 Polyploidy.

of each species and is called *genome*. *Euploidy* is a type of ploidy in which genome contains whole set of chromosomes. Euploidy includes monoploidy, diploidy and polyploidy. When the organism contains more than two sets of chromosomes (genomes), it is called *polyploidy*. The polyploidy occurs in a multiple series of the basic chromosome or genome numbers of 3, 4, 5, 6, 7, 8, etc. and accordingly, it is called triploidy, tetraploidy, pentaploidy, hexaploidy, heptaploidy and octaploidy, respectively (Fig. 6.5).

Polyploidy is caused by physical agents like X-rays, centrifugation, temperature chocks and chemical agents mainly, colchicine, veratrine, sulphanilamide, hexachlorocyclohexane and mercuric chloride. The chemical agents cause disturbance to mitotic spindle of dividing diploid cell and induce nonsegregation of already duplicated chromosome and thus, convert diploids into tetraploid cells. It may cause formation of new species, adaptability to various habitats and mainly accumulation of vitamins.

The ideal chemical agent to induce polyploidy is colchicine, an alkaloid obtained from *Colchicum autumnale, Colchicum luteum, Colchicum speciosum* and other speices of *Colchicum*. The tropolone alkaloid, colchicine prevents sister chromatids from separating into daughter nuclei at anaphase. These chromatids remain attached by their common centromere in C-metaphase. The chromatids eventually separate, but remain in the same nucleus. An interphase occurs, followed by a second C-metaphase, involving a doubled chromosome complement. Hence, the chromatid pairs are doubled in second C-metaphase. Likewise, the cell undergoes one, two or more than two rounds of DNA replication and results into polyploidy. The activity of colchicine is caused due to its interaction with disulphide bonds of spindle protein and by inhibition of conversion of globular proteins into fibrous proteins. The capacity of colchicine to induce polyploidy varies along with its different derivatives. Colchicine treatment to medicinal plants has given promising results in many cases. Chemically induced mutations may lead to variations in biochemical composition of plant. In stramonium, the yield of crop is enhanced by 60–150% in 4n form. Plants such as lobelia, cinchona, belladonna, squill, cannabis, acorus and poppy also show increased yield of respective compound in 4n form. Polyploidy may cause a reduction in total glycoside contents of *Digitalis purpurea* and *Digitalis lanata,* but in the later, it raises slightly the contents of lanatosides A and B. The increase in chemical contents may not be coincidental with phenotype of medicinal plant.

There may be reduction in size in addition to enhancement in the content of active constituents. Some plants do not show any change in chemical contents as a response to polyploidy.

CHEMODEMES (CHEMICAL RACES)

Chemodemes are regarded as a group of plant species having identical morphological characters, differing in their chemical nature. Chemodemes are, therefore, considered as chemically separate groups within species. The chemodemes can be established only by growing different plants of a species in identical conditions preferably, from the seeds and for many generations. By this way, it shows variations either in type or contents of certain constituents, including secondary metabolites of medicinal importance. The chemical characters acquired as a result of chemodemes are hereditary.

Chemical races have been reported in *D. purpurea* and *D. lanata*. These chemical races in *D. purpurea* yield different proportions of glycosides obtained from digitoxin and gitoxin. Based on lanatosides A and C contents, the chemical races in *D. lanata* are *D. lanata* Ehrh., chemovarieties A and C. The chemodemes in *Rheum palmatum* exhibit different rhein–chrysophanol ratio. The different varieties of *Prunus communis* (almond) differ by the presence or absence of amygdalin. Chemical races have also been reported in *Duboisia myoporoides, Duboisia leichhardtii* and *Claviceps purpurea*. Three chemotypes are evident in *Withania somnifera*. The chemotype I contains mainly withaferin; chemotype II has identical compounds, whereas chemotype III contains withanolides (a mixture of steroidal lactones).

The volatile oil-bearing plants like *Cinnamomum zeylanicum, Cinnamomum camphora, Eucalyptus dives, Ocimum menthaefolium* and *Ocimum sanctum* exhibit chemical races, with varied contents of aromatic constituents.

HYBRIDIZATION

A hybrid is an organism that results from crossing of two species or varieties differing at least in one set of characters. The process through which hybrids are produced is called *hybridization*. The resultant hybrids are monohybrids (one pair of different characters), dihybrids (two pairs of different characters) or polyhybrid (more than two pairs of different characters). The process of hybridization helps in yielding a single variety with favourable characters of other varieties or species and some times, producing new and favourable characters that are not present in both the parents (Fig. 6.6).

The hybridization of *Withania somnifera*—Israeli chemotype II and *W. somnifera*—South African chemotype have led to the formation of a new hybrid that contains three new withanolides. The hybrids such as *D. purpurea* × *D. lanata* and *D. purpurea* × *D. lutea* contain lanatoside A as main glycoside, along with lanatosides B and E. However, the hybrids do not contain lanatoside

Figure 6.6 Hybridization.

C or purpurea glycoside A. During the cross of *Solanum incanum* (1.8% solasodine) and *Solanum melongena* (traces of solasodine), first generation bears more fruits (berries) with solasodine content up to 0.5%. However, the second generation has proved to be a high-yielding source of solasodine.

A recent development in hybridization is the protoplast culture. The plant protoplasts are the cells without cell walls. Such protoplasts in cultures can be fused together (protoplast fusion or asexual hybridization). The fusion can be arranged in cells of same origin or between different species.

GENETIC ENGINEERING AND RECOMBINANT DNA TECHNOLOGY IN PLANTS

The technique of artificial synthesis of new genes and their subsequent transplantation in the genome of an organism or the method of correcting defective genes of organism by molecular biological techniques form the discipline known as *genetic engineering*. In recent years, these techniques have gained importance in relation to their utility in drugs and pharmaceuticals. The secondary metabolites from plants that are expected to give promising results by genetic engineering include, cardiac glycosides, morphine alkaloid, stigmasterol or diosgenin (for synthesis of steroid hormones and corticosteroids). In respect of such plants, the transfer of entire biosynthesis route to bacteria by technique of gene technology is expected to yield different economic processes (Fig. 6.7).

Recombinant DNA technology involves gene splitting, so as to change characters of plants and animals by implanting in them genes from other organisms and, in some cases, even from other species. Important genetically engineered products available in market are human insulin, human growth hormone, interferons, hepatitis B vaccine, interleukin-2 (to fight certain types of cancer), β-endorphins (for treatment of pains), tissue plasminogen activator (TPA; for treating heart attack), somatostatin (for treatment of pituitary diseases) and erythropoietin or EPO (for treating anaemia in patients undergoing kidney dialysis).

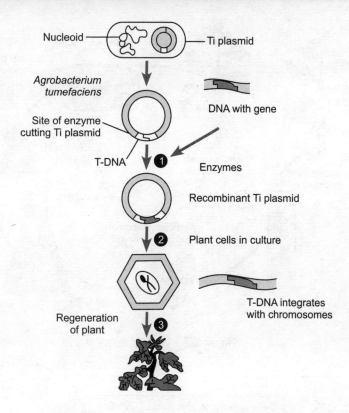

Figure 6.7 Recombinant DNA technology in plants.

Transgenic Plants

Application of genetic engineering techniques for the improvement of plants is a fast-developing area of research. The advent of modern technique of genetic modification has enabled to remove individual genes from one species and insert them into another, without the need for sexual compatibility. Once the new gene has been inserted into the plant, offspring that contains the copies of the new gene can be produced in the traditional manner.

Transgenic plants (Fig. 6.8) also carry the stable integrated foreign genes. The introduction of foreign gene is made by genetic transformation (gene transfer), also called *transfection*. It is necessary that DNA remains stable and can pass the new gene to its offspring. Transgenic plants are useful in causing crop improvement. The herbicide- and insect-resistant species and species with high yields of oil and protein, as well as other enhanced quantities of products (Fig. 6.9), are obtained through this technique.

The complete process of transformation includes the following steps:

1. Gene transfer
2. Selection of transgenic plants
3. Recovery of transgenic whole plants

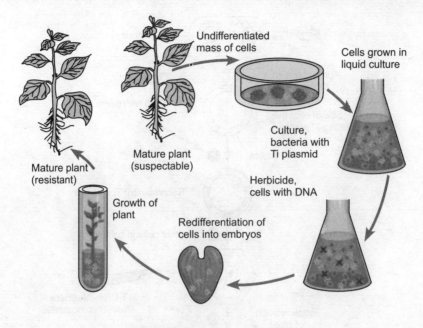

Figure 6.8 Transgenic plant.

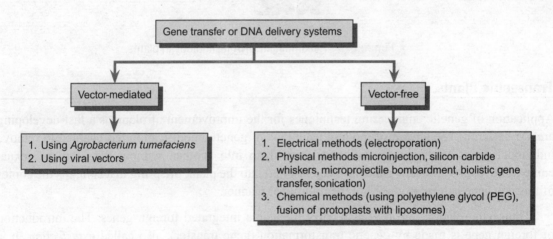

Figure 6.9 Process of transformation.

Gene Transfer

1. **By using *Agrobacterium tumefaciens*:** This Gram-negative bacterium commonly occurs in soil and causes crown gall disease (tumorous outgrowth) in higher plants. It enters the plant through the mechanically wounded tissue site. *Agrobacterium tumefaciens* exhibits capability to transfer some of its DNA into nuclear DNA of an infected host plant mainly, because of functions encoded on its tumour-inducing (Ti) plasmid. The transferred DNA or xDNA of this plasmid encodes substances and acts as plant hormone, causing proliferation of

tissue in infected plant. Ti plasmid of this bacterium has been modified by means of genetic engineering to create 'disarmed' plasmids (also called *vectors*), which can carry any DNA sequence into plants infected with *A. tumefaciens*, without tumorous growth on the host plant. These cloning vectors derived from Ti plasmid have been used for replication in both *A. tumefaciens* and *E. coli* bacterial strains. This makes possible the cloning of foreign DNA sequences into the vectors using *E. coli* host as well as further transfer of the completed vector to an *A. tumefaciens* host (Fig. 6.10).

2. **By electroporation**: In this technique, short electrical pulses of high field strength are used to promote DNA uptake by protoplasts. The optimal electroporation conditions may vary from

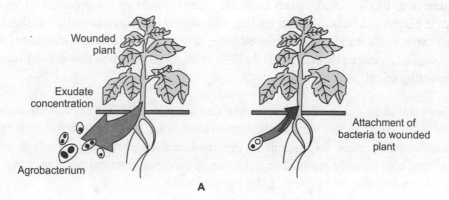

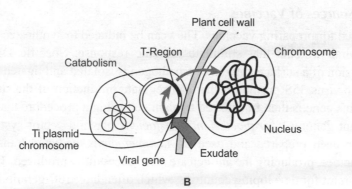

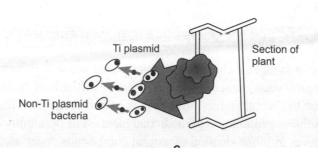

Figure 6.10 Gene transfer.

plant to plant. Buffers greatly influence the gene transfer efficiency and also the protoplast survival rate.
3. **By using microinjection**: In this technique, microcapillaries and microscopic devices are employed to deliver DNA into protoplasts. It is a very skilled and expensive process and requires precision in use of techniques.
4. **Biolistic gene transfer**: In this technique, DNA is bombarded into the intact cell using a biolistic device, such as particle gun. The technique can only be used for intact tissues. It is not suitable for protoplasts, as pressure is required to be applied. This is also a very accurate technique.
5. **Silicon carbide whiskers**: Silicon carbide forms long, needle-shaped crystals (whiskers) in solution (e.g. PEG, 1–3%). When cells are vortex mixed in the presence of whiskers and DNA, the DNA can be introduced into the cells following penetration by whiskers.
6. **Gene transfer by chemical agents**: Various chemical treatments are used to stimulate DNA uptake by protoplasts. Among them, PEG (15–20%) is the most common that acts by increasing the permeability of cell membranes.

The first plant to be genetically transformed was tobacco (1983), followed by cereals (1990). Most cereal food crops in the western countries have been made insect and pest resistant by incorporating suitable genes. As a result of this modification, percentage spoilage of crops per annum has been considerably reduced and the use of harmful chemical pesticides has also come down. It is an eco-friendly technique of plant improvisation.

Transgenic Plants as Sources of Vaccines

The plants can be utilized for preparing vaccines. They can be induced to synthesize appropriately folded bacterial and viral proteins that stimulate immune response. Specific DNA sequence encoding for the expression of a surface antigen of a pathogen is isolated and ligated to a promoter (e.g. cauliflower mosaic virus 35S promoter) that can regulate production of the surface antigen in a transgenic plant. This gene is then transferred to plant cell using a procedure that results in its integration into the plant genome, *Agrobacterium tumefaciens* plasmid vector system. Flanking T-DNA sequences have been prepared and used in *A. tumefaciens*-mediated transformation of tomato. Transgenic tomatoes, producing the antigen are also successfully produced. The transgenic vaccines are more important for developing countries, which often have refrigeration problems.

PLANT TISSUE CULTURES AS SOURCES OF BIOMEDICINALS

Despite recent developments in synthetic chemistry, higher plants are still important sources of medicinal compounds. In recent years, however, it has become difficult to maintain an ample supply of medicinal plants due to several factors, such as lack of conservation of the environment, increasing labour costs and ruthless exploitation of medicinal plants. The possibility of utilizing the technique of plant tissue cultures in bioproduction of natural compounds under aseptic conditions by methods similar to those used to culture microorganisms has opened new approach in the field of biotechnology. The use of tissue culture technique for biosynthesis of secondary metabolites

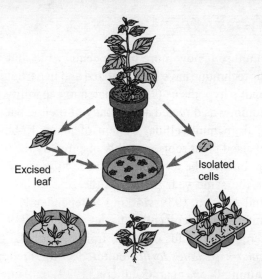

Figure 6.11 Plant tissue culture.

of pharmaceutical significance holds promise for in vitro production of plant constituents. The technique of plant tissue cultures (Fig. 6.11) affords possible solutions to some of the problems of conventional cultivation and procurement of crude drugs. The major advantages expected from plant tissue culture systems over conventional cultivation technique are summarized as follows:

1. It is possible to use plant cell culture technique for synthesis of those medicinal compounds that are too difficult or impossible to synthesize chemically.
2. It offers the prospect of absolutely uniform biomass obtainable at all times and manageable under regulated and reproducible conditions, rarely possible in working with entire living potentials.
3. The cultured cells can be maintained free from any microbial contamination and insect attack.
4. The cells of any tropical or temperate plant may be multiplied to yield specific metabolites produced by them.
5. The useful natural compounds may be produced under controlled environmental conditions, independent of soil conditions and changes in climatic conditions.
6. The technique is useful in study of biogenesis of secondary metabolites. It is possible to feed labelled precursors to cell cultures and deduce interpretations pertaining to metabolic pathway of desired compound.
7. The biotransformation reactions in plant cell cultures can be attempted. Specific modification of chemical structures of certain compounds may be achieved more easily in cultured plant cells rather than in microorganisms or by chemical synthesis.
8. Through plant tissue culture, the immobilized cell systems can be established that may be used for various biotransformation or biochemical reactions. A particular strain of cells obtained from suspension cultures is immobilized by suspending it in sodium alginate solution, precipitating the alginate plus entrapped cells with calcium chloride solution, pelleting and allowing the product to harden.

HISTORICAL DEVELOPMENT

The technique of plant tissue culture is now more than a century old, but it is only in last three decades that implications of this technique have been realized and in particular, its pharmaceutical potential appreciated. Haberlandt's hypothesis (1902) related to capability of isolated plant cells for cultivation on artificial medium was directed at the study of their capacity and characteristics as elementary organisms free from multicellular system of plant. Robbins (1922) and Kotte (1922) successfully cultured excised plant roots, possibly inspired by the hypothesis put forward by Haberlandt. White (1934) published the technique of successful continued culture of excised tomato root tips. Gautheret in the same year published the first of his many classical papers on cambial tissue cultures, and later in 1939 working independently of one another, White et al. reported the first unlimited culture of callus tissue. White (1939) reported continuous cultivation of a similar undifferentiated callus obtained from procambial tissue of young stems of a hybrid *Nicotiana (N. glauca* × *N. langsdorffii)* cultured in the medium he had developed for excised tomato roots, containing 0–5% agar. Gautheret (1939) explained behaviour of carrot explants on a medium compounded of inorganic salt mixture and dextrose containing aneurine, cystein hydrochloride and a low concentration of indole acetic acid. Muir et al. (1954) reported growth of liquid cultures containing single cells and small clumps of cells of *Tagetes erecta* and *Nicotiana tabaccum.* Reinert (1956) obtained evidence for occurrence of cell divisions in a similar suspension of single cells and cell groups from *Picea glauca.* Jones et al. (1960) examined growth in hanging drop culture of separated cells from a callus of hybrid tobaccos *(N. tabaccum* × *N. glutinosa).* George et al. (1961) reported plant propagation by tissue culture. Dougall (1966) reported biosynthesis of protein amino acids in plant tissue culture and isotope competition experiments using protein amino acids. Kaul et al. (1967) reported production of cardioactive substances by plant tissue cultures and their screening for cardiovascular activity. Kokate and Radwan (1979) reported establishment of cell cultures of *Solanum* species known to harbour steroidal alkaloids. Hall and Thomas (1987) explained molecular approaches for the manipulation development process in plants. Jain and Newton (1988) reported micropropagation of selected somaclones of *Begonia* and *Saintpaulia.* Phillipson (1990) established antidiabetic activity of leaf and callus extracts of *Aegle marmelos* in rabbit. Bhojwani and Razdan (1992) reported rapid multiplication of safed musli (*Chlorophytum borivilianum*) through shoot proliferation and again in 1996 established in vitro morphogenesis in zygotic embryo cultures of neem. Lo et al. (1997) described regeneration in African violet *(Saintpaulia ionantha)* using different leaf explants. Nair and Pushpangadan (2003) described in vitro organogenesis and somatic embryogenesis in *Adenia hondala* (an endangered medicinal plant of the Western Ghats). Nahar and Bari (2006) published plant tissue culture studies on two different races of purslane (*Portulaca oleracea).* Kumar and Kanwar (2007) reported direct shoot regeneration, callus induction and plant regeneration from callus tissue in moss rose (*Portulaca grandiflora*). Shih and Doran (2009) described in vitro propagation of plant virus, using different forms of plant tissue culture and modes of culture operation. Jain and Bashir (2010) explained in vitro propagation of a medicinal plant, *Portulaca grandiflora.* Bhojwani and Razdan (2010) reported in vitro propagation of *Aloe vera*—a plant with medicinal properties.

TYPES OF CULTURES

Three main lines of technical developments constituting methodology of plant tissue culture are as follows:

1. The culture of isolated plant organs (particularly of isolated roots, but to a lesser extent, of stem tips, leaf primordia, immature embryos and flower structures)
2. The growth of callus or static mass on solidified nutrient medium
3. The growth of biomass in liquid nutrient medium representing cell-mixed suspensions

The plant tissue culture consists of a number of techniques for growing plant organs and cells. The *organ cultures* are obtained from roots, leaves, stem tips, flowers and fruits. The methods employed for initiation of such cultures depend upon nature of the organ acquired. For example, excised root cultures are initiated by placing sectioned lengths of surface-sterilized root tip in liquid nutrient medium and periodically, subculturing to fresh medium. They demonstrate high growth rate and metabolic activity and a low level of variability and are, therefore, important tools in determining responses to varying conditions (Fig. 6.12).

The plant tissue is cultured on solid agar (*callus cultures*) or in liquid nutrient medium (*suspension cultures*). When grown on agar medium, the tissue forms a callus or a mass of unorganized cells. The technique of callus culture is useful in for starting and maintaining cell lines, as well as, for studies related to organogenesis and meristem culture. The liquid suspension cultures consist of mixtures of cell aggregates, cell clusters and single cells. The growth rate of such culture is generally much higher than the growth on solidified medium. The technique offers better control of the growth of biomass, because the cells are surrounded on all sides by the nutrient medium. For the same reason, the cell material is probably more uniform physiologically. The callus and suspension cultures can be derived from tissues of most of the species, but the ease of starting the cultures varies with the

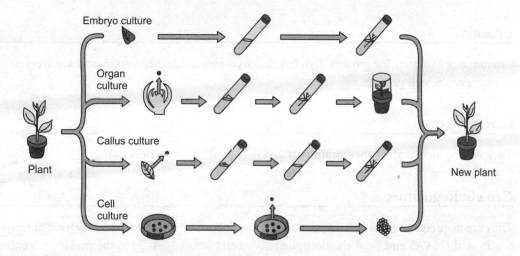

Figure 6.12 Types of plant tissue culture.

type of plant and the tissue origin. Almost any part of a plant can be induced to produce a callus and a suspension culture. The necessary tissue can be obtained from roots, seedlings, pollen, stem and leaf portions. It usually grows as a mass of undifferentiated cells on enriched solidified medium.

CULTURE MEDIUM

An ideal nutrient medium for the growth of cell cultures consists of inorganic salts, a carbon source, vitamins, growth regulators and some organic supplements. The pioneering studies for culture medium were initially carried out by Gautheret (1939) and White (1939). For a long time, it had been the subject of research for many scientists to suggest enriched nutrient medium for optimum growth of biomass. As a result of these studies, several well-defined standard nutrient media have been developed in basic or modified forms and used for tissue culture studies. The media have been developed by Gautheret (1942), White (1943), Hildebrandt et al. (1946), Heller (1953), Nitsch and Nitsch (1956), Murashige and Skoog (1962), Eriksson (1965) and (1968), and are commonly used for establishment of cultures.

An ideal nutrient medium for plant tissue cultures contains five classes of ingredients, as discussed below.

Inorganic Salts

The concentration of potassium and of nitrate is at least 20–25 mM for each, whereas, concentrations of 1–3 mM of phosphate, sulphate and magnesium is adequate. Ammonium is essential, although amounts in excess of 8 mM could be deleterious. The recommended micronutrients are salts of manganese, zinc, molybdenum, cobalt, copper, iron, iodide and boric acid. Iron can be incorporated in chelated form.

Vitamins

Thiamine is essential for growth. Pyridoxine, myo-inositol and nicotinic acid are frequently added to improve the cell growth.

Carbon Source

Sucrose or glucose at a concentration of 2–4% is suitable as carbon source.

Growth Regulators

They are needed to induce cell division. The growth hormones most frequently used are naphthalene acetic acid (NAA) and 2-, 4-dichlorophenoxy acetic acid (2-, 4-D) in the molar concentrations of 10^{-7} to 5×10^{-5}. Both 2, 4-D and NAA are degraded very slowly by plant cells and are stable to autoclaving. Cytokinins such as kinetin or benzyladenine ($10^{-7} - 10^{-5}$ M) are also used in conjunction with 2-, 4-D or NAA to obtain good callus formation (Table. 6.1).

Table 6.1 List of plant growth regulators

Hormones	Product names	Functions in plant tissue cultures
Auxins	Indole-3-acetic acid (IAA) Indole-3-butyric acid Indole-3-butyric acid, potassium salt α-Naphthaleneacetic acid 2,4-Dichlorophenoxyacetic acid p-Chlorophenoxyacetic acid	Adventitious root formation (high concentration) Adventitious shoot formation (low concentration) Induction of somatic embryos Cell division Callus formation and growth Inhibition of axillary buds Inhibition of root elongation
Cytokinins	6-Benzylaminopurine 6-γ, γ-Dimethylallylaminopurine (2iP) Kinetin thidiazuron (TDZ) N-(2-chloro-4-pyridyl)-N'-phenylurea Zeatin Zeatin, riboside	Adventitious shoot formation Inhibition of root formation Promotes cell division Modulates callus initiation and growth Stimulation of axillary bud breaking and growth Inhibition of shoot elongation Inhibition of leaf senescence
Gibberellins	Gibberellic acid	Stimulates shoot elongation Releases seeds, embryos and apical buds from dormancy Inhibits adventitious root formation
Abscisic acid	Abscisic acid	Stimulates bulb and tuber formation and maturation of embryos Promotes the start of dormancy
Polyamines	Putrescine, spermidine	Promotes adventitious root formation Promotes somatic embryogenesis and shoot formation

Organic Supplements

Protein hydrolysates, malt extract, yeast extract and coconut milk (liquid endosperm) are commonly used for enhancement in the growth rate of the cells in biomass.

Preparation of Medium

The chemicals are dissolved in glass distilled water, the stock solutions of vitamins and micronutrients. Moreover, growth hormones are added and the pH of the medium is adjusted from 5.5 to 5.7. The solution is made up to the volume with distilled water. Around 50 and 100 mL quantities of solution are distributed into 250 mL Erlenmeyer flasks. The flasks are stoppered with plugs of cotton wool and autoclaved at 120°C for 15 min. After sterilization, flasks are removed for cooling, as early as possible. The agar medium is autoclaved in lots of 500 mL and subsequently poured into sterile containers while it is hot and allowed to settle. All media are stored at 10°C prior to use.

SURFACE STERILIZATION OF EXPLANTS

The surface sterilization is carried out for the organ from which the tissue is aseptically excised or for spore/seed used for germination to yield the tissue explant. The commonly used surface-sterilizing agents are hydrogen peroxide (10–12%), sodium hypochlorite (1–2%), bromine water (1–2%), mercuric chloride (0.1–1%) and silver nitrate (1%).

The seeds are treated with 70% ethanol for about 2 minutes, washed with sterile distilled water and treated with chosen surface-sterilizing agent for a fixed duration. Thereafter, they are once again rinsed with sterile distilled water and kept for germination under aseptic conditions. The seeds are aseptically placed on double layers of presterilized filter paper in Petri dishes moistened sufficiently with sterile distilled water or on moist cotton plugs in Petri dishes or culture tubes. They are germinated in dark at 26–28°C. A small portion of the seedling is used for the initiation of the callus culture. The aerial portion of plants, such as bud, leaf and sections of stem, are surface sterilized by submerging it in 70% ethanol for 2–3 min followed by 2–3 rinses in sterile distilled water.

ESTABLISHMENT OF CULTURES

Callus Cultures

The surface-sterilized plant part is aseptically transferred onto solidified nutrient medium in flasks, glass jars or culture tubes and incubated in dark at 26–28°C. After 4 weeks, the callus usually grows to a size of about 4–5 times of the explant. Many tissue explants possess some degree of polarity, with the result that callus is formed more easily at one surface. It is observed that callus is formed from stem segment particularly from the surface that is directed in vivo towards the root. The development of callus is more rapid from the tissue not in contact with and particularly not immersed in the solidified culture medium. The maintenance of growth in callus tissue is done by subculturing the transfer on each occasion of a piece of healthy tissue every 4 weeks into the flask containing fresh solidified nutrient medium. At 5–10°C, many cultures can be stored for a longer time in healthy condition. The growth of many cultures (particularly of those that form chlorophyll) is usually stimulated by low-intensity illumination. Light either on a 12-hour cycle or continuously is, therefore, usually provided in the incubation chambers by fluorescent tubes.

Suspension Cultures

They are usually initiated by transferring an established callus tissue to an agitated liquid nutrient medium in Erlenmeyer culture vessels (30–60 mL medium per 250-mL flask). The composition of the medium for suspension cultures is usually identical to that of callus cultures except agar. The soft callus generally forms a suspension culture without much difficulty. The release of cells and tissue fragments from less friable callus masses, and the maintenance of a good degree of cell separation may often be promoted by the presence of a high auxin concentration in the liquid medium, an appropriate balance between yeast extract and auxin or between auxin and kinetin. The suspension cultures are incubated at 25°C in darkness or under low intensity fluorescent light. The continuous agitation of flask cultures is done on a horizontal shaker (100–200 rpm). The aluminium foils or parafilms are used to seal culture flasks, which reduce evaporation during the process of culture growth. A cell suspension is usually formed within 4–6 weeks. The cells grown in cultures are meristematic and generally undifferentiated. The suspension cultures are subcultured by transferring untreated or fractionated aliquots of the suspension to fresh medium at regular intervals.

Growth Measurement

The growth of cultured cells is measured in fresh and dry weight, cell number, packed cell volume, total nitrogen, etc. However, none of these methods of growth measurement takes into consideration growth in all its facets that include cell division, elongation and differentiation. The method of dry weight determination is advantageous as it is a simple technique and gives an acceptable assessment of overall synthetic activity of cells.

The production of useful secondary metabolites and regeneration of plants are two major applications of plant tissue culture. The regeneration of whole, fertile plants from selected or biotechnologically engineered cells has been achieved in many cases. It is achieved mainly with the technique of organogenesis and in few cases, embryogenesis. The technology of plant regeneration is successfully applied for various purposes, like production of virus-free plants, germplasm storage and studies on plant biology.

Organogenesis

The phenomenon of organogenesis refers to the process whereby, tissue callus or cells may be induced to form shoots and complete plants. This process can be initiated by manipulations in the concentrations of growth hormones. Benzyladenine and other cytokinins alone or in combination with naphthalene acetic acid or indole acetic acid and at times, gibberellic acid bring about differentiation and shoot formation. The basic requirement for this technique is either the pre-existing meristematic primordial, which is available through the medium of explant or induced meristematic primordia. The induction of meristem, as well as, the differentiation into root and shoot are achieved supplementing with appropriate levels of plant hormones like auxins, cytokinins and also their relative proportions with each other. The formation of root or shoot is governed by ratio of auxin to cytokinin. Auxin to cytokinin in 4:1 proportion causes shoot formation, whereas the same in 100:1 proportion induces root formation. The shoot formation is favoured by IAA and NAA. High auxin concentrations, however, cause the formation of meristem-like cells.

By means of clonal propagation, technique of regeneration of plants has been commercially exploited for routine multiplication of plant species. The plant meristems are generally virus free as many of plant viruses cannot enter or survive in meristematic tissue. Therefore, meristem tip culture technique is used for the development of virus-free plants. It is propagated from apical shoot tips and used as a source of sterile shoot cultures, which are further regenerated into virus-free plants.

The protoplast fusion is achieved through the medium of protoplast culture. Using this technique, it is possible to induce desired plant yields, disease resistance, etc. It is beneficial for the purpose of somatic hybridization of plant cells. The cell wall may be hydrolyzed by enzymes such as cellulase, hemicellulase and pectinase and isolated protoplasts are mounted in proper medium containing mannitol/sorbitol for maintaining osmolarity. The protoplast fusion is achieved by chemical fusion or electrofusion. It may be either a self–self fusion or binary fusion (for regeneration).

PHYTOPHARMACEUTICALS IN PLANT TISSUE CULTURES

Besides regeneration of plants, the techniques of plant tissue culture have wide applications in production of useful secondary metabolites. A large number of secondary products have been reported to accumulate in tissue cultures. The cell suspension cultures are most widely used mainly due to their potential for scale up to bioreactor levels. Using cell cultures, higher yields of secondary products have been reported—for ginsenosides from *Panax ginseng,* serpentine from *Catharanthus roseus,* shikonin from *Lithospermum erythrorhizon,* etc.

The factors conventionally used for manipulation of yield of secondary metabolites in cell suspension cultures are light, temperature, concentrations of nitrogen, precursors, osmotic stress, plant hormones, carbohydrate and phosphate source and elicitors like fungal extracts.

The exploration of biosynthetic capabilities of cell cultures derived from different higher plants has been systematically investigated in the last four decades by scientists all over the world. In many cases, however, the desired compounds are not produced; whereas in others only low amounts could be detected. The poor yields of secondary metabolites are mainly due to (1) differences between young, dispersed and rapidly dividing cells, akin to meristem tissue; (2) mature slow-growing cells; (3) repression dormancy of desired biosynthetic pathway; and (4) nonexcretion of products from culture by cells. These problems can be significantly overcome by development of hairy root cultures and establishment of immobilized cell systems. The techniques to encourage excretion of desired product into medium are being attempted. The factors limit the success of cell suspension cultures and low accumulation of desired metabolite and slow growth of plant cells. On the other hand, however, examples do exist, showing callus (static) and cell suspension cultures to be capable of synthesizing secondary metabolites with yields comparable to the intact plant. During the last few years, promising findings have been reported for a variety of medicinally important secondary metabolites, some of which are with potential for industrial exploitation.

The secondary metabolites often synthesized after the cell growth in tissue cultures are ceased. It has been possible to establish large-scale production of biomass, containing useful secondary metabolites by defining nutritional requirements and ensuring proper environmental conditions for their in vitro growth. It is achieved by extending the bioreactor time and necessitating a change to production medium with an extended run time. The overall production of a secondary metabolite in cultures depends upon the rate of accumulation within productive cells and proportion of such cells in culture. The productivity of metabolite for commercial considerations is defined as gram product/L culture/day. The yield of biomass, rate of growth, yield of desired product and the number of runs that may be carried out per year are the major factors affecting productivity on an annual basis for a bioreactor process. The cultures capable of producing more than 1 g of compound per litre with a value exceeding US$1000 per kg are considered to be commercially exploitable. The total market size for interesting compound should be large enough to warrant the capital expenditure needed to develop a tissue culture system. In view of these economical constraints, industrial level production by tissue culture technique is limited to a few plant-specific compounds. These include shikonin, digitalis glycosides, rosmarinic acid, diosgenin-derived steroid hormone precursors, opium alkaloids (codeine and morphine), ginsengosides, ajmalicine and other indole alkaloids,

Table 6.2 High levels of biomedicinals in plant cell cultures

Compounds	Species	Yield (% dry wt)
Acetonide	*Syringa vulgaris*	5.0
Ajmalicine	*Catharanthus roseus*	1.8
Anthraquinones	*Morinda citrifolia*	18.0
Benzylisoquinolines	*Coptis japonica*	15.0
Berberine alkaloids	*Berberis wilsoniae*	10.0
Diosgenin	*Dioscorea deltoidea*	7.8
Ginsengoside	*Panax ginseng*	17.0
Nicotine	*Nicotiana tabacum*	5.0
Rosmarinic acid	*Coleus blumei*	21.4
Serpentine	*Catharanthus roseus*	2.2
Shikimic acid	*Galium mollugo*	10.0
Shikonin	*Lithospermum erythrorhizon*	12.4

including vinblastine and vincristine, and possible complex mixture such as rose and jasmine oil. Mitsui Petrochemical Industries of Japan is producing red-coloured phenolic naphthoquinone compound—shikonin from cell cultures of *Lithospermum erythrorhizon*—which is used as red dye and as an astringent. A German pharmaceutical company is producing digitalis glycosides by biotransformation in cell cultures. The anticancer alkaloids of *Catharanthus roseus* are produced on industrial scale using tissue culture technique based on selection of high-yielding cell lines. The selection of a cell line with a high content of desired metabolite and use of specific precursor are prerequisites to the development of cell cultures with high yields. The applicability of enzyme immunoassay and radioimmunoassay techniques for quantitative analysis of bioconstituents has revolutionized the selection of plants and cell clones with high yields. The techniques have been successfully applied in the bioproduction of vindoline, serpentine and ajmalicine. The unusual ether lipids, 1(3),2-diacylglyceryl-3(1)-*O*-4'-(*N,N,N*-trimethyl)homoserine were isolated from algae cultures of *Chlorella fusca*.

The biomedicinals of pharmaceutical significance produced at high levels in plant cell cultures are indicated in Table 6.2.

BIOPRODUCTION OF USEFUL METABOLITES IN HAIRY ROOT AND MULTIPLE SHOOT CULTURES

Hairy cultures are developed by way of incorporating a segment of Ri-DNA from *Agrobacterium rhizogenes* into the plant genome that brings out a phenotypic character called *hairy roots* (Fig. 6.13). These are able to grow rapidly and indefinitely with profuse lateral roots. They grow on a medium

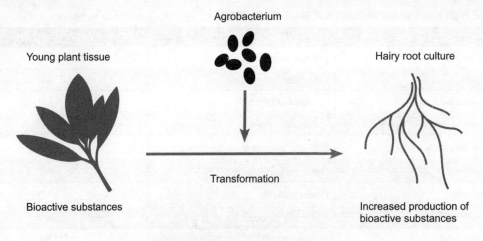

Figure 6.13 Hairy root culture.

without plant hormones and can accumulate levels of secondary metabolites similar to those occurring in natural roots. Many plants have been subjected to these studies. This technique has yielded positive results mostly in nonwoody dicotyledonous plants. It is too early to envisage the commercial impact of genetic manipulation on secondary metabolite production in tissue cultures. The hairy root cultures of *Catharanthus roseus* capable of producing appreciable quantities of ajmalicine, serpentine and catharanthine have been successfully established.

The bioproduction of scopolamine and hyoscyamine in hairy roots developed from *Scopolia japonica* has in some cases found to be lower or greater as compared to the entire plant. The bioproduction of shikonin and betacynin in appreciable quantities has been reported in hairy root cultures of *Lithospermum erythrorhizon* and *Beta vulgaris*, respectively. The inherent ability of a large number of medicinal plants to generate multiple shoots in culture from axillary and shoot tip meristem has opened up new avenues for bioproduction of medicinal agents and preservation/propagation of elite genotype. The axillary meristem and shoot tip develop into shoots on a minimal inorganic basal MS medium both on agar surface and in liquid suspension. They can be propagated by routine periodic subcultures in liquid medium. The type of explant, growth conditions and combination of hormone used play an important role in maintenance of cultures. The bioactive shoot cultures have been established for belladonna, dioscorea, solanum and vinca.

Immobilized Cell Systems

The technique of immobilizing plant cells or enzymes has facilitated the utility of plant cell biotechnology for production of pharmaceuticals. Immobilization by use of matrices such as calcium alginate, agarose, gelatin, carrageenan or polyacrylamide has number of advantages, including reuse of expensive biocatalysts, simplified control with automatic separation of the catalyst from the product and stabilization of the biocatalyst. The plant tissue culture techniques have wide applications in the study of fundamental problems of plant breeding, plant physiology, cytology, genetics, as well as viral, bacterial and nematode plant relationships. The use of these techniques offers a unique approach to the analysis of the metabolic potentialities of higher plant cells, for studies of physiological interrelationships amongst cells, tissues and organs within the multicellular plant; for controlling the expression and differentiation and morphogenesis of the totipotency of such cells. Some of the medicinal compounds localized in morphologically specialized organs or tissues have been produced in culture systems, not only by inducing specific organized structures in submerged cultures, but also by certain culture strains in the absence of organization.

The recent advances in the field of plant tissue cultures have strengthened our hopes towards the realization of full-scale industrial production of biomedicinals in the near future. The increasing awareness of pharmacists in the biotechnological applications of plant tissue culture promises an ever-increasing involvement in a branch of bioscience that could lead to production of new pharmaceutical products to be used in service of the mankind.

QUESTIONS

1. Write a note on mutation in plants.
2. Write a short note on polyploidy, haploid culture, transgenic plants, organogenesis and embryogenesis, callus culture and chemical mutagens.
3. Describe application of transgenic plants.
4. Explain role of various factors in the production of secondary metabolites.
5. Explain role of mutation and polyploidy in improving quality of plants.
6. What is mutation? Write the different types of mutation with its application.
7. Write methods of culturing cells in suspension.
8. Write a short note on embryo culture.
9. Write a short note on factors affecting rate of mutation.
10. Write a note on nutrients incorporated in culture medium of plant tissue culture.
11. Write in brief application of transgenic plants.
12. Explain role of growth regulators in production of secondary metabolites.
13. Explain role of mutation and polyploidy in improving quality of crops.
14. Give an account of pharmaceuticals produced in tissue cultures.

CHAPTER 7
Microbial Culture and Fermentation Process

CHAPTER OUTLINE

- Introduction
- Microorganisms
- Microbial Culture
- Quality Control of Microbial Culture Media
- Questions

INTRODUCTION

Microbial population dominates the biospheres in terms of number and metabolic impact. Amongst the various stages of microbes, *prokaryotes* have a wide spread on the planet, often tolerating extremes of pH, temperature, salts, concentration, etc. Metabolic diversity is greater amongst *prokaryotes* than all *eukaryotes*. The bacterial, yeast and fungal populations are systematically utilised for the manufacture of various chemicals, biochemicals, beverages, antibiotics, etc. Microbes have been involved in the daily affairs of human beings who have been taking their help from time immemorial. The production of alcohol, vinegar, amino acids, vitamins, antibiotics, therapeutics antibodies, acetone and other solvents and recombinant proteins is accomplished by the large-scale cultivation of microbial cell lines of bacteria, algae, yeast and fungi on industrial scale.

The microbiological culture refers to the multiplication of microbial organisms in predetermined culture medium under controlled laboratory conditions. Microbial cultures are used to determine the type of organism, its abundance in the sample being tested, or both. The culture is used for diagnosis of the cause of infectious disease by letting the agent multiply in a predetermined medium. For example, a throat culture is prepared by scraping the lining of tissue in the back of the throat and blotting the sample into a medium for enabling screening for harmful microorganism, such as *Streptococcus pyogenes*, the causative microorganism of strep throat. Implementation of quality measures, compliance with the recommendations of Convention on Biological Diversity (CBD) and adoption of latest bioinformatics tools are essential prerequisites for microbial culture collections to provide resources for the emerging area of the knowledge-based bioeconomy. It should be in association with the deposition of phylogenetically and physiologically diverse microbiological organisms. In view of the fact that the vast majority of microbial isolates do not find their way into public collections, a strategy should be devised to encourage researchers to deposit a higher fraction of strains.

MICROORGANISMS

Types

Out of more than one million microorganisms present in nature, about a few hundred species synthesize the useful products. The examples of some of these important microorganisms are given below.

1. **Algae**: *Chlorella sorokiniana, Spirulina maxima, Gephyrocapsa oceanica, Haematococcus pluvialis.*
2. **Bacteria**: *Acetobacter woodii, Bacillus polymyxa, Bacillus subtilis, Bacillus thuringiensis, Clostridium aceticum, Pseudomonas denitrificans,* etc.
3. **Actinomycetes**: *Micromonospora purpurea, Nocardia mediterranei, Streptomyces aureofaciens, Streptomyces griseus, Streptomyces noursei,* etc.
4. **Fungi**: *Aspergillus niger, Aspergillus oryzae, Aspergillus terreus, Candida utilis, Cephalosporium acremonium, Fusarium moniliforme, Gibberella fujikuroi, Morchella esculenta, Paecilomyces lilacinus, Penicillium chrysogenum, Penicillium notatum, Rhizopus nigricans, Saccharomyces cerevisiae, Saccharomycopsis lipolytica, Sardinella rouxi, Trichoderma harzianum, Trichoderma reesei, Trichoderma viride,* etc.

 MICROBIAL CULTURE

The microorganism isolated from different sources (soil, water and air) or obtained through genetic manipulations are cultured on sterilized growth medium supplemented with sources of carbon, nitrogen, phosphorus, amino acid, trace elements, etc.

Principle of Microbial Growth and Culture System

The growth of microorganism is a highly complex process and is usually expressed in terms of increase in cell number or cell mass. The process of growth depends on the availability of requisite nutrients and their transport into the cells, and the environmental factor, such as aeration, O_2 supply, temperature and pH. Doubling time normally increases with increasing cell size and its complexity. It normally presents in the following ranges for different types of microorganisms: bacteria, 0.30–1 h; yeast, 1–2 h; animal cells, 25–48 h and plant cells, 20–70 h.

Culture Ingredients and Preparation

1. **Carbon source**: The main carbon source used in the culture medium is carbohydrate (sugar). It is required to provide energy and carbon skeleton for the synthesis of various other biological compounds.
2. **Nitrogen source**: Commonly used nitrogen sources in culture medium are ammonium salts, amino acid mixture, urea, animal tissue extracts and plant tissue extracts.

3. **Microelements**: Microelements required in small amounts are Ca, Co, Fe, Mg, Mn, Zn, etc.
4. **Growth factors**: Growth factors are certain organic compounds that are essential for the growth and multiplication of cell, but cannot be synthesized by the cells, e.g. adrenomedullin (AM), erythropoietin (EPO), bone morphogenetic proteins (BMPs), insulin-like growth factor (IGF), etc.
5. **Antifoams**: This is not a nutritive component of the medium. The culture medium rich in nutritive compounds, such as starch, proteins and other organic materials on agitation for aeration can result in excessive foaming. In the production of useful substances by deep-aerobic culture, a large quantity of foam is formed, which causes various problems. In such cases, a fermenter is filled with foam, which lowers a culturing capacity per unit volume and causes the culture solution to overflow from the fermenter. It is necessary to overcome this problem of foaming. This may be arrested by using a suitable antifoaming agent (e.g. Acepol™ 20, Acepol™ 7203, etc.).
6. **Water**: Whether it is liquid or solid, water is the base of any culture medium. In the preparation of the agar nutrient medium, water is comparatively required less than that of liquid nutrient medium. Water is also required for a large number of other services in the laboratory, such as cooling, heating, steaming, etc.

Types of Cultures

The culture vessels and culture medium are sterilized by autoclaving at 120°C for 15–20 min to prevent the growth of unwanted microorganism and possible contamination. If the laboratory-scale experiments are carried out in 100–1000 mL flask, or in lesser volumes of 50 mL or 10 mL, the sterilization may be carried out alternatively in a pressure cooker of convenient size.

Solid or Semisolid Culture

The nutrient medium is supplemented with different concentrations of agar to give solid or semisolid medium. For research purpose, these types of media are regularly used. They are usually not preferred for large-scale production of microbial products, because such media occupy space and are difficult to harvest. However, for the production of amylase from *A. oryzae*, it is used. After the growth of *A. oryzae* for several days at 30°C, mycelia are harvested, dried and ground, which result in crude preparation of amylase.

Batch Culture

Batch fermentation is regarded as a closed system. The sterile nutrient culture medium in the bioreactor is inoculated with microorganism. The incubation is carried out under optimal condition of pH, temperature, oxygen supply, agitation duration, etc.

The following six typical phases of growth are observed in batch fermentation (Fig. 7.1):

1. Lag phase (a)
2. Acceleration phase (b)
3. Logarithmic phase (c)

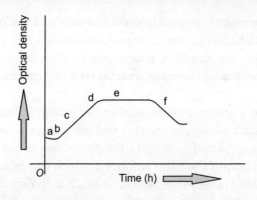

Figure 7.1 Typical growth phases.

4. Deceleration phase (d)
5. Stationary phase (e)
6. Death phase (f)

Lag phase

The initial brief period of culturing after inoculation is referred to as lag phase. During this phase of growth, the microorganism adapts to the new environment. There is no increase in the cell number, although the cellular weight may slightly increase. The length of the lag phase is variable and is mostly determined by the new set of physical and physiological conditions.

Acceleration phase

This is a brief transient period during which cells start growing slowly. In fact, acceleration phase connects the lag phase and log phase.

Log phase

This is the most active phase for growth of microorganism, wherein multiplication occurs rapidly. When the number of cells or biomass is plotted against time on semilogarithmic graph, a straight line is obtained, hence the term log phase. As long as excess substrate is present, growth rate of microbes in log phase is independent of substrate concentration, and there are no growth inhibitors in the medium.

Deceleration phase

With the decline in growth rate of microorganism at the end of log phase, commencement of declaration phase take place.

Stationary phase

As the substrate in the growth medium gets depleted, and the metabolic end products that are formed inhibit the growth, the cells enter into the stationary phase. The microbial growth may either slow down or completely stop during this phase of growth.

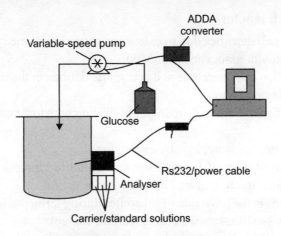

Figure 7.2 Fed-batch culture.

Death phase

This phase is associated with cessation of metabolic activity and depletion of energy reserves. The cell dies at an exponential rate.

Fed-Batch Culture

In this type of culture, the limiting substrate is fed without diluting the culture. The volume of culture can also be practically maintained constant by feeding the growth-limiting substrate in undiluted form, as a very concentrated liquid or gas (e.g. oxygen). Alternatively, the limiting substrate can be added by dialysis. In a photosynthetic culture, radiation may serve the purpose of growth-limiting factor without affecting the culture volume. A certain type of extended fed-batch—*the cyclic fed-batch culture for fixed volume systems*—refers to a periodic withdrawal of a portion of the culture and use of the residual culture as the starting point for a further fed-batch process. Basically, once the fermentation reaches a stage, when aerobic conditions cannot be maintained anymore, the culture is removed and the biomass is diluted to the original volume with sterile water or medium containing the feed substrate. The dilution decreases the biomass concentration and results in an increase in the specific growth rate (Fig. 7.2). Subsequently, as feeding continues, the growth rate declines gradually as biomass increases and approaches the maximum sustainable level in the vessel once more. At this stage, the culture may be diluted again to optimize the process.

Variable volume fed-batch

As the name implies, a variable volume fed-batch is one in which the volume changes with the fermentation time due to the quality of substrate feed.

The feed can be provided according to one of the following options.
1. The same medium used in the batch mode is added.
2. A solution of the limiting substrate at the same concentration as that in the initial medium is added.
3. A very concentrated solution of the limiting substrate is added.

Advantages of fed-batch reactors

1. Production of high cell densities due to extension of working time (particularly, important in the production of growth-associated products).
2. The mode of operation can overcome and control deviations in the organism's growth pattern as found in batch fermentation.
3. Alternative mode of operation for fermentations dealing with toxic substrates or low-solubility compounds.
4. Allows the replacement of water loss by evaporation.
5. Control over the production of by-products or catabolite repression effects due to limited provision of substrates solely required for product formation.
6. Controlled conditions in the provision of substrates during fermentation, particularly regarding the concentration of specific substrates (e.g. carbon source).
7. Increase of antibiotic-marked plasmid stability by providing the correspondent antibiotic during the time span of the fermentation.
8. No additional special piece of equipment is required as compared with the batch fermentation mode of operation.

Limitations of fed-batch reactors

1. Special operational skills are required for development and setting up of the process.
2. It requires previous analysis of the microorganism, its requirements and the understanding of its physiology with the profile of productivity.
3. The special care is required to be taken in the design of the process to ensure that toxins do not accumulate to inhibitory levels.
4. The quantities of the components to control must be above the detection limits of the available measuring equipment.

Continuous Culture

The culture medium is continuous in bioreactor, and it is replaced with a fresh sterile medium.

Homogeneously mixed bioreactor

The culture solution is homogeneously mixed in this type of bioreactor. The bioreactors are of the following two types:

1. **Chemostat bioreactor**: In this case, the concentration of any one of the substrates is adjusted to control the cell growth and maintain a steady state. An important feature of chemostat is that microorganism can be grown in a physiological steady state, wherein growth occurs at a constant rate and all culture parameters—such as culture volume, dissolved oxygen concentration, nutrient and product concentrations, cell density, pH, etc.—remain constant. The microorganisms grown in chemostats naturally strive in the steady state. If a low amount of cells are present in the bioreactor, the cells can grow at growth rates higher than the dilution rate, as growth is not limited by the addition of the limiting nutrient. The limiting nutrient is a nutrient that is essential for growth and present in the media at a limiting concentration (all other nutrients are usually supplied in surplus). If the cell

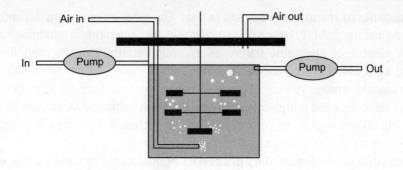

Figure 7.3 Chemostat bioreactor.

concentration becomes too high, the amount of cells that are removed from the reactor cannot be replenished by growth as the addition of the limiting nutrient is insufficient. It results in an equilibrium situation (steady state), where the rate of cell growth is equal to that of cell removal (Fig. 7.3).

2. **Turbidostat bioreactor**: In this case, turbidity measurement is used to monitor the biomass concentration.

Plug flow bioreactor

In this bioreactor, the culture solution is fed through a tubular reaction vessel without back mixing. The composition of medium, the quantity of cells, O_2 supply and product formation vary at different locations in the bioreactor.

Continuous fermentation process is used for the production of antibiotics, organic solvents, single-cell proteins, beer and ethanol, besides its use in wastewater treatment.

Advantage of continuous cultures

1. Continuous fermentation can be run in a cost-effective manner.
2. The size of the bioreactor and other equipment are smaller as compared to batch fermentation.
3. The yield of the product is more consistent.
4. The time required for cleaning and preparing the bioreactor for reuse between two successive fermentations can be avoided.

Limitation of continuous cultures

1. Maintenance of sterile condition for such a long period is difficult.
2. It is difficult to maintain the same quality of the culture medium for all the additions.
3. Recombinant cells with plasmid construct cannot function continuously.

Choice of Fermenter

The selection of the type, size, mode of operation and the pattern of the fermenter to be used in a particular biological process is based on the following characteristics:

1. **Characteristics of the microorganism in use**: The mode of operation depends substantially on the type and the stability of the microbial strain. The operational conditions are determined based on aerobic or anaerobic nature of microbial strain. The size and the shape of the microbial cells also have considerable effect on the design of the fermenter. The spherical cells are usually smaller in size and less sensitive to shear than the filamentous organisms. The spherical cells need a higher degree of dispersion of the air as compared to filamentous mycelia. Small dimensions of cells ensure a high surface to volume ratio and a high rate of uptake of substrate. Therefore, a rapid growth is possible.
2. **Characteristics predetermined by properties of medium**: The choice of the strain generally not only determines the culture medium, but also it exerts a pronounced influence on the choice of the bioreactor. The characteristics of the medium such as physical and biokinetics, difficulty in sterilization and rheological behaviour are the most important properties to be considered for the selection of the reactor. When the substrate or product is responsible for the high viscosity, the medium frequently has a new toning behaviour. A high apparent viscosity due to the morphology of the organism is almost always associated with a non-Newtonian behaviour. The physical properties of the substrate used generally differ like gaseous, liquid and water soluble, solid and water soluble, liquid and water insoluble and solid and sparingly soluble or insoluble in water.
3. **Characteristics predetermined by parameters of biomedical process**: The important factor influencing choice of the fermenter in cultivation of aerobic organism is specific oxygen transfer rate in the medium. The rate of growth and rate of product formation are temperature dependent. Consequently, the cultivation and the formation of the product are usually carried out at controlled temperature.

Some important fermentation-based commercial products are given in Table 7.1.

Table 7.1 Some products produced by microorganisms in fermenter

Products	Microorganisms
Industrial chemicals	
Ethanol	*Saccharomyces cerevisiae*
Citric acid	*Aspergillus niger*
Gluconic acid	*A. niger*
Acetic acid	*Acetobacter* spp.
Lactic acid	*Lactobacillus delbrueckii*
Amino acids	
L-lysine	*Corynebacterium glutamicum*
MSG	*C. glutamicum*
Glutamic acid	*C. glutamicum*
Vitamins	
Riboflavin	*Ashbya gossypii*
Vitamin B_{12}	*Pseudomonas denitrificans*
Ascorbic acid	*Gluconobacter oxydans*

(Continued)

Products	Microorganisms
Enzymes	
Amylase	*Aspergillus oryzae*
Cellulose	*Trichoderma reesei*
Invertase	*Saccharomyces cerevisiae*
Lipase	*Saccharomycopsis lipolytica*
Polysaccharides	
Dextran	*Leuconostoc mesenteroides*
Xanthan gum	*Xanthomonas campestris*
Pharmaceuticals	
Penicillin	*Penicillium notatum*
Cephalosporins	*Cephalosporium acremonium*
Amphotericin B	*Streptomyces nodosus*
Kanamycin	*Streptomyces kanamyceticus*
Neomycin	*Streptomyces fradiae*
Streptomycin	*Streptomyces griseus*

QUALITY CONTROL OF MICROBIAL CULTURE MEDIA

The culture media are widely employed for isolation, identification and sensitivity testing of different pathogenic microorganisms. Most of the laboratories usually prepare their own media for routine diagnostics, as well as research purposes. However, to ensure that the media prepared are of good quality and capable of giving satisfactory results, proper quality management system should be in place. For this purpose, following parameters of media preparation should be thoroughly checked and then taken up for laboratory use.

Raw Materials

The quality of the medium depends directly upon the quality of the raw materials used for its preparation. Water is the most important raw material used for the preparation of culture medium. The parameters to be checked for water are presence of copper ions, conductivity and pH. Ideally, there should be no copper ions present in water, because they are inhibitory to the growth of microorganisms. The quality of Petri dishes used is also an important factor. Normally Petri dishes are ethylene oxide sterilized or gamma irradiated. The maximum permissible limit for residual ethylene oxide is 1 µg/g, and it can be measured by standard gas chromatographic method. Only borosilicate glassware should be used, because soda glass can leach alkali into the medium.

For blood-containing culture medium, sterility, homogeneity, viscosity and colour of the blood should be scrupulously checked. For other additives, the certificate of analysis and sterility conditions should be taken into consideration.

Sterilization Parameter

Sterilization of the medium plays an important role in maintaining the quality of the culture medium. Sterilization by autoclaving is the usual practice in preparation of medium. However, the time of autoclaving and the quantity of medium sterilized should be closely monitored. Excessive heat treatment of complex culture medium may result in destruction of its nutrients either by direct thermal degradation or by reactions between the components. Therefore, it is very important to optimize the heating process to minimize the loss of nutrients.

Physical Parameter

The gross physical appearance of medium often suggests the quality. The culture medium prepared should be screened for physical characteristics, such as excessive bubbles or pits, unequal filling of plates (uniform levelling), cracked medium in plate and effect on freezing or crystallization. All these characters can be checked visually by naked eye. However, in order to ascertain unequal filling of plates, thickness of medium can be checked at two ends of two diameters of the plate, which are at right angles to each other. Thus, all the four sides can be simultaneously checked. The thickness at these four points is noted down, and the mean thickness is determined and reported as mean thickness of the medium in the plate, which must be 4.0 ± 0.2 mm. The pH value can be measured during preparation of the medium, before and after autoclaving, by using the standard pH metre after proper calibration with standard buffers.

Microbiological Parameter

Growth-supporting characteristics are the most important parameters in quality control of culture medium. Standard inoculating procedures should be used. Both qualitative and quantitative results should be examined. While testing new lots, both previous batch and new batch should be simultaneously grown.

Econometric Method

This is a simple and numerical method. Both absolute growth index (AGI) and relative growth index (RGI) can be determined. The method is based on streaking of inoculum to extinction. This method can be used to compare results with previous batches of same medium or between selective and nonselective media. In this method, 5 mL of brain heart infusion broth is inoculated with loopful of chosen test organism and incubated for 4 h.

Productivity Ratio

The productivity ratio (PR) of a culture medium is related to control medium, which is usually a nutritious agar-like tryptone soy agar (TSA). The inoculum used should be the same for both media and PR, and it is calculated by counting the colonies on test and control medium.

Total productivity = Output quantity/Input quantity

$$PR = \frac{\text{No. of colonies on test} \times \text{dilution factor}}{\text{No. of colonies on control} \times \text{dilution factor}}$$

The method employed is modified Miles and Misra technique. A 10-fold dilution of an overnight culture of test organism is prepared in buffered peptone water. The test plates are divided into four quadrants, and each quadrant is marked with the dilution to be used. Starting with highest dilution, each drop of the dilution is placed on the relevant quadrant. Same steps are repeated for control plate. Each drop is spread in the corresponding quadrant and incubated at 37°C for 18 h. The colonies of the lowest dilution are counted for both test and control plates.

Detection of Contamination

It is an important parameter for determination of the quality of the culture medium. It is recommended that the whole batch of the prepared medium be checked for contamination by keeping the plates with medium, at least for 3 days at room temperature. Alternatively, two plates from the test batch can be placed into the incubator set at 37°C for 24 h. After required incubation, the plates are checked for any growth. If the growth is observed, the process is repeated, with two more plates from the same batch. If contamination occurs again, it is inferred that contamination has occurred in the prepared batch, which is required to be discarded.

Gel Strength Parameters

Gel strength is an indication of level of solidification of agar in the culture medium. It is measured by using a tripod stand with a central rod that is used to impart pressure on the agar. The lower end of the rod has a spherical portion, which rests on the surface of the medium. The upper portion of the rod has a platform on which standard weights are placed. The spherical portion of the central rod is placed on the medium, and weights are placed on the upper platform one by one and observed for some time. The process is continued until the agar breaks under the weight of the central rod. While calculating the gel strength, the weight of the central rod should be deducted. The force imparted by the rod on the agar surface can be calculated by the formula

$w/\pi r^2$, where w = weight kept on the platform, r = radius of the spherical portion at the lower end of the central rod and $\pi = 3.14$.

A gel strength of about 300–500 dyne/cm² is usually considered to be satisfactory.

QUESTIONS

1. What are the factors that affect microculture?
2. Differentiate between shake flask and solid-state fermentation. Explain their advantages.
3. Describe the physical and chemical environment for microbial growth.
4. Give some examples of viscous fermentation products.

5. What do you mean by fed-batch fermentation?
6. Write a short note on batch culture.
7. Draw a neat labelled diagram of fermentor and write its uses.
8. Differentiate between batch and continuous culture.
9. Give design and operation of batch fermentation with diagram.

CHAPTER 8: Genetics and Molecular Biotechnology

CHAPTER OUTLINE

- Introduction
- DNA as Segment of Molecular Heredity
- Genes
- Mutations
- Molecular Biotechnology
- Questions

INTRODUCTION

Genetics refers to the study of inheritance and variability of the characteristics of an organism. *Heredity* is concerned with exact transmission of genetic information from parents to their progeny.

Classical genetics: It refers to the study of mutants—study of properties of mutants, mapping and generating hypothesis. The main principle of classical genetics is the function of genes and their products that are observed when the genes are muted.

Reverse genetics: It is the study of isolation of genes with the help of phenotype and its linkage with molecular markers.

Gene comprises the genome of cell that may consist of single DNA molecule or some other small DNA molecules. The genotype variation brought about by changes in genetic information could be because of mutation or loss of plasmid. The genetic material and components, particularly nucleic acid, were first isolated in 1868 by Friedrich Miescher. He established the material isolated containing phosphorus, known as nucleoprotein. Hydrolysis of nucleic acid from thymus yielded purine and pyrimidine bases, 2 deoxyribose and phosphoric acid. The acid has been named as *deoxyribonucleic acid (DNA)*. Another nucleic acid was isolated that contained uracil instead of thymine and ribose in place of deoxyribose. This has been known as *ribonucleic acid (RNA)*. A small portion of DNA that is localized outside the chromosomal region is known as *plasmid*.

History of Genetics

The modern science of genetics originated with the discovery of hereditary characteristics by Gregor Johann Mendel (1822–84). Each such unit of hereditary nature, which is called a genetic unit or *gene*, is a substance that satisfies at least the following two essential requirements:

1. It is inherited between generations in such a fashion that each descendent has a physical copy of this material.
2. It provides information to its carriers in respect to structure, function and other biological attributes.

Although, the science of genetics began with the applied and theoretical work of Gregor Mendel in the mid 1800s, other theories of inheritance preceded Mendel. A popular theory put forward by Jean-Baptiste Lamarck during those days was the concept of *blending inheritance*—the concept that individuals inherit smooth blend of traits from their parents. Mendel's work disapproved this theory proving that traits are composed of combinations of distinct genes rather than a continuous blend, and the experiences of individuals do not affect the genes they pass to their children. Other theories included the pangenesis of Charles Darwin (1809–82) related to both acquired and inherited aspects and Francis Galton's (1822–1911) reformulation of pangenesis as both particulate and inherited.

The modern science of genetics traces its roots to the observations made by Gregor Johann Mendel, a German-Czech Augustinian monk and scientist who studied the nature of inheritance in plants. The importance of Mendel's work did not gain worldwide attention until 1890. Thomas Hunt Morgan (1866–1945) argued that genes are on chromosomes, based on observations of a sex-linked white eye mutation in fruit flies. Alfred Sturtevant (1891–1970), a student of Thomas Hunt Morgan, used the phenomenon of genetic linkage to show that genes are arranged linearly on the chromosome.

Important Contributions

Some of the contributions of modern genetics are as follows:

1. It has revolutionized the approaches of genetic analysis.
2. All possible mutations can be studied by simply changing cloned DNA before reintroducing it into cells.
3. Method of transferring cloned DNA back into living cells has provided the key to producing gene product (e.g. hormones and interferons) in large quantities in cells that are simple and easy to cultivate.
4. Isolation and characterization of genes involved in cancer is possible.
5. Interspecific gene transfer to establish conservation is a reality today.

Problems and Limitations of Genetics

They can be briefly enlisted as follows:

1. Some mutations are lethal in homozygotes.
2. Only mutation that produces readily detectable phenotype can be studied.

3. The genetic analysis is not possible except by retrospective studies of family pedigrees.
4. In dipole organism, such mutation can be detected after appropriate breeding strategy to generate homozygote.

Evaluation and Molecular Biology of Plants

The questions concerning the evaluation of organelles have been a key force driving studies of organelle molecular biology. The endosymbiont origin of cellular organelles was proposed in 1882, and this hypothesis was further developed in early 20th century. Molecular feature of organelle genes and genome provided the confirming data for the endosymbiont hypothesis. Molecular research over the subsequent decades reveals many prokaryotic features in the modern-day plant organelles, including some aspects of organelles division, genome, coding content and organization, transcription, translation, RNA processing and protein turnover. Present-day mitochondrial genomes are extremely diverse, and the mitochondria and the land plants are uniquely expanded over the ancestral type. The functional transfer of genes from organelle to nuclear genes has occurred frequently and recently in the plant lineages.

DNA AS SEGMENT OF MOLECULAR HEREDITY

It has been proved experimentally that DNA is the molecule of heredity. The importance of nucleus that contains DNA is established by the observation that nuclei of male and female germi cells fuse during the process of fertilization.

Structure of DNA

The structure of DNA is the physical basis for inheritance. The replication of DNA duplicates the genetic information by splitting the strands and using each splitted strand as a template for synthesis of a new partner strand. The genes are arranged linearly along long chains of DNA sequence called *chromosomes*. In bacteria, each cell usually contains a single circular chromosome, whereas DNA arranged in multiple linear chromosomes are characteristics of eukaryotic organisms (including plants and animals). These DNA strands are often extremely long. The largest human chromosome, for example, is about 247 million base pairs in length.

The DNA of a chromosome is associated with structural proteins that organize, compact and control access to the DNA, forming a material called *chromatin*. In eukaryotes, chromatin is usually composed of nucleosomes, and segments of DNA wound around cores of histone proteins. The full unit of hereditary material in an organism (usually combined DNA sequences of all chromosomes) is called the *genome* (Fig. 8.1).

Nucleotide

The molecular basis for genes is DNA. It is composed of a chain of nucleotides, of which there are four types (Fig. 8.2): (1) adenine (A), (2) cytosine (C), (3) guanine (G) and (4) thymine (T). Genetic

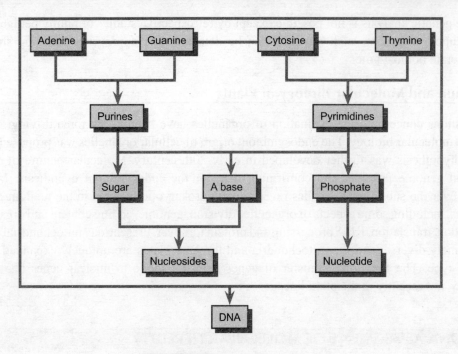

Figure 8.1 Structure of DNA.

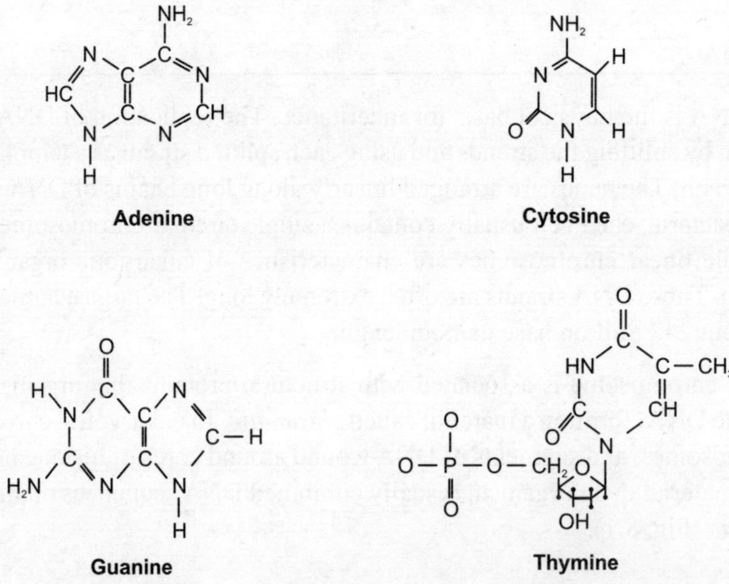

Figure 8.2 Structures of adenine, guanine, cytosine and thymine.

information exists in the sequence of these nucleotides, and genes exist as stretches of sequence along the DNA chain.

Each nucleotide in DNA preferentially pairs with its partner nucleotide on the opposite strand. A pairs with T, and C pairs with G. Thus, in its two-stranded form, each strand essentially contains all necessary information, redundant with its partner strand. Viruses are the only exception to this rule. Sometimes viruses utilize the very similar molecule RNA instead of DNA as their genetic material.

Double Helix

The structure of DNA was put forward by Watson and Crick in 1953 using X-ray diffraction picture taken by Franklin and Wilkins. The X-ray diffraction picture of the double helix shows repeated pattern of bands that reflects the regularity of the DNA. Pitch of the helix is 34 Å, and the spacing between bases is 3.4 Å. The diameter of the helix is 20 Å. The double helix is said to be antiparallel. One of the strands runs in the $5'\rightarrow 3'$ direction, and the other in the $3'\rightarrow 5'$ direction. Only antiparallel polynucleotide forms a stable helix. The double helix is not absolutely regular and when viewed from the outside, a major groove and a minor groove can be seen. These are important for interaction with proteins for replication of the DNA and for expression of the genetic information. The double helix is right handed. DNA normally exists as a double-stranded molecule, coiled into the shape of a double helix (Fig. 8.3).

DNA Replication and Inheritance

The *cell division* is a process by which a single cell divides into two usually identical daughter cells. The growth, development and reproduction of organism depend upon cell division. It requires first making a duplicate copy of every gene in the genome in a process called *DNA replication*. The copies are made by specialized enzymes known as DNA polymerases, which read one strand of the double-helical DNA (template strand) and synthesize a new complementary strand. Because the DNA double helix is held together by base pairing, the sequence of one strand completely specifies the sequence of its complement. Therefore, only one strand needs to be read by the enzyme to produce a faithful copy.

The process of DNA replication is semiconservative, wherein the copy of genome inherited by each daughter cell contains one original and one newly synthesized strand of DNA. After completion of DNA replication, the cell must physically separate the two copies of the genome and divide into two distinct membrane-bound cells. In prokaryotes (bacteria and archaea), this usually occurs through relatively a simple bioprocess known as *binary fission*, wherein each circular genome attaches to the cell membrane. It is separated into the daughter cells as the membrane invigilates to split the cytoplasm into two membrane-bound portions. The process of binary fission is extremely fast as compared to the rates of cell division in eukaryotes. The cell division of eukaryotic is a more complex process referred to the *cell cycle*. The DNA replication occurs during a phase of this cycle

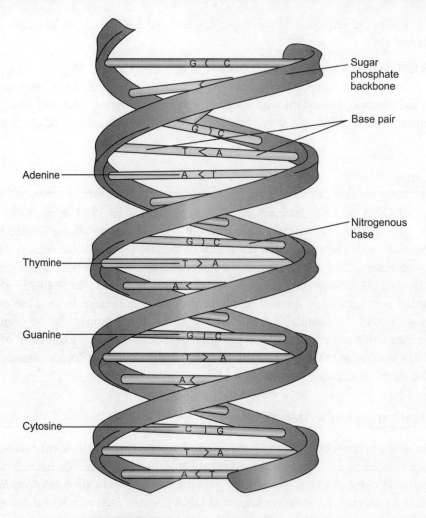

Figure 8.3 Structure of double helix of DNA.

known as *S* phase, whereas the process of segregating chromosomes and splitting the cytoplasm occurs during *M* phase. In many single-celled eukaryotes, e.g. yeast, reproduction by budding is a common phenomenon that results in asymmetrical portions of cytoplasm in the two daughter cells (Fig. 8.4).

GENES

A *gene* is a unit of heredity in a living organism. It is normally a stretch of DNA that codes for a type of protein or for an RNA chain that has a function in the organism. Genes hold the information to build and maintain organism's cells and pass genetic traits to offspring. All living things depend on genes, as they specify all proteins and functional RNA chains. The *gene* is a locatable region of genomic sequence, corresponding to a unit of inheritance, which is associated with regulatory

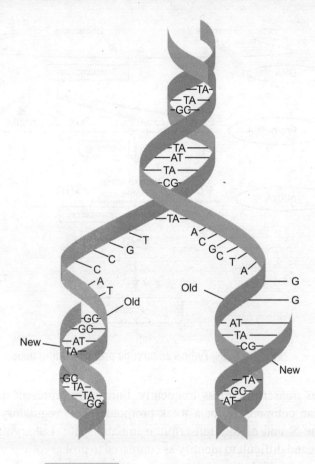

Figure 8.4 DNA replication and inheritance.

regions, transcribed regions and or other functional sequence regions. Colloquial usage of the term *gene* (e.g. good genes, hair colour genes) may actually refer to an *allele*. A gene is the basic instruction and a sequence of nucleic acid (DNA or, in case of certain viruses RNA), whereas an allele refers to one variant of that instruction (Fig. 8.5).

In addition to possessing regulatory regions (promoters), all genes have regions that explicitly code for a protein or RNA product. The promoter provides a position that is recognized by the transcription machinery when a gene is about to be transcribed and expressed. Promoters and enhancers determine what portions of the DNA shall be transcribed into the precursor mRNA (pre-mRNA). The pre-mRNA is then spliced into messenger RNA (mRNA) that is later translated into protein.

A gene may contain more than one promoter, resulting in RNAs that differ in how far they extend in the 5′ end. Although, promoter regions have a consensus sequence (most common sequence at this position), some genes contain 'strong' promoters that bind the transcription machinery well, and others have 'weak' promoters that bind poorly. The weak promoters usually permit a lower rate of transcription as compared to the strong promoters, because the transcription machinery binds

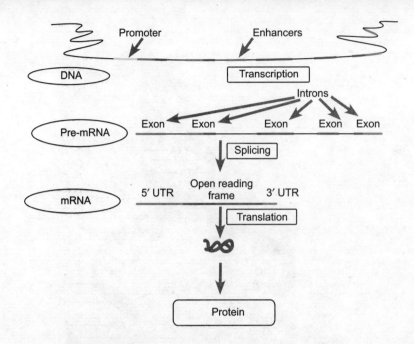

Figure 8.5 Typical eukaryotic protein-coding gene.

to them and initiates transcription less frequently. Enhancers represent other possible regulatory regions, and they can compensate for a weak promoter. Most regulatory regions are 'upstream' before or toward the 5' end of the transcription initiation site. Eukaryotic promoter regions are much more complex and difficult to identify as compared to prokaryotic promoters. The prokaryotic genes are usually organized into *operons*, or groups of genes whose products have related functions and which are transcribed as a unit. The eukaryotic genes are transcribed only one at a time, but may include long stretches of DNA known as *intones*, which are transcribed, but never translated into protein. They are spliced out before translation.

Chromosomes

The total complement of genes in an organism or cell is known as its *genome*. It may be stored on one or more chromosomes. The region of the chromosome at which a particular gene is located is called its *locus*. A chromosome consists of a single, very long DNA helix, whereupon thousands of genes are encoded. Prokaryotes—bacteria and archaea—typically store their genomes on a single large circular chromosome, sometimes supplemented by additional small circles of DNA called *plasmids*. These usually encode only a few genes and are easily transferable between individuals. For example, the genes for antibiotic resistance are usually encoded on bacterial plasmids and can be passed between individual cells (even those of different species), via horizontal gene transfer. The majority of eukaryotic genes are stored on multiple linear chromosomes. However, some simple eukaryotes also possess plasmids with small numbers of genes. The eukaryotic genes are packed within the nucleus in complex with storage proteins referred to as *histones*. The manner in which

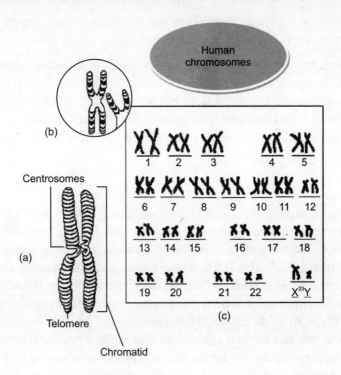

Figure 8.6 Chromosomes of the human male.

DNA is stored on the histone, as well as chemical modifications of histone itself, are regulatory mechanisms indicative of whether a particular region of DNA is accessible for gene expression. The ends of eukaryotic chromosomes are capped by long stretches of repetitive sequences known as *telomeres*, which do not code for any gene product, but are present to prevent degradation of coding and regulatory regions during DNA replication. The length of the telomeres tends to decrease each time the genome is replicated in preparation for cell division. The loss of telomeres has been proposed as an explanation for the loss of the ability to divide, and by extension for the aging process in organisms (Fig. 8.6).

Advancement in technology related to human artificial chromosomes: The technology of *human artificial chromosome (HAC)* has developed rapidly over the past one decade. Recent reports show that HACs are useful gene transfer vectors in expression studies and important tools for determining human chromosome function. HACs have been used to complement gene deficiencies in human cultured cells by transfer of large genomic loci also containing the regulatory elements for appropriate expression. They offer the possibility to express large human transgenes in animals especially, in mouse models of human genetic diseases.

Gene Expression

Genetic information is encoded as a series of genes in the base sequence of DNA molecules. *Gene expression* is the term commonly used to explain how cell decodes the information to synthesize proteins required for cellular function. It involves the synthesis of a complementary of RNA molecules whose sequence is specific to the amino acid sequence of a protein.

In all organisms, there are two major steps involved separating a protein-coding gene from its protein. First, the DNA on which the gene resides must be transcribed from DNA to messenger RNA (mRNA) and then it must be translated from mRNA to protein. RNA-coding genes must go through the first step, but are not translated into protein. The process of producing a biologically functional molecule of either RNA or protein is called *gene expression*, and the resulting molecule itself is known as a *gene product*.

Transcription

This is the first stage of gene expression. It involves the synthesis of RNA from a DNA template by polymerase enzyme. The process of genetic transcription yields a single-stranded RNA molecule known as *messenger RNA*. The nucleotide sequence in messenger RNA is complementary to the DNA from which it is transcribed. The DNA strand whose sequence matches with that of the RNA is referred to as the *coding strand*, and the strand from which the RNA is synthesized is known as *template strand*. Transcription is performed by an enzyme RNA polymerase, which reads the template strand in the 3' to 5' direction and synthesizes the RNA from 5' to 3'. To initiate transcription, the enzyme polymerase first recognizes and binds a promoter region of the gene. Therefore, the major function of gene regulation is blocking or sequestering of promoter region, either by tight binding through repressor molecules that physically block the polymerase, or by organizing DNA so that the promoter region is not accessible.

In prokaryotes, transcription occurs in the cytoplasm. For very long transcripts, translation may begin at the 5' end of the RNA, whereas the 3' end is still being transcribed. In eukaryotes, transcription necessarily occurs in the nucleus, where the cell's DNA is sequestered. The RNA molecule produced by polymerase enzyme is known as *primary transcript*. It must undergo posttranscriptional modifications before being exported to the cytoplasm for translation. The splicing of introns present within the transcribed region is a specific modification related to eukaryotes. Alternative splicing mechanisms can result in mature transcripts from the same gene with different sequences and thus, coding for different proteins is possible. This is a major form of regulation in eukaryotic cells.

Analysis of transcription

The levels of gene expression are measured by comparison among three groups of blastocysts (in vivo, in vitro culture [IVC] in KSOM and IVC in KSOM+FCS). Different patterns of gene expression and development are found between embryos cultured in vitro or in vivo. Moreover, when the embryos produced in KSOM are compared with those of KSOM+FCS, it was observed that the presence of FCS affected the expression of 198 genes. The processes of metabolism, proliferation, apoptosis and morphogenetic pathways are the most common processes affected by IVC. However, the presence of FCS during IVC preferentially affects genes associated with certain molecular and biological functions related to epigenetic mechanisms. These results indicate that culture-induced alterations in transcription (at the blastocyst stage) related to epigenetic mechanisms help in foundation for understanding the molecular origin at the time of preimplantation development of the long-term consequences of IVC in mammals.

Translation

It is a process in which a mature mRNA molecule is used as a template for synthesizing a new protein. Translation is carried out by ribosome, large complexes of RNA and protein responsible for chemical reactions related to addition of new amino acids to a growing polypeptide chain by formation of peptide bonds. The genetic code is read through interactions with specialized RNA molecules known as *transfer RNA (tRNA)*. Each tRNA has three unpaired bases known as *anticodons* that are complementary to the *codons* it reads. The tRNA is also covalently attached to the amino acid specified by the complementary codon. When the tRNA binds to its complementary codon in mRNA strand, the ribosome legates its amino acid cargo to the new polypeptide chain, which is synthesized from amino terminus to carboxyl terminus. During and after its synthesis, the new protein must fold to its active three-dimensional structure, before it carries out its cellular function.

Genetic Code

Genetic code refers to the set of rules by which a gene is translated into a functional protein. Each gene consists of a specific sequence of nucleotides encoded in a DNA (or sometimes RNA) strand. A link between nucleotides (the basic building blocks of genetic material) and amino acids (the basic building blocks of proteins) is required to be established for genes to be successfully translated into functional proteins.

A total of 64 base triplets called *codons* encode around 20 amino acids. Most amino acids have more than one codon. The correspondence between codons and amino acids is nearly universal among all living organism. It is known as the *degeneracy* of the genetic code and it helps to minimizing the effect of mutation (Fig. 8.7).

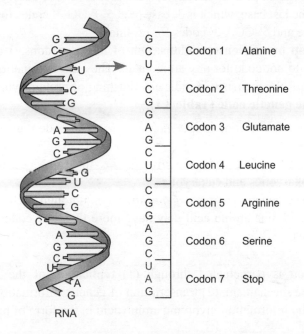

Figure 8.7 Single-stranded RNA molecule illustrating the position of three-base codons.

Table 8.1 Characters of the genetic code

First base (5' end)		Second base				First base (3' end)
		U	C	A	G	
U	UUU UUC	Phe	UCU UCC UCA UCG	Ser	UAU UAC	Tyr	UGU UGC	Cys	U C
	UUA UUG	Leu			UAA Stop UAG Stop	UGA Stop UGG Trp	A G
C	CUU CUC CUA CUG	Leu	CCU CCC CCA CCG	Pro	CAU CAC	His	CGU CGC	Arg	U C
					CAA CAG	Gln	CGA CGG		A G
A	AUU AUC AUA	Ile	ACU ACC ACA ACG	Thr	AAU AAC	Asn	AGU AGC	Ser	U C
	AUG	Met or start			AAA AUG	Lys	AGA AGG	Arg	A G
G	GUU GUC GUA GUG	Val	GCU GCC GCA GCG	Ala	GAU GAC	Asp	GGU GGC	Gly	U C
					GAA GAG	Glu	GGA GGG		A G

1. **Translation of codons**: It is through the table or dictionary used to translate codon sequence. Each triplet is read from 5' to 3' direction so the first base is 5' base, followed by the middle base and finally the last base, which is 3' base (e.g. 5'-AUG-3' codes for methionine, 5'-UCU-3' codes for serine and 5'-CCA-3' codes for proline).
2. **Termination (stop or nonsense) codons**: Out of the 64 codons, 3 codons, namely UAA, UAG and UGA, do not code for any amino acid. They are termination codes. When one of them appears in mRNA sequence, it indicates finishing of protein synthesis.
3. **Characters of the genetic code (Table 8.1)**
   a. *Specificity:* The genetic code is *specific*, i.e. a specific codon always codes for the same amino acid.
   b. *Universality:* The genetic code is *universal*, i.e. the same codon is used in all living organisms, prokaryotics and eukaryotics.
   c. *Degeneracy:* The genetic code is *degenerative*, i.e. although each codon corresponds to a single amino acid, one amino acid may have more than one codon, e.g. arginine has six different codons.

Genetic information is perpetuated through (1) replication of the DNA and its genetic information by template mechanism, (2) transcription of genetic information into RNA molecules and (3) translation of the information involving amino acid in the form of proteins.

MUTATIONS

The replication of DNA is usually accurate with an error rate per site of around 10^{-6} to 10^{-10} in eukaryotes. Rarely, spontaneous alterations in the base sequence of a particular gene arise from a number of sources, e.g. errors in DNA replication and the aftermath of DNA damage. These errors are known as *mutations*. Mutation is a random undirected, heritable variation caused by an alteration in the nucleotide sequence at some point of the DNA of the cell. A cell or an organism that shows the effect of the mutation is known as *mutant*. Each gene undergoes mutation with a fixed frequency mutation rate of individual gene in bacteria from 10^{-2} to 10^{-4} per bacterium per division.

During the process of DNA replication, errors occasionally occurring in polymerization of second strand result into *mutations*. The mutations have an impact on the phenotype of an organism especially, if they occur within the protein coding sequence of a gene. The errors in alignment during meiosis in organisms that use chromosomal crossover to exchange DNA and recombine gene can also cause mutations. Errors in crossover are especially likely when similar sequences cause partner chromosomes to adopt a mistaken alignment; this makes some regions in genomes more prone to mutating than other.

The mutation occurring at different frequencies are discussed below.

Induced Mutation

They are mutations resulted by mutagens that are chemical or physical agents capable of increasing the rate of mutation. The mutation rate is defined as the average number of mutations per cell per division, and it is expressed as a negative exponent per cell division. A variety of mutagens are known to increase the rate of mutation in microorganism. These include both physical and chemical agents.

Physical agents

The commonly used physical agents to induce mutation are ultraviolet light, X-ray, γ-radiation, heat, etc. All these agents can cause nonselective mutation. Irradiation of DNA with UV rays generally results in formation of covalent bonds between thymine molecules on the same stand of DNA yielding thymine-thymine dimers. The mutation due to UV is nonsense type of mutation due to change in one or few bases in the DNA.

Chemical agents

These are nitrous acid, base analogues, hydroxylamine, alkylating agents such as ethyl ethane sulphonate, ethyl methane sulphonate, sulphur mustard, nitrosoguanidine, etc.

Spontaneous Mutation

Mutation that occurs in the absence of all mutagenic agents is known as *spontaneous mutation*. It was first provided by Salvador Luria and Max Delbruck in 1943 by the fluctuation test, wherein a

Table 8.2 Types, mechanism and frequency of mutation

Types of mutation	Mechanism	Frequency per cell division
Point mutation	1. Mistakes in DNA replication 2. DNA damage by chemical mutagens (or by radiation)	~10^{-10}/base pair ~10^{-5}/gene ~0.5/cell
Submicroscopic deletion or insertion	1. Unequal crossing over 2. Misalignment during DNA replication 3. DNA damage by chemical mutagens (or by radiation) and misrepair 4. Insertion of mobile element	Included in the above
Microscopically visible deletion, translocation or inversion	1. DNA damage by chemical mutagens (or by radiation) and misrepair 2. Unequal crossing over	6×10^{-4}

series of tubes containing 0.5 mL of cells were incubated without phage until a certain population size was reached. The cultures were then exposed to phage by pouring the contents of each tube into an agar plate containing phage. The colony count from series of similar cultures were then compared with the result of a series of samples taken from one culture started with a similar density of cells/mL and allowed to reach a similar population. The type, mechanism of action and frequency of mutation are given in Table 8.2.

MOLECULAR BIOTECHNOLOGY

Molecular biotechnology is an exciting and revolutionary scientific discipline that is based on the ability of research to transfer specific units of genetic information from one organism to another.

Today, with the acquisition of sufficient knowledge of the biochemistry, genetics and molecular biology of microorganism, it is possible to accelerate the development of useful and improved biology products and processes and to create new products that would not have otherwise occurred. The functional unit of inheritance and the discipline that is concerned with the manipulation of genes for the purpose of providing useful product services using living organism is known as *molecular biotechnology*.

The molecular biotechnology laboratory techniques include chemical synthesis of genes, the polymer chain reaction and DNA sequencing. The molecular biotechnologist makes use of a number of different biological systems for genetic manipulation and for production of commercial product of these replicate biological units.

Techniques of Molecular Biology

Since early 1960s, molecular biologists have learnt to characterize, isolate and manipulate the molecular components of cells and organisms. These components include DNA, the repository of genetic information, RNA, proteins and the major structural and enzymatic types of molecule in cells.

Expression Cloning

It is one of the most basic techniques of molecular biology for the study protein function. In this technique, DNA coding for a protein of interest is cloned (using PCR and/or restriction enzymes) into a *plasmid* (known as an expression vector). This plasmid may contain special promoter elements to drive production of the protein of interest, and may also contain antibiotic resistance markers. Such a plasmid can be inserted into either bacterial or animal cell. The techniques of transformation (by uptake of naked DNA), conjugation (by cell–cell contact) or by transduction (by viral vector) may be used for introduction of DNA into bacterial cells. Introduction of DNA into eukaryotic cells such as animal cells by physical or chemical means is referred to as *transfection*.

Different transfection techniques used are calcium phosphate transfection, microinjection, electroporation and liposome transfection. DNA can also be introduced into eukaryotic cells using viruses or bacteria as carriers, the latter is known as *bactofection*, and in particular uses *Agrobacterium tumefaciens*. The plasmid may be integrated into the genome, resulting in a *stable transfection*, or remain independent of the genome, called *transient transfection*. In both cases, DNA coding for a protein of interest is inside a cell and the protein can now be expressed. Different systems, such as inducible promoters and specific cell-signalling factors are available for facilitating expression of the protein of interest at high levels.

Large quantities of protein can then be extracted from the bacterial or eukaryotic cells and tested for enzymatic activity under a variety of situations. The proteins may be crystallized, its tertiary structure can be studied, or in the pharmaceutical industry, the activity of new drugs against the protein may be determined.

Polymerase Chain Reaction

The polymerase chain reaction (PCR) is a powerful and most widely used technique in molecular biology. It is a quick, inexpensive and simple tool. The technique amplifies specific DNA fragments from minute quantities of source DNA material, even if the source of DNA is of relatively poor quality. The PCR is an extremely versatile technique for copying DNA, as it allows a single DNA sequence to be copied millions of times, or altered in predetermined ways. The technique of PCR is useful for introducing restriction enzyme sites, or for mutating particular bases of DNA, the latter is referred to as *quick change*. PCR can also be used to determine whether a particular DNA fragment is found in a cDNA library.

PCR indicates different variations such as reverse transcription PCR (RT-PCR) for amplification of RNA, or real-time PCR (QPCR) for quantitative measurement of DNA or RNA molecules. Different steps involved in PCR procedure are given below.

1. **Denaturation**: The DNA fragments are heated at high temperatures for reducing the DNA double helix to single strands. These strands are accessible to primers.
2. **Annealing**: The reaction mixture is cooled down. Primers anneal to the complementary regions in the DNA template strands. The double strands are formed again between primers and complementary sequences.
3. **Extension**: The enzyme DNA polymerase synthesizes a complementary strand, which reads the opposing strand sequence and extends the primers by adding nucleotides in the order in which they can pair.

Gel Electrophoresis

Gel electrophoresis is an important tool of molecular biology. The basic principle is that DNA, RNA and proteins can all be separated by means of an electric field. Proteins may be separated on the basis of size by using an SDS-PAGE gel, or on the basis of size and electric charge by using 2D gel electrophoresis. The DNA and RNA can be separated on the basis of size by running the DNA through an agarose gel.

Macromolecule Blotting and Probing

The terms 'Northern', 'Western' and 'Eastern' blotting are derived after the technique of Southern blotting was described by Edwin Southern for the hybridization of blotted DNA. Patricia Thomas, developer of the RNA blot, which then became known as the *Northern blot* actually did not use the term. Further, combinations of these techniques referred to as South-Westerns (protein–DNA hybridizations), North-Westerns (to detect protein–RNA interactions) and Far-Westerns (protein–protein interactions) are presently mentioned in the literature.

1. **Southern blotting**: Named after its inventor, biologist Edwin Southern, the Southern blot is a method for probing for the presence of a specific DNA sequence within a DNA sample. DNA samples before or after restriction enzyme digestions are separated by gel electrophoresis and then transferred to a membrane by blotting via capillary action. The membrane is then exposed to a labelled DNA probe that has a complement base sequence. Majority of the original protocols were based on use of radioactive labels. However, nonradioactive alternatives are now available. Southern blotting is less commonly used in laboratory due to better capacity of other techniques, such as PCR, to detect specific DNA sequences from DNA samples. These blots are still used for measuring transgene copy number in transgenic mice or in the engineering of gene knockout.

2. **Northern blotting**: The technique is a combination of denaturing RNA gel electrophoresis and a blot. In this technique, RNA is separated based on size and is then transferred to a membrane that is probed with a labelled complement of a sequence of interest. The results may be visualized by different ways, depending on the label used. The revelation of bands represents the sizes of the RNA detected in sample. The intensity of these bands is related to the amount of the target RNA in the samples analysed. The technique is used to study the expression patterns of a specific type of RNA molecule in relation to a set of different samples of RNA. The procedure is commonly used to study when and how much gene expression is occurring by measuring the content of RNA present in different samples. It is one of the basic tools for determining the time and conditions certain genes are expressed in living tissues.

3. **Western blotting**: In cell culture, monoclonal antibodies can be used for a variety of analytical and preparative techniques. In Western blotting technique, the proteins are first separated by size in a thin gel sandwiched between two glass plates in a technique known as *sodium dodecyl sulphate polyacrylamide gel electrophoresis (SDS-PAGE)*. The proteins in the gel are then transferred to a PVDF, nylon, nitrocellulose or other support membrane. This membrane can then be probed with solutions of antibodies. The antibodies that specifically bind to the protein of interest can be visualized by a variety of techniques, including coloured products, chemiluminescence or autoradiography. The antibodies are usually labelled with enzymes.

When a chemiluminescent substrate is exposed to the enzyme, it allows detection. Antibodies to most proteins can be generated by injecting small amounts of protein into an animal such as a mouse, rabbit, sheep or donkey (polyclonal antibodies) or produced in cell culture. These antibodies are used in different analytical or preparative techniques. The Western blotting techniques are useful, not only in detection but also, in quantitative analysis. Analogous methods to Western blotting can be employed to directly stain specific proteins in live cells or tissue sections. However, these immunostaining methods, such as FISH, are used more often in cell biology research.

4. **Eastern blotting**: Eastern blotting technique is used to detect posttranslational modification of proteins. The Proteins blotted on PVDF or nitrocellulose membrane are probed for modifications using specific substrates.

QUESTIONS

1. What are restrictions of endonuclease and DNA ligase?
2. Write a short note on mutation, mutagenesis, mutants, cloning vectors, endonuclease and restriction endonuclease.
3. Describe different DNA sequencing techniques.
4. Describe gene mapping.
5. Write a note on plant genome.
6. Write a note on DNA replication.
7. Discuss various steps of DNA replication and write a note on DNA repair mechanisms.
8. What is genetic code? Why is there only three letter code?
9. What are the factors that affect mutation of microbes? How is mutation isolated?

CHAPTER 9: Recombinant DNA Technology

CHAPTER OUTLINE

- Introduction
- Outline of Recombinant DNA Technology
- Introduction of the Recombinant DNAs into a Suitable Host Cell and Their Identification
- Applications of Recombinant DNA Technology
- Production of Human Follicle-Stimulating Hormone
- Disease Diagnosis by Recombinant DNA Methods
- Recombinant DNA Technology for Improving
- Milk Production
- Applications in the Pulp and Paper Industry
- Applications in Antibiotic-Producing Microorganisms
- Genetic Engineering as a Part of Recombinant DNA Technology
- Questions

INTRODUCTION

Recombinant DNA (rDNA) is a form of artificial DNA that is produced by combining two or more sequences that would not usually occur together. Through genetic modification, it is created with the introduction of relevant DNA into an existing organismal DNA, such as the plasmids of bacteria, to code for or alter different traits for a specific purpose, e.g. antibiotic resistance. In a nutshell, recombinant DNA technology refers to a set of techniques for manipulating DNA and making recombinant molecules for use in basic and clinical research.

Recombinant DNA is a tool that facilitates understanding the structure, function and regulation of genes and their products, including identifying genes and isolating, modifying and re-expressing genes in other hosts or organisms. The advent of new molecular biological techniques has enabled scientists to determine the nucleotide sequence of DNA and map parts of the genome of eukaryotic cells. The advent of classical recombinant DNA technology provided opportunities for large-scale production of therapeutics or human-derived proteins and peptide. In addition, it offers potential for remodelling of protein drug for site specificity, reducing immunogenicity, stability and improving pharmacokinetics.

In 1970s, a new technology emerged allowing molecular biologists to isolate and characterize genes and their protein products with unprecedented power and precision. One important discovery was the ability to recombine segments of DNA from diverse sources into new composite molecules or recombinants. The recombinant DNA technique was first proposed by Peter Lobban, a graduate student, with Dale Kaiser at the Stanford University, Department of Biochemistry. The technique was then realized by Lobban and Kaiser; Jackson, Symons and Berg and Stanley Norman Cohen, Chang, Herbert Boyer and Helling. The present-day DNA technology has its roots in the experiments performed by Boyer and Cohen (1973). They successfully recombined two site-specific plasmids, pSC101 and pSC102 and cloned the new plasmid in *Escherichia coli*. The second set of experiment of Boyer and Cohen was more organized. A gene-encoding protein (required to form rRNA) was isolated from the cell of African clawed frog *Xenopus laevis*, by use of a restriction endonuclease enzyme.

OUTLINE OF RECOMBINANT DNA TECHNOLOGY

Principle of Gene Cloning

The basic principles for production of the recombinant proteins include isolation of a gene from a cell, placing the gene in a different cell and fermenting a culture of the recombinant cells to produce the desired protein.

Following five essential steps are involved in DNA cloning:

1. Generation of DNA fragments and selection of the desired piece of DNA
2. Insertions of the selected DNA in to a cloning vector (*plasmid*) to create a *recombinant DNA* or *chimeric DNA*
3. Introduction of the recombinant vector into host cell
4. Multiplication and selection of clones containing the recombinant molecules
5. Expression of the gene to produce the desired product

Genomic DNA Libraries

Construction

Gene library is prepared from fragmented genomic DNA synthesized from cellular mRNA. The genomic library is the best course of action when the source of the protein is either unknown or is suspected to be synthesized by a small subset of the cells in a tissue. These libraries are of special significance in the recombinant-based production of protein, as they provide an easy opportunity for selection of the gene of interest.

Preparation

The strategies for preparation of the genomic library are well established in the literature. Genomic library derived from commonly used animal species are available within the scientific community. Generally, the genomic library could be constructed in bacteriophages λ cloning vector, in order to obtain enough independent yields (a 99% probability) in the presence of any one sequence in a mammalian genomic. The method involves insertion of genomic DNA fragments into a restrictive site of the vector, replacing a segment of vector DNA.

Isolation of Gene Sequence from the Libraries

The choice of best screening method is the most difficult hurdle, as no single method can give full assurance of successful gene isolation in a reasonable period of time. As all methods have potential pitfalls and limitations, choice can only be made on the basis of assaying the quality of the resource at hand and the nature of the protein being stalked.

Plasmid as Cloning Vector

A vector is a DNA molecule with ability to replicate in appropriate host cells within which the DNA insert is integrated for cloning. A vector must have an origin of DNA replication with functions in the host cell. The most useful cloning vehicles are *plasmids*. They are defined as genetically homogeneous constant monomeric units with ability to replicate independent of the chromosomes. The plasmid exists as a double-stranded circular DNA molecule. When both the strands of DNA are intact circles, the molecules are referred to as *covalently closed circles*. Phenotypic traits exhibited by plasmid carry the genes. The basic property that makes plasmids highly acceptable as best possible vector systems is that they are the replicons, which are stably inherited in the extrachromosomal state.

Recently, a number of linear plasmids have been described in bacteria such as *Streptomyces* species and *Spirochete borrelia*. Many different *E. coli* plasmids are also used as vectors. The natural plasmids have been modified, shortened, reconstructed and recombined both in vitro and in vivo to create plasmids of enhanced utility and also with specific functions. The important genes that allow easy selection of the recombinant DNA are given in Table 9.1.

The plasmid for cloning should possess the following properties:
1. Low molecular weight
2. Easy to isolate and purify
3. Ability to replicate autonomously
4. Single site for a large number of restriction endonucleases preferably, in gene with a readily scorable phenotype

Table 9.1 Antibiotic resistance genes found in R plasmid, their proteins and mechanism of resistance

Antibiotic	Enzyme produced by the gene	Mechanism of resistance
Amplicillin	Penicillinase	Hydrolysis of C–N bond in β-lactam ring
Kanamycin	Kanamycin acetyltransferase	N-acetylation of antibiotic
Neomycin	Aminoglycoside phosphotransferase	O-phosphorylation of antibiotic
Streptomycin	Streptomycin phosphotransferase	Phosphorylation of OH– on antibiotic

5. Ability to confer readily selectable phenotypic trait on host cells
6. Ability to integrate either itself or with DNA insert it carries into the genome of the host cell

Role of DNA Ligase

The vector used in rDNA experiment can be processed by splicing and cleaving vector DNA at a single specific site with a restriction enzyme. (e.g. the plasmid pSC101 is split at a unique site by the *Eco*RI restriction enzyme). Any DNA fragment may be inserted into this plasmid, if it has the same cohesive ends. The DNA fragments and the cut plasmid may be annealed and joined by DNA ligase (the enzyme that catalyzes the formation of phospodiester bonds between the two DNA chains). The cohesive end joining process for DNA molecules can be accomplished by using a short, chemically synthesized DNA that can be cleared by restriction enzymes.

INTRODUCTION OF THE RECOMBINANT DNAs INTO A SUITABLE HOST CELL AND THEIR IDENTIFICATION

The recombinant DNA is constructed in vitro and introduced into *E. coli*. Different steps involved in construction of recombinant DNA are described below in Figure 9.1.
1. The plasmid is cut with restriction endonuclease to produce DNA with sticky ends. The plasmid with sticky ends can combine with any DNA, which has similar stick.
2. A desired gene with same sticky ends is obtained from human DNA (by cutting human DNA with same restriction endonuclease).
3. Two fragments are annealed and covalently linked by DNA ligase to produce recombinant DNA.
4. The recombinant DNA, thus generated, is inserted into bacteria by a process known as *transformation*.
5. The colonies of bacteria containing recombinant DNA, as well as normal bacteria are grown on agar medium.
6. Clones are transferred to nitrocellulose disc.
7. By treating with sodium dodecyl sulphate (SDS), DNA of clones is liberated. The DNA is then denatured and made single stranded by treatment with alkali.
8. The recombinant DNA is identified by using DNA probe. cDNA is a single-stranded molecule. cDNA can be also prepared from mRNA using reverse transcriptase. This type of cDNA synthesis is possible when amino acid sequence of gene product is known. Usually, base composition of cDNA probe is complementary to base composition of desired gene.
9. After washing excess probe, the disc is dried and exposed to X-ray film.
10. On development of X-ray film, a faint image of disc containing dark spot is seen.
11. Finally, colonies containing recombinant DNA are separated by comparing image with original plate and cultured.
12. In the culture medium, gene product accumulates due to expression of inserted human gene by bacteria (Fig. 9.1).

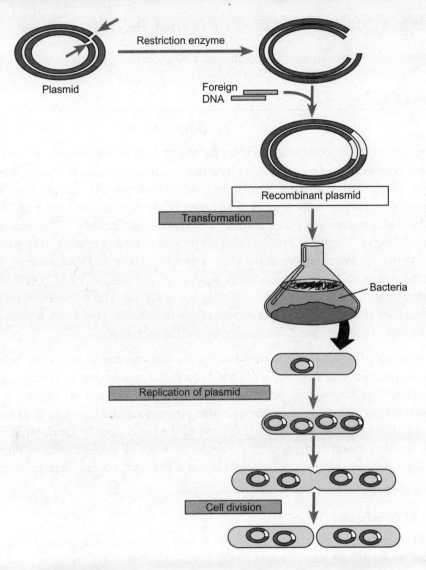

Figure 9.1 General procedure for recombinant DNA technology.

Identification of Clones with Recombinant DNAs

In case the vector has two selectable markers, the DNA insert may be placed in one of these markers. It is observed that some vectors contain a gene or sometimes only part of a gene, which complements a function missing in their host cells. When the DNA inserts codes for a gene product that is defective in the autotrophic host cells, a direct selection for the recombinant DNA is possible. Similarly, selection by suppression of nonsense mutation present in the host also permits a direct selection for the recombinant DNA.

Some λ vectors retain the lysogenic function as well, e.g. λgt 10. In such vectors, the DNA insert may be placed within the lyses repressor gene cell.

APPLICATIONS OF RECOMBINANT DNA TECHNOLOGY

The wide spectrum of applications of recombinant DNA technology is discussed below.

Production of Biologicals

Insulin

Insulin is the first (1982) recombinant product for use in human, produced in *E. coli*. The human insulin protein is processed extensively after translation. The processing of insulin is accomplished in two steps. The primary product known as *preproinsulin* is 110 amino acid long. During translocation of membranes of the protein, the prepart of the protein (the stretch of 24 amino acids serving as the leader sequence for membranes translocation) is cleaved off. The remaining protein (86 amino acid long) is known as *proinsulin*. This protein is further processed in pancreatic cells, whereas an internal fragment known as the C or connecting chain of 33 amino acid, together with a few assorted amino acids are enzymatically separated. The chains A and B left are associated through S bond to form the mature and biologically active insulin. The information for the fragment A is synthesized by linking a set of appropriate oligonucleotides. This DNA is then fused by the ligation to the end of the gene lac Z, controlled by the lac promoter.

In the plasmid pBR322 (a very well-known *E. coli* cloning vector), the information for fragment B is synthesized in two steps: (1) the N-terminal coding part is synthesized by linking oligonucleosides. This fragment is fused to the plasmid pBR322 and propagated in *E. coli*, (2) the C-terminal coding part is synthesized and also propagated after ligating it to pBR322. Both these parts are linked together and fused at the end of the lac Z gene in the plasmid pBR322. The treatment of isolated fusion proteins with cyanogen bromide allows isolation of the fragments A and B. The final step consists of mixing A and B and allowing the S bonds to form spontaneously (Fig. 9.2).

Production of Interferons *(Fig. 9.3)*

The synthesis of interferon, based on molecular biology, was successful when DNA sequence for human leukocyte interferon was attached to the yeasts, alcohol dehydrogenase gene in a plasmid and was introduced into cells of *Saccharomyces cerevisiae*. These yeast cells were capable of synthesizing about one million molecules of interferon. In *E. coli*, the plasmid could also successfully replicate. However, the production was relatively slow in *E. coli* as compared to yeast. In yeast, it is easy to grow and replicate glycoprotein derived from mammalian cells.

Growth Hormones

Human growth hormones secreted from the anterior lobe of the pituitary gland are responsible for regulating and monitoring the growth throughout the body (Fig. 9.4).

Erythropoietin

Erythropoietin (EPO) is a glycoprotein hormone produced in kidney and is responsible for regulating production of red blood cells. It works via EPO-specific receptor, situated in the erythroid progenitor

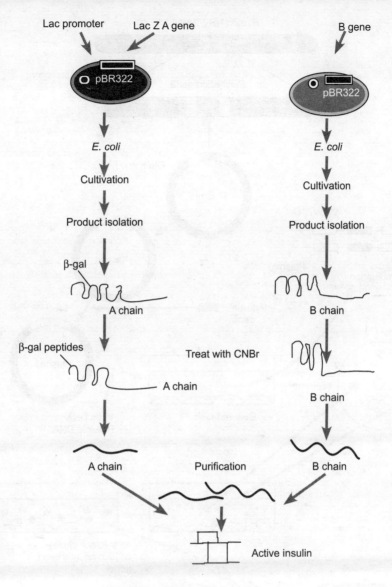

Figure 9.2 Production of insulin.

cell in the bone marrow. The recombinant DNA technology has offered convenient method for production of EPO. The conventional method of obtaining molecules from sources like sheep is very costly. With the development of the recombinant technology, a cost-effective and improvised method is available for the production of EPO. Now, the peptide is produced by effectively cloning the production gene in *E. coli* expression host, using the SV40 virus as the vector.

Cytokines

Some biologically active components capable of regulation, differentiation, proliferation and maturation of various lymphoid and accessory cells were discovered in culture supernatant of

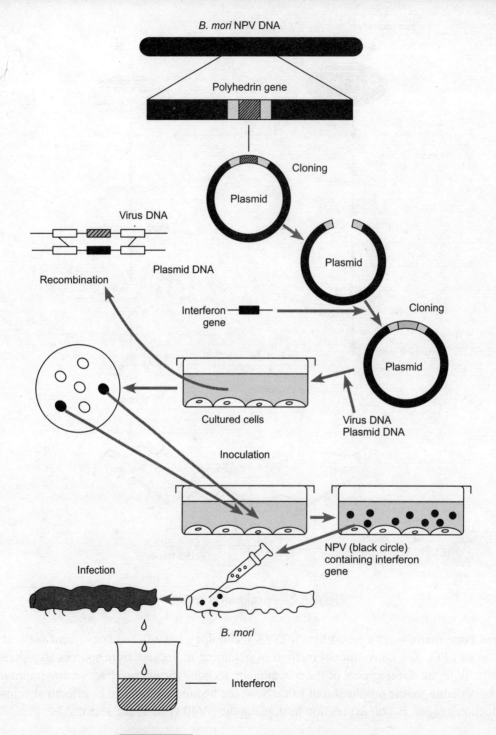

Figure 9.3 Method of production of interferon.

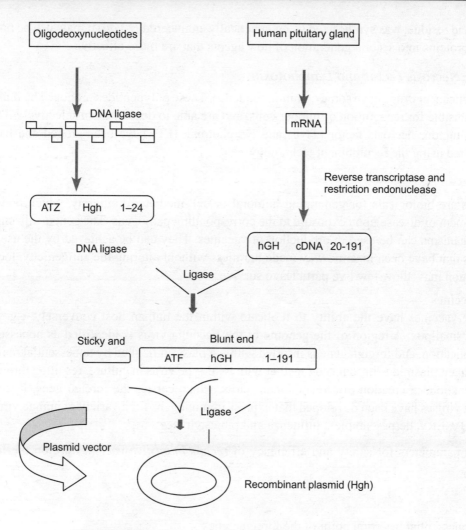

Figure 9.4 Strategy in cloning of growth hormone genes.

allergenic lymphocytes. It was the beginning of cytokines in 1960s. The gene responsible for production of cytokines could be cloned. The general approaches are based on generation of cDNA library from appropriate cytokine-producing cell lines. Subsequently, the gene of interest could be transected to monkey kidney cell line, which contains an integrated SV40 genome.

Interleukins

These are natural immunomodulators capable of mediating communication between leukocytes and other cells. A complete set of cell–cell interaction is mediated by lymphokines or cytokines.

Interleukin 1 and interleukin 2 are also important for antibody production by lymphocytes. With the application of recombinant DNA technique, it has become possible to clone *IL2* genes from a number of different species. An analogue of the human *IL2* genes, a continuous chain of 133 amino acid residue, was synthesized using genetically engineered *E. coli*. It may also be possible to modify proteins into second generation of new agents that are more effective.

Tumour Necrosis Factor and Lymphotoxin

These anticancer drugs have some common sequence. These polypeptides produced by human cells are responsible for recognition of tumour cells, and are able to destroy them selectively. The genes for both tumour necrosis factor (TNF) and lymphotoxin (LT) have been cloned and bacterially synthesized using the recombinant technology.

Vaccines

Vaccines are biologicals for generating humoral or cell-mediated immunity, which prevents the development of disease upon exposure to the corresponding pathogens. The genetically engineered microorganisms can be used in production of vaccines. They can be achieved by the use of virus particles that have been in some way made harmless without altering the antigenicity. Incomplete inactivation may allow few live particles to survive.

Live vaccines

The live vaccines have the ability to replicate within the human host conversely, e.g. Vaccinia virus of smallpox. A region of the genome of the Vaccinia virus is identified as nonessential for viral replication, and foreign gene is inserted within a plasmid flanked by nonessential region. This recombinant plasmid is introduced together with wild-type virus in culture, resulting through DNA recombination, in creation of a recombinant virus, which carries the foreign gene. Recombinant vaccinia viruses have been developed that express immunogens for a variety of human viruses, e.g. HIV, HBV, ERV, herpes simplex, influenza and rabies viruses.

The parameters for safety and efficiency of *recombinant live vaccines* are on the following lines:

Safety

1. Nonessential insertion point of the foreign genes
2. Defective host range of the recombinant virus
3. Stable attenuation of the parental virus
4. Extensive clinical testing of the parental virus

Efficacy

1. High level expression of the foreign protein
2. Multiple foreign genes in a single viral gene

Killed vaccines

The killed vaccines are unable to replicate or infect the host. Through recombinant DNA technology, a process is initiated by identification of that protein component of a virus or a microbial pathogen that itself can elicit the production of antibodies processing the capacity to neutralize infectivity and protecting the host against attack by the pathogens.

PRODUCTION OF HUMAN FOLLICLE-STIMULATING HORMONE

Human follicle-stimulating hormone (FSH) is now produced in vitro by recombinant DNA technology. FSH, being a complex heterodimeric protein, eukaryotic cell line has been selected for expression work (Chinese hamster ovary cells). The pharmaceutical preparation of recombinant human FSH (rFSH) differs from that of human menopausal gonadotrophin (HMG) and the first generation of urinary human FSH (uFSH) in terms of

1. source of bulk materials,
2. purity and specific activity,
3. batch to batch consistency and
4. complete absence of luteinizing hormone (LH) activity.

Pharmacokinetic profile of rFSH has shown an absolute bioavailability of 75% after both SC and IM administration and an apparent terminal half-life of 37 ± 25 h. This profile is very similar to that of uFSH. Clinical efficacy and safety are currently demonstrated through several randomized and well-controlled studies, comparing rFSH-administered SC with uFSH-administered IM for stimulating follicular development prior to assisted reproduction treatment. In World Health Organization (WHO) group II anovulation, to date, 1000 patients have been treated with rFSH. Moreover, rFSH has recently been used successfully in association with recombinant human LH for inducing ovulation and pregnancy in WHO group I anovulatory patients. It is likely that rFSH may replace all urinary-derived FSH preparations for stimulating ovarian follicular development.

DISEASE DIAGNOSIS BY RECOMBINANT DNA METHODS

The recombinant DNA technology has rapidly expanded our ability to diagnose diseases. Substantial advances in simplification of procedures for diagnostic purposes have been made, and the physicians have been immensely benefited as a consequence of these developments. Recombinant DNA procedures are useful for identification of molecular defects in men accounting for heritable diseases, somatic mutations associated with neoplasia and acquired infectious disease. The wide applications of recombinant DNA diagnostics depend on simplicity, speed of results and cost containment and can be briefly summarized as described below.

1. **Plasminogen activator** produced by recombinant DNA technology is useful in clinical studies in the patients with evolving myocardial infarction of tissue type.
2. **Diagnostic of AIDS**: The recombination technology is used to exploit potential for development of AIDS vaccines. Several recombinant vaccines for human immunodeficiency viruses are presently being developed and assessed in volunteers as potential vaccines for AIDS.
3. **Detection of genetic disorder**: It has been recognized that genetic disease in man results from single recessive mutation and attributed either to defective or absence of protein product. Antenatal diagnosis of haemoglobinopathies by rDNA techniques is a prototype for other genetic disorder.

4. **Diagnosis of affected and carrier states of hereditary diseases**: The tests have been developed to determine if human beings are carriers of the cystic fibrosis gene, Huntington's disease gene, Tay-Sachs disease gene or the Duchenne muscular dystrophy gene. This has been made possible by recombinant DNA technology.

RECOMBINANT DNA TECHNOLOGY FOR IMPROVING MILK PRODUCTION

Although the galactopoietic properties of bovine somatotropin (bST) were known for several decades, systematic experiments to analyse these properties were not initiated until the advent of recombinant bST (rbST) and awareness of the potential economic impact of rbST on the dairy industry. The administration of rbST to lactating animals significantly increases milk production. The development of transgenic animals carrying and expressing fusion genes coding for hormones that may increase milk production efficiency is also possible. Several such fusion genes including metallothionein-I gene promoter sequence have been identified and used in most studies. Possible alternative gene promoter sequences include those from prolactin, phosphoenolpyruvate carboxykinase and β-lactoglobulin genes.

These new developments shall lead to more cost-efficient animal production in general. However, implementation of such new techniques requires acceptance by both consumers and producers.

APPLICATIONS IN THE PULP AND PAPER INDUSTRY

New gene selection techniques (recombinant DNA) are currently available to explore and exploit useful properties of various biological systems hitherto regarded as interesting but of little or no immediate commercial value. The applications of genetic engineering techniques in the pulp and paper industry are diverse in nature. These techniques are used to provide much needed fundamental information on the cellular and molecular mechanisms involved in expression of extracellular enzymes that degrade lignocellulosic pulping wastes. The information obtained from the studies on cellulolytic fungi and bacteria can be effectively used to genetically engineer a yeast or bacterium capable of converting pulping wastes into ethanol and other useful by-products.

APPLICATIONS IN ANTIBIOTIC-PRODUCING MICROORGANISMS

The current yields of industrially important products are almost entirely due to mutation and selection of producing microorganisms and media development. However, recombination has also helped in improvisation of strains for production of certain antibiotics (Table 9.2).

Table 9.2 Some examples of strain improvement and isolation of new products by recombination

Organism	Product	Genetic manipulation	Strain improvement novel product
Aspergillus nidulans	Penicillin	Sexual recombination	Improved yield
A. nidulans	Cephalosporins	Protoplast fusion	Improved yield, growth rate and sporulation
Nocardia mediterranei	Rifampicin	Conjugation	Range of novel ansamycins
Streptomyces hygroscopicus	Turimycin	Conjugation	Iremycin

GENETIC ENGINEERING AS A PART OF RECOMBINANT DNA TECHNOLOGY

The steps involved in synthesis of recombinant proteins through genetic engineering are as follows:

1. Recombinant technology begins with the isolation of a gene of interest. The gene is then inserted into a vector and cloned. A vector is a piece of DNA that is capable of independent growth. Commonly used vectors are bacterial plasmids and viral phages. The gene of interest (foreign DNA) is integrated into the plasmid or phage, and this is referred to as recombinant DNA.
2. The cloning is done before introducing the vector containing foreign DNA into host cells. Cloning is necessary to produce numerous copies of the DNA, since the initial supply is inadequate to insert into host cells.
3. Once the vector is isolated in large quantities, it can be introduced into the desired host cells such as mammalian, special bacterial cells or yeast. The host cells then synthesize the foreign protein from the recombinant DNA. When the cells are grown in vast quantities, the foreign or recombinant protein can be isolated and purified in large amounts.

Genetic engineering produces proteins that offer advantages over proteins isolated from other biological sources. These advantages include the following:

a. Batch-to-batch consistency
b. Steady supply
c. High purity
d. High specific activity
e. Recombinant DNA technology is not only an important tool in scientific research, but also useful in the diagnosis and treatment of diseases and genetic disorders in many areas of medicine.

QUESTIONS

1. What is genetic engineering?
2. Explain the production of insulin by rDNA technology.
3. Give the application of rDNA technology.
4. Discuss the concept, methodology and pharmaceutical application of rDNA technology.
5. Write a brief note on disease diagnosis by rDNA methods.

CHAPTER 10: Microbial Transformation

CHAPTER OUTLINE

- Introduction
- Types of Reactions Mediated by Microorganisms
- Design for Biotransformation
- Biotransformation Process and Its Improvements with Special Reference to Steroids
- Questions

INTRODUCTION

Microbial transformation is a routine economically and ecologically competitive technology for organic chemists who are in search of new production routes for fine chemicals, pharmaceuticals and agrochemical compounds. Biotransformation deals with the use of natural and recombinant microorganisms (bacteria, yeast, fungi), enzymes, whole cells, etc., as catalysts in organic synthesis. It is an attractive approach to generate structural diversity in a chemical library, which may be utilized to synthesize chemical structures that are otherwise difficult to obtain by other means. The major difficulty for a specific biotransformation of a certain substrate is to find the appropriate microorganism. Biotransformation plays a key role in the areas of foodstuff, chiral drug industry, vitamins, specialty chemicals and animal feedstock.

Transformation refers to the basic mechanisms for genetic exchange in bacteria. It could be either a natural process, that is, one that has evolved in certain bacteria, or it may be an artificial process whereby, the recipient cells are forced to take up DNA by a physical, chemical or enzymatic treatment. In either case, exogenous DNA (DNA that is outside the host cell) is taken into a recipient cell where it is incorporated into the recipient genome, changing the genetic set-up of the bacterium.

The bioremediation and biotransformation methods harness the naturally occurring microbial catabolic diversity to degrade, transform or accumulate a huge range of compounds, including hydrocarbons (e.g. oil), polyaromatic hydrocarbons (PAHs), polychlorinated biphenyls (PCBs), pharmaceutical substances, radionuclides and metals. Major methodological breakthroughs in recent years have enabled detailed genomic, metagenomic, proteomic, bioinformatic and other high-throughput analyses of environmentally relevant microorganisms, providing unprecedented insights into biotransformation and biodegradative pathways. Moreover, this also demonstrates the ability of organisms to adapt to changing environmental conditions.

Biotransformation of various pollutants is a sustainable way to clean up contaminated environments. Biological processes are important in the removal of contaminants and pollutants from the environment. Some microorganisms possess an astonishing catabolic versatility to degrade or transform such compounds.

New methodological breakthroughs in sequencing, proteomics, genomics, bioinformatics and imaging are producing wide range of information. In the field of environmental microbiology, genome-based global studies open a new era, providing unprecedented views of metabolic and regulatory networks as well as clues to the evolution of biochemical pathways relevant to biotransformation.

Biotransformation is the process in which a substance is changed from one chemical to another (transformed) by a chemical reaction within the body. *Metabolism* or *metabolic transformations* are terms frequently used for the biotransformation process. Biotransformation is vital to survival as it transforms absorbed nutrients (food, oxygen, etc.) into substances required for normal body functions. For some pharmaceuticals, it is a metabolite that is therapeutic and not the absorbed drug. For example, phenoxybenzamine (dibenzyline), a drug given to relieve hypertension, is biotransformed into a metabolite, which is the active agent. Biotransformation also serves as an important defence mechanism in the toxic xenobiotics and body wastes that are converted into less harmful substances and which can be excreted from the body.

Chemical reactions catalysed by microorganisms or by enzyme preparations derived from biomass are referred to as *biotransformations*. With a multiplicity of constitutive or inducible enzymes, microorganisms are capable of performing a vast number of chemical reactions that are essential for maintaining the life functions of the cell, including growth and reproduction. In fact, there exists an enzyme-catalysed equivalent for almost every type of chemical reaction. Microorganisms employ such enzyme-catalysed reactions that are well organized in metabolic pathways for the degradation or synthesis of a great variety of chemical compounds.

Transformation: Discovery and Types

The first report of transformation was an example of natural transformation. Dr Frederick Griffith, a public health microbiologist—while studying bacterial pneumonia during the 1920s—discovered that the bacterial colonies from the lungs of animals with pneumonia grown on the agar plates were of reasonable size and had a glistening, mucoid appearance. When he transferred these colonies repeatedly from one agar plate to another, the mutant colonies were smaller in size and chalky in appearance. Dr Griffith named the original strains as 'smooth' strains and the mutants as 'rough' strains. When he injected mice with smooth strains, they contracted pneumonia, and smooth strains of the bacterium could be reisolated from the infected mice. However, when he infected the mice with rough strains, they did not develop the disease.

The smooth strains were capable of causing disease or were virulent, whereas the rough strains did not cause disease or were nonvirulent.

Natural Transformation

It is a physiological process that is genetically encoded in a wide range of bacteria. There appear to be two basic mechanisms by which bacteria can become competent for transformation. Most bacteria shift their physiology in order to transform DNA, that is, they must become 'competent' for taking up exogenous DNA. In some bacteria, including *Streptococcus pneumoniae* and *Bacillus subtilis*, competence is externally regulated. These bacteria produce and secrete a small protein called *competence factor* that accumulates in the growth medium.

Artificial Transformation

While a wide variety of bacteria can transform naturally, many other species are unable to take up DNA from an outside source. In some cases DNA can be forced into these cells by physical, chemical or enzymatic treatment. This is especially important in genetic engineering, as artificial transformation is essential for the introduction of genetically altered sequences into recipient cells. One of the most common methods is a chemical process whereby cells are heat shocked, followed by treatment with the DNA and a high concentration of calcium ions. The calcium ions precipitate the DNA on the surface of the cell, where the DNA is forced into the recipient.

TYPES OF REACTIONS MEDIATED BY MICROORGANISMS

Different types of cells can be used to convert an added compound into another compound, involving many forms of enzymatic reaction including oxidation, dehydration, hydroxylation, amination, isomerization, etc. These bioconversions have advantages over chemical processes in that the reaction can be very specific, and it can be produced at moderate temperature. Examples of successful transformations using enzymes include the production of steroids, conversion of antibiotics and prostaglandins, etc.

Industrial transformation requires production of large quantities of enzyme. The half-life of enzymes may be improved by immobilization and extraction simplified by the use of whole cells. In any bioprocess, the bioreactor is an essential component in an integrated process with upstream and downstream components. The upstream consists of storage tanks, growth and media preparation, followed by sterilization. Also, seed culture for inoculation and sterilized raw material (mainly, sugar and nutrients) are required for the bioreactor to operate. The sterilization of the bioreactor is carried out by steam at 15 pounds per square inch gauge (psig) at 121°C. Alternatively, any disinfectant chemical reagent such as ethylene oxide may be used. The downstream processing involves extraction of the product and purification as normal chemical units of operation. The solids are separated from the liquid, and the solution and supernatant from separation unit may go further for purification after the product is concentrated.

The specific oxidizing reactions and oxidizing enzymes are numerous. The examples of different types of oxidizing reactions are illustrated below.

$$-\underset{H}{\overset{H}{C}}-\underset{H}{\overset{H}{C}}- + \tfrac{1}{2}O_2 \xrightarrow{\text{Oxygenation}} -\underset{H}{\overset{H}{C}}-\underset{OH}{\overset{H}{C}}-$$

$$-\underset{H}{\overset{H}{C}}-\underset{H}{\overset{H}{C}}- \xrightarrow{\text{Dedydrogenation}} -\overset{H}{C}=\overset{H}{C}- + H_2$$

$$Fe^{3+} + Cu^+ \xrightarrow{\text{Electron transfer}} Fe^{2+} + Cu^{2+}$$

The microbial systems can carry out the following types of transformation reactions:

1. Oxidation (including dehydrogenation)
2. Reduction
3. Esterification
4. Hydrolysis

In addition, the following are reported in literature for microbial modification of antibiotics:

1. Adenylylation and ribonucleotide formation
2. Phosphorylation
3. Demethylation and deamination

Oxidative Biotransformations

Cyclohexanone monooxygenase from *Acinetobacter* sp. NCIB 9871 is expressed in Baker's yeast to create a general reagent for asymmetric Baeyer–Villiger oxidations. The behaviour of 3-substituted cyclohexanones is based upon the size of other substituents. The engineering yeast strain cleanly converted the antipodes of 3-methyl and 3-ethylcyclohexanone to divergent regioisomers. On the other hand, for larger substituents, both antipodes were oxidized by the enzyme to a single regioisomer (Stewart, 1998; Stewart et al., 1998).

The microbial transformation of the D-enantiomer of 13-ethyl-17-B-hydroxy-18,19-dinor-17 alpha-pregn-4-en-20-yn-3-one was investigated by Hu et al. (1998). Transformation of the named compound by *Cunninghamella blakesleeana* gave 6-beta-, 7-beta-, 15-alpha-hydroxy derivatives, 4-beta-, 5-beta-, 6-beta-, 7-beta-, 15-alpha-dihydroxy derivatives and 6-beta-, 10-beta-dihydroxy derivative. Biotransformation of the DL-1 by *Cunninghamella* species usually gave 10-beta-hydroxy products with the low enantiomeric excess or as racemic form.

Chloroperoxidase was capable of catalysing a broad spectrum of stereoselective hydroxylation, sulphonation and epoxidation reactions. The chiral epoxidation of alkenes was published by Hagel et al. (1998).

The first evidence of stimulation of an immediate early gene product, specifically prostaglandin G/H synthase-2 (PGHS-2), by an arachidonic acid epoxygenase metabolite, 14,15-epoxyeicosatrienoic acid, as well as of a heterologous regulation of PGHS-2 synthesis by these monooxygenase products was demonstrated by Peri et al. (1998).

Hydrolysis of Oxiranes, Nitriles and Amides

A review in the field of enzyme-catalysed epoxide hydrolysis has been published and cited by Roberts (1998). Yeast strains from different genera were screened for the hydrolysis of 1,2-epoxyoctane. Asymmetric hydrolysis of 1,2-epoxyoctane was reported for eight yeast strains belonging to the genera *Trichosporon, Rhodotorula* and *Rhodosporidium*. All these strains preferentially hydrolysed (R)-1,2-epoxyoctane to (R)-1,2-octanediol.

The recombinant epoxide hydrolase from *Agrobacterium radiobacter* AD1 was used to obtain enantiomerically pure epoxides by means of a kinetic resolution. Epoxides such as styrene oxide and various derivatives thereof and phenyl glycidyl ether were obtained in high enantiomeric excess and in a good yield (Spelberg et al., 1998).

The biohydrolysis with fungal epoxide hydrolases can be carried out in a preparative scale for different types of epoxides. Good yields were obtained by using an efficient reactor. Asymmetric biocatalytic hydrolysis of (±)-2,3-disubstituted oxiranes leading to the formation of vicinal diols was accomplished by using the epoxide hydrolase activity of various bacterial strains. The mechanism of this deracemization was elucidated with partially purified epoxide hydrolase Nocardia EH1 (Kroutil et al., 1998).

Esterification and Ester Hydrolysis

Several microbial lipases were screened for their property to catalyse the synthesis of citronellyl and geranyl esters in organic media using vinyl esters as acylating agents. On the basis of these results, an immobilized lipase, SP435 from *Candida antarctica* was selected for the study of the effect of reaction parameters on terpinyl acetate synthesis (Akoh and Yee, 1998).

Dry mycelium of *Rhizopus delemar* MIM catalysed the formation of geranyl acetate using geraniol and acetic acid in heptanes. It also catalysed the direct acetylation of different primary alcohols with molar conversions ranging from 65 to 98% (Molinari et al., 1998).

The synthesis of geranyl acetate and citronellyl acetate by alcoholysis reaction catalysed by immobilized lipase from *Mucor miehei* was studied in a solvent-free system. Reactions were carried out at a terpene alcohol/acyl donor molar ratio of 1:5 for ethyl and butyl acetates as acyl donors with excellent yields (Chatterjee and Bhattacharya, 1998).

Investigations on aminolysis of esters catalysed by different lipases from different origins were carried out. Lipases from *Rhizopus niveus, Candida antarctica* B and PPL gave the best enantioselectivities in the resolution of chiral esters. The *Candida rugosa* and *Pseudomonas cepacia* lipases were less interesting lipases. The aminolysis is interesting for the resolution of racemic amines, but not for the resolution of racemic esters. The immobilization does not alter the enantiopreference of the lipases (De Castro and Gago, 1998).

Using the ability of lipase to distinguish between enantiomers and enantiotopic group, the syntheses of several classes of enantiomerically pure compounds such as g- and d-lactones, nucleosides, carba analogues of phospholipids and myo-inositol phosphates were described (Andersch et al., 1998).

Lipase from *Humicola lanuginosa* and *Candida antarctica* lipase B were adsorbed onto methylated-controlled glass bends with varying degrees of hydrophobicity. The effect of support hydrophobicity on the amount of lipase adsorbed was studied. The immobilized lipases were examined for their activity in an alcoholysis reaction of cod liver oil with ethanol in supercritical carbon dioxide (Gunnlaugsdottir et al., 1998).

Alcoholysis involving triolein and stearyl alcohol for production of wax esters with lipases from *Alcaligenes* sp. and *Chromobacterium viscosum* were studied. Lower yields with *Mucor javanicus, Mucor miehei* and *Pseudomonas fluorescens* (35%) were obtained. Other lipases from extracts of *Porcine pancreas*, *C. rugosa*, *Rhizopus javanicus* and *Rhizopus niveus* showed only low catalytic activity (less than 2%; Decagny et al., 1998). *C. rugosa* lipase-catalysed ester hydrolysis on carboxy-funtionalized sulphoxides showed the highest enantioselectivity as compared with those from *Aspergillus niger, Candida antarctica* B and *Pseudomonas* sp. (Lowendahl and Allenmark, 1998).

In order to use lipase in an economically suitable and inexpensive way, a simple medium without organic nitrogen must be considered. Lipase production by Brazilian strain of *Penicillium citrinum* was attempted (Pimentel et al., 1997), which was totally free of mycotoxin (Pimentel et al., 1996). This lipase can reversibly catalyse the esterification reactions in a free form (Pereira et al., 1997).

Some of the examples of changes in antibiotics are given in Table 10.1.

Table 10.1 Microbial transformations of antibiotics

Antibiotic	Changes noted
Actinomycin	Hydrolysis of lactones
Cephalosporin	Hydrolysis of lactam, hydrolysis of peptide bond, hydrolysis of ester
Chloramphenicol	Hydrolysis of amide, reduction of nitro group, acylation of hydroxyl
Circulin	Hydrolysis of peptide ring
Clindamycin	Sulphoxide formation, phosphorylation, demethylation ribonucleotidation
Colistin	Hydrolysis of peptide chain
Cordycepin	Deamination
Cycloheximide	Acylation
Echinomycin	Hydrolysis of lactone
Etamycin	Hydrolysis of lactone
Formycin	Beta deamination, oxidation
Fusidic acid	Oxidation (hydroxyl to ketone), hydroxylation

(Continued)

Antibiotic	Changes noted
Gentamicin A	Phosphorylation 40
Gramicidin S	Hydrolysis of peptide ring
Griseofulvin	Demethylation, hydroxylation, reduction
Kanamycin	Phosphorylation, acylation
Lincomycin	Sulphoxidation, demethylation, phosphorylation
Mannosidostreptomycin	Hydrolysis, adenylylation
Mycophenolic acid	Oxidation
Nisin	Hydrolysis
Paromamine	Phosphorylation
Penicillin	Hydrolysis of lactam, hydrolysis of peptide
Rifamycin S	Acylation, esterification
Polymyxin	Hydrolysis of peptide
Rifamycin B	Acetylation
Staphylomycin S	Hydrolysis of lactone
Stendomycin	Hydrolysis of lactone
Spectinomycin	Adenylylation
Spiramycin	Acylation
Streptomycin	Adenylylation, phosphorylation
5-alpha-6-anhydrotetracycline	Rehydration
12-alpha-deoxytetracycline	Hydroxylation
Toyocamycin	Hydrolysis
Tylosin	Reduction
T-2636 antibiotics	Deacetylation, dehydrogenation, acylation

DESIGN FOR BIOTRANSFORMATION

Production of Vanillin by One-Step Biotransformation Using Fungus *Pycnoporus cinnabarinus*

One-step biotransformation process for vanillin production from ferulic acid using the wild fungal strain *P. cinnabarinus* has been reported. The effects of medium composition variables (i.e. carbon, nitrogen) and environmental factors such as pH on vanillin production were observed. Subsequently, concentrations of medium components were optimized using an orthogonal matrix method. After primary screening, it was observed that the glucose as carbon source and corn steep liquor and ammonium chloride as organic and inorganic nitrogen source, respectively, supported maximum biotransformation of ferulic acid to vanillin. Under statistically optimized conditions, vanillin production from ferulic acid by *P. cinnabarinus* was 126 mg/L with a molar yield of 54%. The overall molar yield of vanillin production could be increased by four times.

Characterization of Competent Cells and Early Events of *Agrobacterium*-Mediated Genetic Transformation in *Arabidopsis thaliana*

The foreign DNA is inserted through a complex interaction between *Agrobacteria* and host plant cells. The marker gene beta-glucuronidase of *Escherichia coli* and cytological methods are used to characterize competent cells for *Agrobacterium*-mediated transformation. In cotyledon and leaf explants, competent cells are the mesophyll cells that are dedifferentiating, a process induced by wounding and/or by phytohormones. The cells are located either at the cut surface or within the explant after phytohormone treatment. In root explants, competent cells are present in dedifferentiating pericycle, and these are produced only after phytohormone treatment. Irrespective of their origin, the competent cells are small, isodiametric with thin primary cell walls, small and multiple vacuoles, prominent nuclei and dense cytoplasm. In both cotyledon and root explants, histological enumeration and beta-glucuronidase assays have shown that the number of putatively competent cells is increased by preculture treatment, indicating that cell activation and cell division following wounding are insufficient for transformation without phytohormone treatment. Exposure of explants to *Agrobacterium tumefaciens* for 48 h produced neither characteristic stress response nor any gradual loss of viability, nor did it induce cell death. However, in the competent cell, association between the polysaccharide of the host cell wall and that of the bacterial filament was frequently observed, indicating that transformation required polysaccharide-to-polysaccharide contact. Flow cytofluorometry and histological analysis have revealed that abundant transformation required not only cell activation but also cell proliferation. Noncompetent cells can be converted into competent cells with the appropriate treatments of phytohormones before bacterial infection. This certainly aids analysis of critical steps in transformation procedures and facilitates developing new strategies to transform recalcitrant plants.

BIOTRANSFORMATION PROCESS AND ITS IMPROVEMENT WITH SPECIAL REFERENCE TO STEROIDS

Biotransformation of Steroids

Steroids are complex chemical entities with tetracyclic carbon ring. They are important therapeutic agents as steroid hormones regulate various metabolic processes in humans, including human sexuality. The chemical synthesis of steroids is very complex and expensive. Therefore, use of microorganisms such as fungi and bacteria in their production by biotransformation is an important commercial venture.

The biotransformation of steroids differs from conventional fermentation processes in that the end products are not biosynthesized from the ingredients added into the nutrient medium. The steroid precursors are added into the culture medium towards the end of the growth phase, and these are modified by microorganisms resulting in the formation of steroids.

In a typical biotransformation process of a steroid, a high biomass of the microorganism is ensured in a fermentation tank using an appropriate culture medium and incubation conditions.

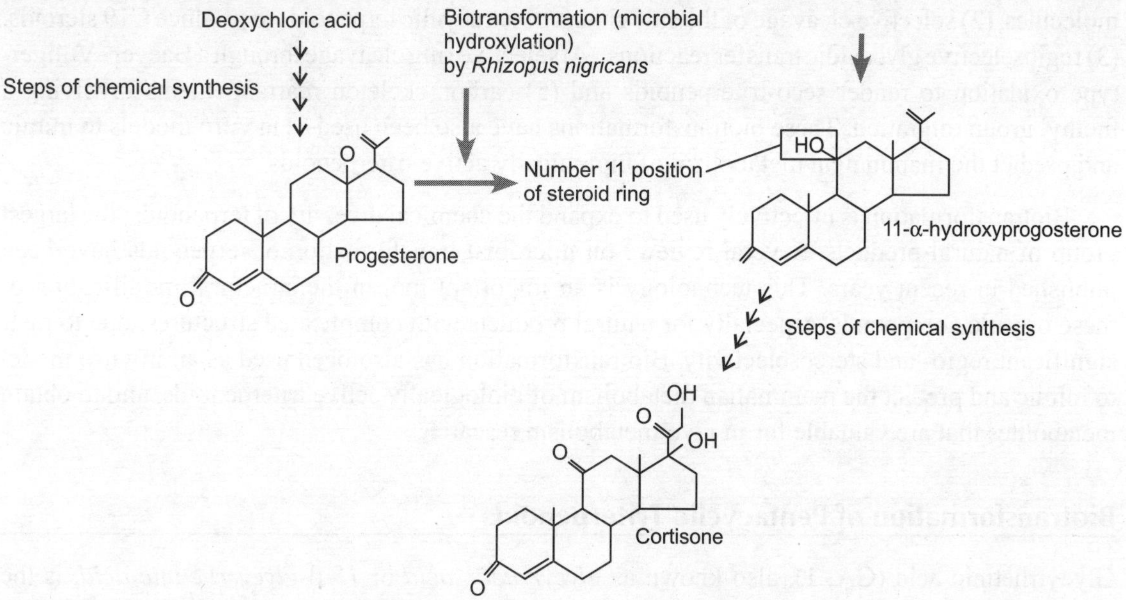

Figure 10.1 Biotransformation of cortisone.

After this, the steroid to be biotransformed is added in the fermentation with solvent, purified chromatographically and recovered by crystallization.

Biotransformation (Microbial Hydroxylation) of Cortisone from Deoxychloric Acid by *Rhizopus nigricans*

Cortisone is a steroid hormone used for rheumatic arthritis in humans. It is synthesized from deoxychloric acid in 37 different steps, many of which require extreme conditions of temperature and pressure. The resulting product is very expensive. The introduction of an oxygen atom at number 11 position of the steroid ring of progesterone, an intermediate, is a difficult reaction, which is easily solved by microorganisms. When progesterone is added to a fermentation tank containing *Rhizopus nigricans* growing for approximately a day, it is hydroxylated at the number 11 position of its steroid ring to yield 11-α-hydroxyprogesterone. The product is recovered by extraction with methylene chloride or other suitable solvents, purified chromatographically and recrystallized. The intermediate 11-α-hydroxyprogesterone is subjected to chemical synthesis, and finally, cortisone is obtained (Fig. 10.1). The biotransformation process for cortisone production has made the whole process easy and very economical from commercial point of view.

Microbial Transformation of Triterpenoids

Several reactions that are difficult to achieve by chemical routes have been accomplished by microbial transformation: (1) introduction of hydroxyl groups into remote positions of the molecules, (2) selective cleavage of the side chains of tetracyclic terpenoids to produce C19 steroids, (3) regioselective glycosidic transfer reactions, (4) selective ring cleavage through a Baeyer–Villiger-type oxidation to render seco-triterpenoids and (5) carbon skeleton rearrangements involving a methyl group migration. These biotransformations have also been used as in vitro models to mimic and predict the mammalian metabolism of biologically active triterpenoids.

Biotransformation is effectively used to expand the chemical diversity of terpenoids, the largest group of natural products. Several reviews on microbial transformation of terpenoids have been published in recent years. This technology is an important tool in the structural modification of these organic compounds, especially for natural products with complicated structures, due to their significant regio- and stereoselectivity. Biotransformation has also been used as an in vitro model to mimic and predict the mammalian metabolism of biologically active triterpenoids, and to obtain metabolites that are valuable for in vivo metabolism research.

Biotransformation of Pentacyclic Triterpenoids

Glycyrrhetinic acid (GA, 1), also known as *glycyrrhetic acid* or *18-β-glycyrrhetinic acid*, is the aglycone of the saponins isolated from *Glycyrrhiza glabra* and other *Glycyrrhiza* species. This triterpenoid and its derivatives have anti-inflammatory, antiulcer and anti–tumour-promoting activities. The major product of the microbial transformation of (GA, 1) by *Curvularia lunata* (ATCC 13432) and by *Mucor spinosus* (AS 3.3450) is identified as 7-hydroxyglycyrrhetinic acid. *Mucor polymophosporus* also produces 7-hydroxyglycyrrhetinic acid as a major metabolite (26.8%) from (GA, 1), along with 15-hydroxyglycyrrhetinic acid (18.2%). The minor metabolites 24-hydroxyglycyrrhetinic acid (2.7%), 6-hydroxyl glycyrrhetinic acid (1.3%), 7-hydroxyglycyrrhetinic acid (1.4%), 3-oxo-7-hydroxyglycyrrhetinic acid (1.1%) and 3-oxo-15-hydroxyglycyrrhetinic acid (0.9%). The main action of *Trichothecium roseum* (ATCC 8685) on (GA, 1) is also directed to C-7 and C-15, yielding 7-, 15-hydroxy, and 7-, 15-dihydroxy derivatives, which is also methylated in the carboxylic acid group by the microorganism. The formation of a methyl ester is also detected in the incubation of (GA, 1) with *Nocardia* sp. (NRRL 5646), which is converted to its methyl ester (methyl glycyrrhetinate). (GA, 1) is also incubated with the mycelium of *Streptomyces* sp. to give three conversion products. *Chainia antibiotica* (IFO 12246) converts (GA, 1) into two new 3,4-seco-oleanane-type compounds.

Microbial Transformation of Cholesterol by *Mycobacterium smegmatis*

M. smegmatis PTCC 1307 is obtained from the Persian Type Culture Collection. The required chemicals and culture media are obtained from Sigma Chemicals and Oxoid, respectively. For analytical thin layer chromatography, 250-μm-thick silica gel precoated plates with 254-nm fluorescent indicator (Sigma Chemical Co.), and for preparative thin layer chromatography, Kieselgel 60 HF (Art. 7741, Merck) are used. The microorganism is grown in 100 mL of a preculture medium (nutrient broth 1.3 g, myo-inositol 0.1 g per 100 mL) in shake flasks by incubating at 30°C with 150 rounds per minute for 24 h. A total of 10 mL of the preculture medium is inoculated and incubated at 30°C with 150 rounds per minute to 100 mL of the production medium (containing nutrient broth 1.3 g, myo-inositol 0.1 g, cholesterol 0.2 g per 100 mL at pH 7). The enzyme inhibitor is added after 24 h. Five chemicals, 1 mM 2,2′-dipyridyl, 1 mM 8-quinolinol, 0.1 mM 1,10-dihydroxy phenanthroline, 1 mM $NiSO_4$ and 10 mM $CoSO_4$ are tested for their enzyme inhibition activity by adding in separate shake flasks. After the incubation period of around 11 days, the broth is extracted three times chloroform. Complete removal of the solvent is followed by thin layer chromatography to detect the products. Separation and purification of the products are carried out with preparative silica gel plates (20 × 40 cm) using ethyl acetate:hexane (3:7) as solvent system. Mass, FTIR, 1H NMR and 13C NMR spectra of compounds are recorded with Finnigan mat TSQ-70, Magna IR Spectrometer 550 (Nicolet) and V-NMR 400 (Varian), respectively. Melting points are taken on a Kofler hot-stage apparatus and are not corrected.

Microbial Transformation of Progesterone In Vitro

M. smegmatis (strain ATCC 14468 from Presque Isle Cultures, Erie, PA) is grown in 200 mL Bacto nutrient broth (3 g beef extract plus 5 g peptone/L) at 24–25°C to yield a cell density of 4×10^6/mL. To determine *M. smegmatis* density, dilution plates are prepared using the pour plate technique, and Quebec colony counter is used for viewing and counting the isolated colonies.

Progesterone is added to each culture to yield a 1-mM solution. Before addition of progesterone and on days 0, 2, 4, 6, 8, 12, 15, 20 and 36 after progesterone is added, triplicate 3-mL samples of the slurried media are extracted in 12 mL 100% methanol (HPLC grade). The vortexed extract is centrifuged and passed through a 0.3 mL washed C-18 solid-phase extraction cartridge (Varian Instrument Co, Walnut Creek, CA). The eluant is dried under nitrogen gas before fractionation.

Each extract is fractionated on a 4.6 × 15 cm C-18 column (Varian Instrument Co.) using HPLC methanol linear gradient with 0.25% *o*-phosphoric acid and 100% methanol. The gradient increased is from 20% methanol at 0 min to 100% methanol at 20–35 min. Ultraviolet detection of column eluants is done at 235 nm. The identification is based on comigration with standards. Quantification of steroids is done based on peak area related to standards. The data is presented as the mean of triplicates (±SE; micromolar).

The percent recovery of extracted steroids in all bacterial sample fractions is determined by sequentially extracting the samples, a second and third time in methanol, followed by solid-phase extraction and quantification by HPLC. The percent yield of progesterone, 17-α-hydroxyprogesterone (17-α-OHP), androstenedione (AED) and androstadienedione (ADD) at each level is 86.7 ± 2.9% (mean ± SE).

QUESTIONS

1. Write a note on biotransformation of pentacyclic triterpenoids.
2. What is biotransformation? Write its application in plant tissue culture.
3. What are microbial conversions? Discuss important bioconversions.
4. Discuss various reactions mediated by microorganisms and give suitable examples.
5. Write a note on biotransformation of steroids.

CHAPTER 11

Hybridoma Technology and Monoclonal Antibodies

CHAPTER OUTLINE

- Introduction
- Principle for Creation of Hybridoma Cells
- Large-Scale Production of Monoclonal Antibodies
- Applications of Monoclonal Antibodies
- Questions

INTRODUCTION

Antibodies are glycoprotein molecules present in the serum. They are produced in response to antigens that are protein or polysaccharide molecules and may be foreign to the body. Antibodies are secreted by a class of blood cells known as *B lymphocytes*, which form a clone of cells known as *plasma cells*. The plasma cells produce proteinaceous molecules known as *antibodies*. Antibodies are produced when the body comes in contact and is invaded by foreign particles. Different types of antibodies may be synthesized depending upon different types of stimulants involved.

Antibodies can be classified into five major classes: (1) IgG, (2) IgM, (3) IgA, (4) IgE and (5) IgD. Because of the multiple epitose on the antigen that could induce proliferation and differentiation of a variety of B cell clones, the antibodies synthesized in response to an antigen are usually heterogeneous. The heterogeneity of antibodies that increases immune protection in vivo often reduces the efficacy of an antiserum for various in vitro uses. Alternatively, it is also possible to generate pure clones of plasma cell in vitro from which monoclonal antibodies (mAbs) with a single antigenic specificity may be isolated.

Antibodies are a part of the defence system to protect the body against the invading foreign substances, namely antigens. In response to an antigen, B lymphocytes gear up and produce different antibodies. These types of antibodies that can react with the same antigen are designated as *polyclonal antibodies* (*PoAbs*). mAbs are a single type of antibody that is directed against a specific antigenic determinant.

The evaluation of technology of mAbs is an important milestone in biotechnology, representing cumulative research efforts in immunology and cell culture methodologies. G. Kohler and C. Milstein (1975) were the first to report on the production of mAbs. The first production of mAbs is an amalgamation of immunochemistry, in vitro cultivation of cancer cells and the molecular biology of malignant transformation, representing three disciplines of basic medical research.

A new era in the immunology was launched in 1975 with the discovery of the *hybridoma technique*, a method of creating pure antibodies against a specific target (antigen). Kohler and Milstein (1975) succeeded in fusing a normal B cell with a myeloma cell, known as a *hybridoma*. Hybridomas are the hybrid cells of myeloma (cancer) with antibody production (lymphocytes). This fusion has an immunized donor (animal). The hybrid cells secrete immunoglobulins, which consist of several types of polypeptides. Hybridoma is an important biotechnological tool and indicates new path for achieving the goal of immunization of human body from infection.

PRINCIPLE FOR CREATION OF HYBRIDOMA CELLS

The myeloma cells used in hybridoma technology must not be capable of synthesizing their own antibodies. The mammalian cells can synthesize nucleotides by two pathways de novo synthesis and salvage pathway. The de novo synthesis of nucleotides requires tetrahydrofolate that is formed from dihydrofolate. The formation of tetrahydrofolate can be blocked by the inhibitor aminopterin.

The establishment of hybridomas and production of mAbs involves the following steps:

1. Immunization
2. Cell fusion
3. Formation and selection of hybrid cells
4. Screening of products
5. Cloning and propagation
6. Characterization and storage

It can be explained as depicted in Figure 11.1.

Immunization

The very first step in hybridoma technology is to immunize an animal, with appropriate antigen. The antigen along with an adjuvant like, Freund's complete or incomplete adjuvant is injected subcutaneously. When the serum concentration of the antibodies is optimal, the animal is killed. The spleen is aseptically removed and disrupted by mechanical or enzymatic method to release the cells.

Cell Fusion

Splenocytes are then mixed with plasmocytoma cells in an appropriate medium. This mixture is exposed to a high concentration of polyethylene glycol (PEG) for short time, and fusion is allowed to take place over a period of time. Mouse is used in the production of mAbs secreting hybridoma with splenocytes, because these animals are inexpensive, and several billion cells can be obtained from a mice spleen. The use of hypoxanthine-guanine phosphoribosyltransferase (HGPRT) cells assures that only hybridomas are selected. After 10 days of culture in the hypoxanthine-aminopterin-

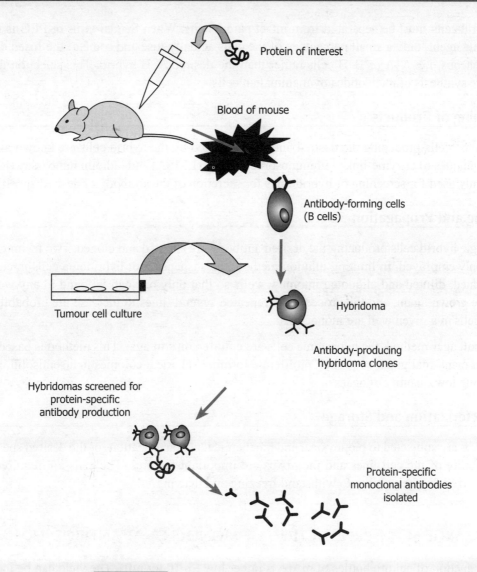

Figure 11.1 Production of monoclonal antibodies.

thymidine (HAT) medium, most of the wells contain dead cells, but a few wells contain small cultures of viable cells. The wells containing viable clusters are then screened for production of antibody, and such positive clones are subcultured at low cell densities in order to ensure clonal purity in each microwell.

Formation and Selection of Hybrid Cells

Important feature of heterokaryon research is its ability to select hybrids of two parent cell lines. When the cells are cultured in HAT medium, only hybridoma cells grow, whereas the rest slowly disappear. Selection of a single antibody-producing hybrid cells is very important. After fusion, the hybrid cells must be separated from intact parent cells. When Sendai virus or PEG is used as the fusing agent, only a small percentage of the cells actually fuse and some of the fused cells are homogeneous, i.e. A–A or B–B cells rather than the desired A–B hybrid. This selection depends upon the synthesis of nucleotides by mammalian cells.

Screening of Products

Hybridoma cells producing the desired antibody secreted by the hybrid cells are known as *mAbs*. The techniques of enzyme-linked immunosorbent assay (ELISA) and radioimmunoassay (RIA) are commonly used for screening of hybridomas for secretion of the antibody of desired specificity.

Cloning and Propagation

The single hybrid cells producing the desired antibody are isolated and cloned. Two techniques are commonly employed. In limiting dilution method, the suspension of hybridoma cells in culture is enumerated, diluted and aliquoted into new wells so that only one cell is found in any well. The cells are grown again, and the procedure is repeated several times to increase the probability that all the cells in a given well are monoclonal.

In soft agar method, the hybridoma cells are cultured on soft agar. This method is based on the fact that many malignant cells shall proliferate forming spherical colonies in a semisolid medium containing low quantity of agar.

Characterization and Storage

The mAbs are subjected to biophysical and biochemical characterization for the desired specificity. The stability of the cell lines and the mAbs are important features. The cells are required to be characterized for their ability to withstand freezing and thawing.

LARGE-SCALE PRODUCTION OF MONOCLONAL ANTIBODIES

The production of culture bottles of mAbs is rather low (5–10 µg/mL). The yield can be increased by growing the hybrid cells as ascites in the peritoneal cavity of mice. In vitro techniques rather than the use of animals are preferred, because several animals are required to be sacrificed to produce mAbs in good concentration.

Human mAbs

The mAbs produced using mice are suitable for in vitro use. For human clinical trials, the use of human mAbs is preferred avoiding any anti-isotype response. The technical difficulties faced in production of human mAbs are as follows:

1. One cannot immunize a human volunteer with the range of antigen that is given to mice.
2. It is difficult to obtain antigen prone B cells in humans.
3. Proper method is selected for assaying mAbs in supernatant hybridomas.

In vitro culturing of the cloned cell lines for production of antibodies and selecting proper method for assaying mAbs in supernatant hybridomas are important prerequisites. The following evaluations are required to be undertaken:

1. Enzyme-linked immunosorbent assay
2. Radioimmunoassay
3. Immunofluoroscence
4. Cytotoxicity
5. Flow cytometry

Advantages and Limitations of mAbs

The advantages and limitations of mAbs are given below.

Advantages

1. In vitro and in vivo production is possible.
2. High reproducibility.
3. One antigenic determinant.
4. High avidity can be produced.
5. Production of cell lines to individual components of a mixture is possible.
6. Determined to very high value of antiserum titre.
7. Cross-reaction can be determined.
8. Radiolabelling and conjugation fluorescent is possible.
9. The immunization and bleeding maintenance of farm or animals are not required.

Limitations

1. There are no suitable myeloma cells in human that can replace mouse myeloma cells.
2. The fused human lymphocytes mouse myeloma cells are very unstable.
3. The method is time consuming and initial cost involved is more.
4. The energy of binding to an antigen is slow in case of mAbs, and also average in case of conventional antiserum.
5. It is not a standard double immunodiffusion method.
6. For ethical reason, humans cannot be immunized against antigens.

Fusion and Culture of Hybridomas

Various steps involved in fusion and culturing of hybridomas are briefly summarized as follows:

1. The mouse is boosted with an intravenous injection of antigen, 72 h before use.
2. On the day of fusion, spleen is collected and a free cell suspension is made by injecting and flushing the spleen with sterile free medium.
3. The plasma cells are maintained in cell culture and fed into fresh flask medium for 16 h before fusion. It ensures that they are in early phase of growth at the time of fusion.
4. Both myeloma and spleen cells are counted and then mixed in appropriate ratio.

5. The cells are centrifuged into a loose pellet by spinning at 1000 rpm for 10 to 15 min.
6. The supernatant is removed and the pellet is overlayed with 1 mL of polyethylene glycol.
7. Following fusion, the cell is diluted with 30 mL of serum-free medium with the first 10 mL medium being added and mixed at 1 mL per min (slow dilution reduces the risk of osmotic disruption of the fused cells).
8. The cell is centrifuged and resuspended in complete medium containing HAT and then dispensed into 96 well tissue culture plates, 10 plates for each 10^8 splenocytes used in the fusion.
9. Added 10^6 thymocytes to each well to serve as feeder cells.
10. The culture is incubated for 3–4 days in carbon dioxide incubator with high humidity.
11. Half of the medium is replaced with fresh HAT medium every fourth day.
12. After each test for antibody transfer positive cell lines to 24 well plates, $3–5 \times 10^6$ feeder cells are added to each well to promote rapid cell growth.
13. The cells are maintained in static culture for a minimum of 2 weeks by removing 50–75% of the hybrid cell after 2–3 days interval.
14. The duplicate cultures are made after 2 weeks and allowed the cells to overgrow and die.
15. The supernatant is collected and assayed for the presence of antibody.

Cloning of Preservation of Hybridomas

After preliminary selection of hybrids, following steps are involved for further processing and preservation of antibodies:

1. The final supernatant is screened in detail to identify antibodies of immediate interest.
2. The parent cell lines are taken from the freezer and cloned by limiting dilution.
3. When viability is good, the cell lines are cloned immediately.
4. The remainders of the excess cells are cultured in a T75 flask for 1–2 days. After 12–14 days, supernatant from the marked wells is assayed.
5. After noting which cloned lines are stable, 4–6 clones of each cell line are expanded into 6 well plates for cell preservation and production of antibody.

APPLICATIONS OF MONOCLONAL ANTIBODIES

mAbs with specificity and high purity have a wide range of applications, which can be broadly categorized as follows:

1. Diagnostic applications
2. Investigational and analytical applications
3. Miscellaneous applications

Diagnostic Applications

mAbs are important diagnostic reagents used in biomedical research, microbiological research in diagnosis of hepatitis, influenza, AIDS, herpes simplex, chlamydia infections and in treatment of cancer and other infections. The worldwide clinical diagnostics industry is valued at approximately US$19 billion, with a growth rate of nearly 5% per year. More than 100 different mAb diagnostic products are currently available in the world market. mAb diagnostic kits are increasingly in use for identification of communicable diseases including transfusion transmissible infections.

1. **In cardiovascular disease**: It is possible to detect the location and the degree of damage to the heart by using radio labelled antimyosin mAb. The antimyosin mAb is specific for human myosin and binds to intracellular myosins exposed as a consequence of myocardial necrosis. Atherosclerosis mAb radiolabelled against platelets can be used to localize the atherosclerotic lesions by imaging techniques.
2. **Site of bacterial infection**: In recent years, attempts are made to detect the sites of infection by using mAb. This is made possible by directing mAb against bacterial antigens.
3. **In cancer**: mAbs with varying degrees of specificity have already been produced against a variety of human and animal tumours. mAbs have important clinical applications in the detection and early diagnosis of cancer, in staging procedures and in therapeutic trials. In addition, many of these reagents have the capacity to distinguish between different types of human tumours and shall be useful in the refinement of histopathologic classification scheme. The mAbs against the antigens on surface of cancer cells are useful for the treatment of cancer. This is brought out by antibody-dependent cell-mediated cytotoxicity, complement-mediated cytotoxicity and phagocytosis of cancer cells by reticuloendothelial system.

    Limitations
    a. The mAbs produced in mice and directly used for therapeutic purpose may lead to the development of anti-mouse antibodies and hypersensitivity reactions.
    b. All the cancer cells may not carry the same antigen for which mAbs have been produced.
    c. The free antigens present in the circulation may bind to mAbs and prevent them from acting on the target cells.
4. **In treatment of AIDS**: Chang et al. (1994) identified a new herpes virus sequence in human immunodeficiency virus (HIV). In addition to its occurrence predominantly in HIV-positive patients, it is also occasionally present in elderly HIV-negative patients. Immunosuppressant is the hallmark of AIDS. This is caused by reduction in CD4 cells of T lymphocytes. The HIV binds to specific receptor on CD4 cells by using surface membrane glycoprotein. Genetic engineering has been successful to attach fragment crystallizable portion (Fc portion) of mouse mAb to human CD4 molecules.
5. **In malaria**: Currently, malarial infection is diagnosed by microscopic examination of blood smears, which is an effective, but labour-intensive and time-consuming process. Although, immunological procedure for detection of plasmodium such as ELISA is rapid and amenable to automation, it does not discriminate to detect antiplasmodium antibodies in the blood of affected individuals. A diagnostic protocol using a DNA probe as a mean of detection is developed for the detection of *Plasmodium falciparum*, parasite causing malaria.
6. **In diabetes**: Hybridoma-produced mAb against insulin is useful for insulin assays because of its specificity and plentiful supply. The spleen cells of male BALB/c mice immunized against monocomponent porcine insulin are hybridized with mouse myeloma cells (P3-X63-Ag8-U1).

The resulting anti-insulin antibody (Ab) is purified and characterized by RIA. The proper concentration of this mAb for RIA is proved to be 1:15,00,000 and the smallest detectable level of porcine insulin by RIA using this Ab is 0.5 ng/mL. These levels are similar to those obtained with PoAb. The binding activity of this mAb to human insulin is quite similar to that of porcine insulin.

7. **Rheumatic arthritis**: Arthritis is an autoimmune disease. Some success has been achieved in trials of rheumatoid arthritis patients by using mAbs directed against T lymphocytes and B lymphocytes.

Some important mAbs along with the names of their manufacturers and other details are given in Table 11.1.

Table 11.2 includes the list of approved and investigational drugs as well as drugs that have been withdrawn from market. Consequently, the column *Use* in Table 11.2 does not necessarily indicate clinical usage.

Investigational and Analytical Applications

Identification and Isolation of Lymphocytes Phenotyping

mAbs specific for various lymphocytes subpopulation express characteristic pattern of membrane protein. This characteristic can be utilized for identification of various stages of lymphocytes differentiation, e.g. human and mice in both CD8 membranes expressed by T helper cells.

Purification of Proteins

Antibodies can also be used in protein purification. When a purified antibody is added to a crude mixture of proteins, the specific protein being sought selectively combines with the antibody and gets precipitated from solution. High efficiency in purification technique can be attained by using antibodies. It is possible to achieve more than 5000-fold purification of interferon-$\alpha 2$.

Miscellaneous Applications

Application of Catalytic mAbs

Catalytic mAbs have wide-ranging applications in biology, medicine, chemistry and biotechnology. In the last couple of years, mAbs with enzyme-like catalytic properties have been successfully raised. They catalyse a variety of reactions (hydrolysis of esters, amides and carbonates, cyclization reactions, bimolecular reactions, elimination reactions, asymmetric synthesis, etc., with rate enhancements of up to 6×10^6-fold and high specificity, including enantiomer specificity. Hybrid antibodies containing covalently-linked catalytic or probe groups have been prepared and antibody-mediated peptide hydrolysis catalysed by metal–ion complexes are reported.

Table 11.1 Generic name, brand name, year, manufacturer and indications of some mAbs

Generic name	Brand and year of FDA approval	Manufacturer	Indications
Infliximab	Remicade (1998)	Johnson & Johnson, Merck	Crohn's disease, ulcerative colitis, ankylosing spondylitis (AS), rheumatoid arthritis, psoriasis (PS), psoriatic arthritis (PSA)
Bevacizumab		Roche	Colon mCRC/metastatic colorectal cancer (MCRC), non–small-cell lung cancer (NSCLC), metastatic breast cancer (MBC), metastatic renal cell carcinoma (mRCC), glioblastoma
Adalimumab	Humira (2002)	Abbott	Rheumatoid arthritis, juvenile idiopathic arthritis, psoriatic arthritis, psoriasis, ankylosing spondylitis (AS), Crohn's disease
Trastuzumab	Herceptin (1998)	Roche	Breast cancer
Ranibizumab	Lucentis (2006)	Novartis, Roche	Macular degeneration
Palivizumab	Synagis (1998)	Astra Zeneca	Coffee ringspot virus (CoRSV)
Cetuximab	Erbitux (2004)	Bristol Myers Squibb, Merck	Head and neck cancer (HNC)
Omalizumab	Xolair (2003)	Roche, Novartis	Allergic asthma
Abciximab	ReoPro (1994)	Johnson & Johnson, Lilly	Thrombosis inhibitor
Panitumumab	Vectibix (2006)	Amgen	Colon cancer
Certolizumab	Cimzia (2008)	Union Chimique Belge	Crohn's disease, rheumatoid arthritis
Golimumab	Simponi (2009)	Johnson & Johnson	Rheumatoid arthritis, psoriatic arthritis, ankylosing spondylitis
Canakinumab	Ilaris (2009)	Novartis	Periodic syndrome
Ofatumumab	Arzerra (2009)	GlaxoSmithKline	Chronic lymphocytic leukaemia
Tocilizumab	Actemra EMA (2009), FDA (2010)	Roche	Rheumatoid arthritis
Rituximab CD20	Rituxan (1997)	Roche	Leukaemia, chronic lymphocytic leukaemia
Natalizumab	Tysabri (2004/2006)	Biogen Idec, Elan	Multiple sclerosis, Crohn's disease
Ustekinumab	Stelara (2008)	Johnson & Johnson	Psoriasis

mAbs in Drug Targeting

The objective of drug targeting is to deliver drugs to a specific site of action through a carrier system. Of all the carrier systems available, mAbs are gaining importance because of their high specificity. In cancer therapy, cytotoxic agents used to kill malignant cells also damage normal cells. mAbs generated against specific antigens, when conjugated to cytotoxic drugs, can selectively deliver drugs to cancer cells while minimizing damage to normal cells.

Table 11.2 Generic name, trade name, source and clinical uses

Generic name	Trade name	Source	Uses
Abciximab	ReoPro	Chimera	Platelet aggregation inhibitor
Adalimumab	Humira	Human	Rheumatoid arthritis
Alemtuzumab	Campath, MabCampath	Human	Chronic lymphocytic leukaemia, cutaneous T cell lymphoma (CTCL)
Altumomab pentetate	Hybri-ceaker	Mouse	Colorectal cancer (diagnosis)
Arcitumomab	CEA-Scan	Mouse	Gastrointestinal cancers (diagnosis)
Atlizumab	Actemra	Human	Rheumatoid arthritis
Basiliximab	Simulect	Chimera	Prevention of organ transplant rejections
Bectumomab	LymphoScan	Mouse	Non-Hodgkin's lymphoma (detection)
Belimumab	Benlysta, LymphoStat-B	Human	Non-Hodgkn's lymphoma
Besilesomab	Scintimun	Mouse	Inflammatory lesions and metastases (detection)
Bevacizumab	Avastin	Human	Metastatic cancer
Biciromab	FibriScint	Mouse	Thromboembolism (diagnosis)
Canakinumab	Ilaris	Human	Rheumatoid arthritis
Capromab pendetide	ProstaScint	Mouse	Prostate cancer (detection)
Catumaxomab	Removab	Rat/mouse hybrid	Ovarian cancer, malignant ascites, gastric cancer
Cetuximab	Erbitux	Chimera	Metastatic colorectal cancer
Daclizumab	Zenapax	Human	Prevention of organ transplant rejections
Denosumab	Prolia	Human	Osteoporosis, bone metastases, etc.
Eculizumab	Soliris	Human	Paroxysmal nocturnal haemoglobinuria
Edrecolomab	Panorex	Mouse	Colorectal carcinoma
Efalizumab	Raptiva	Human	Psoriasis (blocks T cell migration)
Ertumaxomab	Rexomun	Rat/mouse hybrid	Breast cancer
Efungumab	Mycograb	Human	Invasive *Candida* infection
Etaracizumab	Abegrin	Human	Melanoma, prostate cancer, ovarian cancer, etc.
Fanolesomab	NeutroSpec	Mouse	Appendicitis (diagnosis)
Fontolizumab	HuZAF	Human	Crohn's disease
Gemtuzumab ozogamicin	Mylotarg	Human	Acute myelogenous leukaemia
Golimumab	Simponi	Human	Rheumatoid arthritis, psoriatic arthritis, ankylosing spondylitis
Ibritumomab tiuxetan	Zevalin	Mouse	Non-Hodgkin's lymphoma
Igovomab	Indimacis-125	Mouse	Ovarian cancer (diagnosis)
Imciromab	Myoscint	Mouse	Cardiac imaging
Infliximab	Remicade	Chimera	Rheumatoid arthritis, ankylosing spondylitis, psoriatic arthritis, psoriasis, Crohn's disease, ulcerative colitis

QUESTIONS

1. Write a note on production and application of mAbs.
2. Describe hybridoma technology.
3. What are mAbs? Write their applications.
4. Define the terms mAbs, hybridoma technique and hybridization.
5. Write a note on formation and selection of hybrid cells.

CHAPTER 12 Antibiotics

CHAPTER OUTLINE

- Historical Development of Antibiotics
- Antimicrobial Spectrum
- Standardization of Antibiotics
- Screening of Soil for Organisms Producing Antibiotics
- Fermenter or Bioreactor
- Design of Industrial Fermentation Process
- Fermenter Design and Control
- Isolation of Mutants
- Factors Influencing Rate of Mutation
- Isolation of Fermentation Products
- Questions

HISTORICAL DEVELOPMENT OF ANTIBIOTICS

The term *antibiotic* was coined by Selman Waksman in 1942 to describe any substance produced by a microorganism that is antagonistic to the growth of other microorganisms in high dilution. This definition excluded substances that kill bacteria, but are not produced by microorganisms (such as gastric juices and hydrogen peroxide). It also excluded synthetic antibacterial compounds, such as the sulphonamides.

The mixtures with antimicrobial properties that were used in treatments of infections were described over 2000 years ago. Many ancient civilizations, including Indian, Egyptian and Greek used specially selected moulds and plant materials and extracts to treat infections. In ancient days, the treatments for infections were primarily based on medicinal folklore.

The earliest evidence of humans using plants or other natural substances for therapeutic purposes comes from the Neanderthals, who lived over 50,000 years ago. In northern Iraq, archaeologists uncovered evidence of human remains that had been buried with a range of herbs, some of which are now known to be antibacterial. Many of these herbs are still used by the inhabitants of this region.

The first prescription for treating infections has come from the Egyptians around 1550 BC. Written as 'mrht', 'byt' and 'ftt', it was a mixture of lard, honey and lint, and was used as an ointment for dressing wounds. Honey is antibacterial, as it kills bacterial cells by drawing water out of them. In addition, the enzyme inhibin, which is found in honey, converts glucose and oxygen into hydrogen peroxide, a well-known disinfectant.

The use of moulds dates back to the ancient Egyptians, and perhaps even earlier. An Egyptian physician, quoted in the Ebers Papyrus around 1550 BC, stated that if a wound rots, then bind on it spoiled barley bread. The Egyptians used all kinds of moulds to treat surface infections. The ancient Chinese also used moulds to treat boils, carbuncles and other skin infections.

Inorganic substances have been used to treat infections since ancient times. Copper was widely used by the Greeks, Egyptians and Romans, often in combination with honey. Modern scientific tests have proved that copper is, indeed, antibacterial. A skin infection known as *impetigo*, which is caused by the *Staphylococcus aureus* bacteria, is being treated in France with Eau Dalibour, a combination of zinc and copper. This prescription dates from the time of Jacques Dalibour, surgeon general of the army of Louis XIV.

Wine and vinegar have been popular treatments for infected wounds since the time of Hippocrates. Vinegar is an acid and a powerful antiseptic. The antibacterial properties of wine cannot be fully attributed to its alcohol content, as this is very low. Recent chemical analysis of wine has shown the presence of an antibacterial substance called *malvoside*.

Antagonistic activities by fungi against bacteria were first described in England by John Tyndall in 1875. Ehrlich from Germany (1880) noted that certain dyes coloured human, animal or bacterial cells, whereas others did not. He then proposed the idea that it might be possible to create chemicals that would bind to and kill bacteria without harming the human host. After screening hundreds of dyes against various organisms, he discovered a medicinally useful drug, the synthetic antibacterial, salvarsan.

In 1928, Alexander Fleming observed antibiosis against bacteria by a fungus of the genus *Penicillium*. Fleming postulated that the effect was mediated by an antibacterial compound named penicillin, and that its antibacterial properties could be exploited for chemotherapy. Prontosil, the first commercially available antibacterial antibiotic, was developed by Gerhard Domagk and colleagues in 1932 (he received Nobel Prize for medicine for his discovery in 1939). The work was carried out at the Bayer Laboratories of the IG Farben conglomerate in Germany. Prontosil exhibited relatively broad effect against gram-positive cocci but not against enterobacteria. The discovery and development of this first sulphonamide drug opened the era of antibacterial antibiotics.

In 1939, Rene Dubos reported the discovery of gramicidin from *Bacillus brevis*. It was one of the first commercially manufactured antibiotics in use during World War II treating wounds and ulcers.

The scientists in the 19th century began to closely examine various curative substances and discovered that there were good or beneficial bacteria. This could be isolated and encouraged for their growth in the laboratory. The bacterial agents from such bacteria were then tested for their ability to treat disease. This pattern of research and clinical trials continued into the 20th century (and, in fact, continues even today), leading to the production of the antibiotic drugs.

Late 1800s: The growing acceptance of the germ theory of disease, a theory which linked bacteria and other microbes to the causation of a variety of ailments. As a result, scientists began to search for drugs that would kill these disease-causing bacteria.

1871 Joseph Lister (a surgeon) began researching the phenomenon that urine contaminated with mould would not allow the successful growth of bacteria. In 1890, German doctors, Rudolf Emmerich and Oscar Low, were the first to make an effective medication named pyocyanase from microbes. It was the first antibiotic to be used in hospitals.

1928 Sir Alexander Fleming observed that colonies of the bacterium *S. aureus* could be destroyed by the mould *Penicillium notatum*, demonstrating antibacterial properties.

1935 Prontosil, the first sulpha drug, was discovered by German chemist Gerhard Domagk.

1942 The manufacturing process for penicillin G procaine was invented by Howard Florey and Ernst Chain. Penicillin could now be sold as a drug. Fleming, Florey and Chain shared the 1945 Nobel Prize for medicine for their work on penicillin.

1943 American microbiologist Selman Waksman isolated streptomycin from soil bacteria, the first of a new class of drugs called *aminoglycosides*. Streptomycin could treat diseases like tuberculosis (TB). However, the side effects were often too severe.

1955 Tetracycline was patented by Lloyd Conover, which became the most prescribed broad-spectrum antibiotic in the United States.

1957 Nystatin was patented and used to cure many disfiguring and disabling fungal infections.

1981 SmithKline Beecham patented amoxicillin or amoxicillin/clavulanate potassium tablets. The antibiotic was first sold in 1998 under the trade names of Amoxicillin, Amoxil and Trimox. Amoxicillin is a semisynthetic antibiotic.

Discovery of Penicillin and Other Antibiotics

In 1928, while attempting to grow the bacteria *Staphylococcus* spp. on an agar plate, Fleming noticed that the growth of this bacterium was inhibited by a mould that had accidentally contaminated the plate. He decided to identify the mould that was eventually called *P. notatum*. Fleming was excited by this discovery. He cultured the mould in a special broth and injected the broth into some of his patients, who had various infectious diseases. The results were encouraging, and the broth proved to be nontoxic. Unfortunately, Fleming had not made enough of this broth, making his experiment rather limited. When he presented a paper on his findings in 1929, his colleagues in the medical profession were not particularly impressed or interested. It took two other gifted researcher doctors—Florey and Chain, working at Oxford University in the late 1930s and early 1940s—to realize the importance of Dr. Fleming's findings. It was their pioneering work that brought penicillin into clinical use. Florey, an Australian doctor, had gone to Oxford on a scholarship to study pathology. Chain was a German chemist who had fled from the Nazis in the 1930s and had come to rest in England. In 1945, Fleming, Florey and Chain shared Nobel Prize for their work on penicillin.

Penicillin was later produced in oral form and was added to many products, including salves, throat lozenges, nasal ointments and cosmetic creams. Prior to 1955, its sale was not controlled, so anyone could buy it over the counter without a prescription. This excessive and uncontrolled use led to the overgrowth of resistant bacteria, and the damage had been done. Resistance had become a major problem, and epidemics of staphylococcus-resistant infections began to emerge in hospitals.

A systematic search of antibiotic-producing microorganisms from soil was undertaken. In 1939, Dubos isolated from soil a culture of *B. brevis* and produced two valuable antibacterial substances, now known as gramicidin and tyrocidin that killed many gram-positive bacteria.

The study of soil bacteria and the reasons why they are not more capable of causing disease was the lifelong work of Selman Waksman, a research scientist at Rutgers University in New Jersey. In 1939, Merck and Company provided him with financial assistance to mount a search for antibiotics in soil microorganisms. This search culminated in the isolation of streptomycin, the first antibiotic to offer hope to patients with TB. This antibiotic is still used today in the treatment of TB.

In 1947, the antibiotic chloramphenicol was used in a clinical trial to treat an epidemic of typhus in Bolivia. Its success in curbing the epidemic led to its use on the other side of the world– treating scrub typhus in Malaysia.

Modifications of the basic cephalosporin chemical structure led to the development of a whole range of these antibiotics for clinical use. Research into the development of new cephalosporins continues today.

In 1947, chlortetracycline was isolated from a Missouri River mud sample by Benjamin M. Duggar. Chlortetracycline was the first tetracycline introduced. Duggar's discovery has led to the isolation and subsequent development of a large number of very powerful antibiotics, which now rank second only to the penicillins in their use worldwide.

They are active against a broad range of bacteria and are relatively inexpensive to produce. The tetracyclines quickly gained favour and are now used to treat a long list of infections. However, they are also known to cause a number of toxic side effects. The tetracyclines form calcium complexes in growing bone, which may lead to lifelong discolouration and enamel defects in teeth, as well as, reduced bone growth. They are prohibited in the treatment of children below the age of 7, as they may inhibit small children's growth. Tetracyclines also cross the placenta and have a greater toxicity in the fetus. Therefore, they are not administered to pregnant women.

The tetracycline antibiotics form complexes with calcium, magnesium and iron. Therefore, they should not be taken with dairy products or any mineral and vitamin supplements containing calcium, magnesium or iron.

Further research took place during the 1960s that led to the development of the second generation of antibiotics. Among these was methicillin, a semisynthetic derivative of penicillin produced specifically to overcome the problem of penicillin resistance. Methicillin was hailed as a major breakthrough in the fight against bacterial resistance to penicillin, and scientists believed that they could now win this battle. Unfortunately, bacteria had the last word, and we now have bacteria that are resistant to methicillin.

Amoxicillin is another widely used penicillin derivative. Like ampicillin, it has a broad range of activity, as it can treat both gram-positive and gram-negative bacteria.

Ampicillin is also a derivative of penicillin. It was developed to broaden the range of infections that penicillin could treat and has now replaced penicillin to a great extent. It is often the first choice in the treatment of a whole range of infections, including respiratory and urinary tract infections.

Gentamicin is in the same family of antibiotics as streptomycin. It is generally reserved for serious infections, as it can have severe toxic side effects on the ears and kidneys.

Recently, a new family of antibiotics called the fluoroquinolones has been developed by pharmaceutical laboratories. In addition to being effective against a broad range of bacteria, these antibiotics can reach a high concentration in the bloodstream when taken orally. This means that many more infections that may once have required a hospital stay can now be treated at home. The fluoroquinolones are often used for cases in which long courses of antibiotics (weeks to months) are required. A whole range is now available, and is proving effective against bacteria that were once difficult to treat, such as the leprosy bacteria.

The search for new and more effective drugs, which began with Florey, Chain and Waksman, continues today. The pace, however, has slowed remarkably, as it is now much more difficult for pharmaceutical companies to get approval for new drugs. The time gap between the discovery of an antibiotic in the laboratory and the approval to produce it commercially is so great that it has led many companies to abandon the efforts completely. The resistance offered by bacterial strains is also another reason for discontinuation of research undertaken by drug companies.

ANTIMICROBIAL SPECTRUM

Antibacterial spectrum refers to the range of bacteria susceptible to a particular antimicrobial or class of antimicrobials.

Spectrum of activity: A specific antibiotic has a range of microorganisms whose growth it is capable of inhibiting.
1. *Broad spectrum:* The term *broad-spectrum antibiotic* refers to an antibiotic with activity against a wide range of disease-causing bacteria. It acts against both gram-positive and gram-negative bacteria, e.g. levofloxacin, tetracycline.
2. *Narrow spectrum:* An antibiotic that affects only limited number of microorganisms is said to have a narrow spectrum of activity. The spectrum of activity might include most gram-positive bacteria, but few, if any, gram-negative bacteria, e.g. penicillin.

The advantage with broad-spectrum antibiotic is that there is less of a need (as compared with narrow-spectrum antibiotic) to identify the infecting pathogen with real certainty before commencing treatment. On the other hand, a broad-spectrum antibiotic will have a more profound effect on your normal flora, thus interfering with microbial antagonism.

Site of action and organisms from which antibiotics are produced are given in Table 12.1.

Table 12.1 Some clinically important antibiotics

Antibiotic	Producer organism	Site or mode of action	Activity against
Amphoterecin B	*Streptomyces nodosus*	Cell membrane	Fungi
Bacitracin	*Bacillus subtilis*	Wall synthesis	Gram-positive bacteria
Cephalosporin	*Cephalosporium acremonium*	Wall synthesis	Broad-spectrum
Erythromycin	*Streptomyces erythreus*	Protein synthesis	Gram-positive bacteria
Gentamicin	*Micromonospora purpurea*	Protein synthesis	Braod-spectrum
Griseofulvin	*Penicillium griseofulvum*	Microtubules	Dermatophytic fungi
Neomycin	*Streptomyces fradiae*	Protein synthesis	Broad-spectrum
Penicillin	*Penicillium chrysogenum*	Wall synthesis	Gram-positive bacteria
Polymyxin B	*Bacillus polymyxa*	Cell membrane	Gram-negative bacteria
Streptomycin	*Streptomyces griseus*	Protein synthesis	Gram-negative bacteria
Tetracycline	*Streptomyces rimosus*	Protein synthesis	Broad-spectrum
Vancomycin	*Streptomyces orientalis*	Protein synthesis	Gram-positive bacteria

STANDARDIZATION OF ANTIBIOTICS

Antibiotics are not equally effective against all types of microorganisms. Some antibiotics are inhibitory to many species (broad spectrum) while others are inhibitory to only a few species of microorganisms (narrow spectrum). Sometimes, a resistant strain of a microbial species arises as a result of overexposure to an antibiotic. To determine the effectiveness of an antibiotic, sensitivity test is performed against different microbial species. These tests are commonly done by placing small filter paper discs containing antibiotics on the surface of an agar plate heavily seeded with test organisms. The plate is incubated. A zone of inhibition surrounding the filter paper disc indicates that the test organism is sensitive to the antibiotic.

The assay and standardization of antibiotic is done in a similar fashion. An agar plate is seeded with a standard test organism. Measured volume of an antibiotic is placed in glass or metal cylinders on the surface of the inoculated agar. The diameter of the zone of inhibition is a measure of the antibiotic concentration in any given cylinder.

Minimal Inhibitory Concentration (MIC)

Minimal inhibitory concentration (*MIC*) is defined as the lowest concentration of antimicrobial agent required to inhibit growth of the organism. The agar plates, tubes or microtitre trays with two-fold dilutions of antibiotics are inoculated with a standardized inoculum of the bacterium and incubated under standardized conditions following National Committee for Clinical Laboratory Standards (NCCLS) guidelines. The next day, the MIC is recorded as the lowest concentration of antimicrobial agent with no visible growth. The MIC provides information about degree of resistance and might provide some important information about the resistance mechanism and the resistance genes involved. MIC determination performed as agar dilution is regarded as the gold standard for susceptibility testing.

Agar Diffusion Tests

It is often used as a qualitative method to determine whether a bacterium is resistant, intermediately resistant or susceptible to an antibiotic. However, the agar diffusion method can be used for determination of MIC values provided. Moreover, necessary reference curves for conversion of inhibition zones into MIC values are available. After an agar plate is inoculated with the bacterium, a tablet, disk or paper strip with the antimicrobial agent is placed on the surface. During incubation the antimicrobial agent diffuses into the medium and inhibits growth of the bacterium if susceptible. Diffusion tests are cheaper in comparison to most MIC-determination methods. Although E-test is a diffusion test, it has been developed to give an approximate MIC value.

The agar diffusion methods are also influenced by factors like depth of agar medium, diffusion rate of the antimicrobial agent and growth rate of specific bacterium.

The MIC-determination and disk diffusion methods described are in accordance with the international recommendations given by the National Committee for Clinical Laboratory Standards (NCCLS). The NCCLS describes how to perform the tests and sets international guidelines for interpretation of the results. It may be noted that the WHO does not prescribe any specific method for performance and interpretation of susceptibility tests.

Internal quality control should be regularly performed as recommended by NCCLS.

MIC Determination by Broth Dilution (Using Sensititre)

1. **Materials used**
   a. McFarland standard 0.5
   b. Disposable loops (1 and 10 µL)
   c. Microtitre trays with dehydrated antibiotics in two-fold concentrations
   d. Sensititre plates from Trek Diagnostic System Ltd., England
   e. Microtitration reader with mirror
   f. Multichannel pipette
   g. Disposable reservoir for reagents
   h. Graduated pipettes (20–1000 µL)
   i. Nephelometer or white paper with black lines

2. **Media**
   a. Nutrient agar plates for purity control of inoculum suspensions
   b. Sterile normal saline, 4 mL volumes in tubes for nephelometer
   c. 10 mL cation adjusted Mueller-Hinton II broth in sensititre tubes

3. **Bacterial strains**
   a. *Salmonella* strains on nonselective agar
   b. Four strains for quality control: *S. aureus* ATCC 29213, *Pseudomonas aeruginosa* ATCC 27853, *Enterococcus faecalis* ATCC 29212 and *Escherichia coli* ATCC 25922

4. **Safety**: The procedures are carried out in accordance with the local codes of safe practice.

5. **Standardization of inoculum**: The material is picked up from at least 3 to 4 colonies. It is totally suspended in 4 mL saline in tubes and mixed. It is adjusted to McFarland 0.5 (nephelometer).

The nephelometer is calibrated before use and gently the suspensions are turned upside down before measuring. The turbidity of inoculum is adjusted to match that of standard.

If a nephelometer is not available, it is compared visually with the McFarland 0.5 standard using white paper with black lines as background. The McFarland 0.5 suspension is diluted as follows for the species tested at this course:

a. Gram-negative: 10 µL McFarl. 0.5 into 10 mL broth
b. Gram-positive: 50 µL McFarl. 0.5 into 10 mL broth

The suspension should be used for inoculation within 15 min.

6. **Inoculation and incubation**: The microtitre trays are inoculated with 50 µL of the inoculum suspension using a multichannel pipette or Sensititre autoinoculator. The plates are sealed and incubated at 37°C for 18–22 h. Do not stack plates more than two.

   Purity control: About 10 µL of the inoculation suspension is spread on a nutrient agar plate and incubated at 35°C overnight. The quality control strains are run in parallel to the test strains. This is done to avoid picking bacteria that have lost their resistance. The standardization of inoculum is essential because the interpretation of the results is based on a certain inoculum. The dilution procedure varies with the bacterium species and has to be calculated for each. NCCLS recommends that each well contain approximately 5×10^5 CFU/mL after inoculation. Many gram-negative bacterial species grow faster than gram-positive species.

   Each well is inoculated with approximately 55×10^3 gram-negative cells or 2.5×10^4 Gram-positive cells.

   The incubation time is extremely important to obtain reliable endpoints.

7. **Reading MIC/interpretation of results**: The purity of inoculum suspension is checked. It has to be satisfactory for favourable interpretation. The plates are read as follows:

   a. The record sheet is used for orientation of the plates.
   b. The growth is checked in three positive control wells.
   c. The MIC is read as the lowest concentration without visible growth.

The antibiotic trimethoprim and the sulphonamides allow growth of the bacteria for several generations before inhibition occurs. In case of trimethoprim and sulphonamides, the MIC is recorded as the lowest concentration where a growth reduction of 80–90% can be seen. Further interpretation of the MIC is done according to the NCCLS recommendations. The breakpoints are also shown on the record sheet. The acceptable MIC ranges for the quality control strains are as recommended by the NCCLS. Strange patterns of growth in the microtitre trays are often caused by contamination. The MIC is determined from two-fold dilutions of the antimicrobial agent. The true MIC could be anywhere between the observed MIC and the dilution step below. The NCCLS standard does not include breakpoint recommendations for all the compounds tested.

SCREENING OF SOIL FOR ORGANISMS PRODUCING ANTIBIOTICS

The common sources for isolation of microorganism are soils, lakes and river muds. The common techniques used for isolation of industrially useful microorganisms include the following:

1. Direct sponge of the soil

2. Soil dilution
3. Gradient plate method (pour plate and streak plate techniques)
4. Aerosol dilution
5. Flotation
6. Centrifugation

The technique of isolating microorganisms varies according to the nature and physiological properties of the microbe to be isolated. For example, isolation of fungi from a mixture of fungi and bacteria can be achieved easily by incorporating in the growth medium antibiotics to which the bacteria are sensitive. Similarly, from a mixture of sporulating (spore-forming) and nonsporulating bacteria, spore-forming bacteria can be isolated by heating the mixture to 70–80°C for 5–10 min leading to the death of all vegetative forms, so that the remaining spores shall belong to only sporulating bacteria.

Small samples of microbial cultures can also be obtained from permanent cultures, which are maintained by government, as well as private agencies in most countries. Similarly, from a mixture of sporulating (spore-forming) and nonsporulating bacteria, spore forming bacteria can be isolated by heating the mixture to 70–80°C for 5–10 min leading to the death of all vegetative forms, so that the remaining spores belong to only sporulating bacteria.

Isolation of Microorganisms

The soil samples collected from several habitats in different areas are selected for the isolation of *Streptomyces* strains. These habitats include the rhizosphere of plants, agricultural soil, preserved areas and forest soils. The samples are taken up to a depth of 20 cm after removing approximately 3 cm of the soil surface. The samples are placed in polyethylene bags, closed tightly and stored in a refrigerator. The following screening procedure is adopted for the isolation of *Streptomyces*. The soil is pretreated with $CaCO_3$ (10:1 w/w) and incubated at 37°C for 4 days. It is then suspended in sterile Ringer solution (1/4 strength). The test tubes containing dilution of the samples are placed in a water bath at 45°C for 16 h, so that the spores are separated from the vegetative cells. The dilutions are inoculated on the surface of the Actinomycete Isolation Agar plates. The plates are incubated at 28°C until the sporulation of *Streptomyces* colonies occur. *Streptomyces* colonies (where the mycelium remained intact and the aerial mycelium and long spore chains are abundant) are then removed and transferred to the yeast extract—malt extract agar (ISP 2) slants. Pure cultures are obtained from selected colonies for repeated subculturing. After antimicrobial activity screening, the isolated *Streptomyces* strains are maintained as suspensions of spores and mycelial fragments in 10% glycerol (v/v) at −20°C in the Mugla University Collection of Microorganisms (MU).

The *Streptomyces* colonies are characterized morphologically and physiologically following the directions given for the International Streptomyces Project (ISP). General morphology is determined using the Oatmeal Agar plates, incubated in the dark at 28°C for 21 days and then by direct light microscopy examination of the surface of the crosshatched cultures. The colours are determined according to the scale adopted by Prauser (1964). Melanin reactions are detected by growing the isolates on peptone-yeast extract-iron agar (ISP 6). All strains are cultivated on an ISP 2 medium. Some diagnostic characters of highly active *Streptomyces* strains are determined following the directions given in the *Bergey's Manual of Systematic Bacteriology*.

FERMENTER OR BIOREACTOR

The term *fermentation* is derived from the Latin word *fevere* meaning 'to ferment'. Fermentation is an ancient process dating back to thousands of years. It was the means by which bread, wine, beer and cheese were made. Egyptians found that uncooked dough left standing became lighter and softer. Around 4000 BC, wine was made from grape juice through a fermentation process. Beer was made by soaking barley in water. During the process of fermentation, complex organic substances are broken down into simpler ones. The cell (microbial or animal) obtains energy through glycolysis, the splitting of a sugar molecule to extract its electrons. The by-product of this process is excreted from the cell in the form of substances such as alcohol, lactic acid and acetone. With advances in the science of microbiology and technologies like biotechnology, microorganisms are exploited to produce a wide variety of products using fermentation. These include the following:

1. Beverages—beer, wine
2. Dairy products—cheese, yoghurt
3. Chemicals—citric and acetic acid, amino acids, enzymes and vitamins
4. Single cell proteins (SCP)—it is monoculture of bacteria, fungi and algae containing large amounts of protein. It is a cheap source of dietary protein.
5. Antibiotics—they are one of the most important compounds produced by fermentation. Alexander Fleming in 1929 was the first to discover penicillin, an antibiotic.
6. Fuels—ethanol, methanol, methane

Types of End Products of Fermentation

Various types of end products are as follows:

1. Microbial cells (e.g. bacteria, yeast, fungal spores)
2. Microbial enzymes (e.g. milk-clotting enzymes or rennets, recombinant fungal and bacterial rennets for cheese manufacture)
3. Microbial metabolites (e.g. alcohols—ethanol, butanol, 2,3-butanediol, isopropanol; chemicals—lactate, propionate, proteins, vitamins, antibiotics and fuels—methane)
4. Recombinant products (e.g. hormones)

Fermentation is widely used in the pharmaceutical and food industries. It requires the cultivation in submerged culture of an identified microorganism (mainly bacterial) as a monoculture under defined environmental conditions. The incubation regime imposed is designed to maximize the productivity of the organism of interest by providing optimal conditions for population growth (biomass). The product of interest might be a bioactive metabolite or recombinant protein. During an incubation cycle a nutrient energy source (e.g. glucose) is added. The biomass and end product shall increase as this is depleted.

The bioreactor should be capable of the following as a minimum:

1. Aseptic production for extended periods of time
2. Meeting the local containment regulations
3. Monitoring and controlling pH (by acid or base addition), temperature and sterile sampling capability

A fermenter is the setup to carry out the process of fermentation. The fermenters vary from laboratory experimental models of 1–2 L capacity to industrial models of several hundred litres capacity, which refers to the volume of the main fermenting vessel.

A bioreactor differs from a fermenter in that the former is used for the mass culture of plant or animal cells, instead of microorganisms. The chemical compounds synthesized by these cultured cells, such as therapeutic agents, can be extracted easily from the cell biomass.

The design engineering and operational parameters of both fermenters and bioreactors are identical. With the involvement of microorganisms as elicitors in some situations, the distinction between the two concepts is being gradually eroded.

Some important products obtained from naturally occurring microorganisms with their applications are given in Table 12.2. The products derived from genetically engineered microorganisms with their applications are enlisted in Table 12.3.

Table 12.2 Fermentations by naturally occurring microorganisms

Product	Microorganism	Application
Bacitracin	*B. subtilis* (bacterium)	Antiobiotic
Chloramphenicol	*Streptomyces venezuelae* (bacterium)	Antiobiotic
Citric acid	*Aspergillus niger* (fungus)	Food flavouring
Erythromycin	*S. erythraeus* (bacterium)	Antibiotic
Invertase	*Saccharomyces cerevisiae* (fungi)	Candy
Lactase	*Escherichia coli* (bacterium)	Digestive aid
Neomycin	*S. fradiae* (bacterium)	Antibiotic
Pectinase	*A. niger* (fungus)	Fruit juice
Penicillin	*P. notatum* (fungus)	Antibiotic
Riboflavin	*Ashbya gossypii* (fungus)	Vitamin
Streptomycin	*S. griseus* (bacterium)	Antibiotic
Subtilisins	*B. subtilis* (bacterium)	Laundry detergent
Tetracycline	*Streptomyces aureofaciens* (bacterium)	Antibiotic

Table 12.3 Fermentations by genetically engineered microorganisms

Product	Microorganism	Application
Growth hormone B	*E. coli*	Milk production (cows)
Cellulase	*E. coli*	Cellulose
Growth hormone H	*E. coli*	Growth deficiencies
Human insulin	*E. coli*	Diabetics
Monoclonal antibodies	Mammalian cell culture	Therapeutics
Ice-minus	*Pseudomonas syringae*	Prevents ice on plants
Snomax	*P. syringae*	Makes snow
tPA	Mammalian cell culture	Blood clots
Tumour necrosis factor	*E. coli*	Dissolves tumour cells

Abbreviations: B, bovine; H, human; tPA, tissue plasminogen activator.

DESIGN OF INDUSTRIAL FERMENTATION PROCESS

The fermentation process requires the following:
1. A pure culture of the chosen organism, in sufficient quantity and in the correct physiological state
2. Sterilized and carefully composed medium for growth of the organism
3. A seed fermenter, a mini-model of production fermenter to develop an inoculum to initiate the process in the main fermenter. Items 1 and 3 constitute upstream.
4. A production fermenter, the functional large model
5. Equipment for the following:
   a. Drawing the culture medium in steady state
   b. Cell separation
   c. Collection of cell-free supernatant
   d. Product purification
   e. Effluent treatment

This (4 and 5) constitutes the downstream of the fermentation process. Other equipments for fermenting are autoclave, ovens, incubators and pumps.

Fermenters or bioreactors are equipped with an aerator to supply oxygen in aerobic processes, a stirrer to keep the concentration of the medium uniform, and a thermostat to regulate temperature, a pH detector and similar control devices.

Types of fermenters: Airlift fermenters, fixed bed fermenters, tower fermenters, fed-batch culture, batch culture fermenter and continuous culture fermentation

Solid-State Fermentation

It involves the growth of microorganisms on moist solid substrates in the absence of free water. The absence of free water makes the system quite different from submerged liquid fermentations and makes solid-state fermentation (SSF) superior for the production of some products. However, compared with submerged liquid fermentation, relatively little is known about how to design and operate bioreactors for large-scale SSF processes. Despite this, there are commercially successful large-scale SSF processes for the production of soy sauce koji and other traditional fermented foods, citric and gluconic acids, and fungal enzymes, such as cellulases, amylases, lipases and pectinases. In addition, fungal spores have been produced by SSF for use in steroid transformations and as inocula for production of blue vein cheeses.
1. In production of enzyme such as amylase, cellulose, lipase and pectinase.
2. Upgradation of solid wastes from agriculture and food processing for use as fermented animal feeds.
3. Production of fungal spores for use as mycopesticides. SSF is superior for the production of fungal spores because most fungi do not sporulate well in liquid culture.
4. Biopulping of woodchips during production of paper.
5. Production of dehairing enzymes, fungal rennets, β-glucanases and xylanases.
6. Production of gibberellic acid, aroma and flavour compounds, antibiotics and ethanol. There is no systematic framework guiding the design and operation of large-scale bioreactors for

SSF as in case of submerged fementers. As a result, there is limited success in developing successful large-scale processes for these newer SSF products, although large-scale bioreactors have been used to make soy sauce koji. Intermediate-scale processes can be carried out using tray fermentations. Such operations are however, quite labour intensive.

Submerged Fermentation

It is preferred for large-scale fermentations and is used extensively for industrial production of amino acids, antibiotics, organic acids, ethanol, Baker's and distiller's yeasts. While many fungi may not sporulate well in submerged cultures, there is evidence that most non fastidious plant pathogenic fungi can be produced using submerged fermentation. Some fungi are even seemingly restricted to submerged cultures for production of conidia, requiring high moisture contents or aeration levels. Submerged fermentation is considered more readily available, economical and practical than other methods for mass production of biopesticides. Submerged cultures are homogenous and easier to operate aseptically compared to solid media and may be more readily applied on a large scale. It is generally believed that liquid fermentation is preferred or required to produce low-cost bioherbicide agents.

Batch Fermentation

The production of high lysine content animal feed supplement from methanol using the methylotroph *Bacillus methanolicus* strain MGA3 is depicted in Figure 12.1. The inoculum is taken from a 1-mL vial of frozen vegetative rod. The seed build-up is accomplished by using shake flasks (4 L) in minimal salt medium containing methanol to give a 5% v/v starter culture for the next stage. New

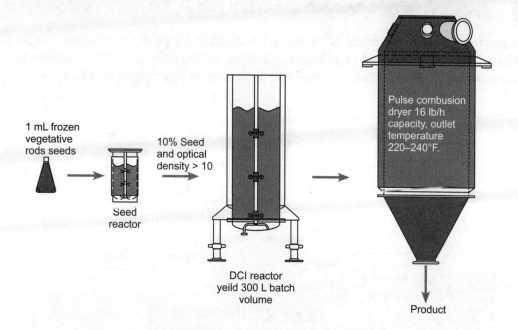

Figure 12.1 Batch fermentation process.

Brunswick Scientific (NBS) fermenter run overnight with dissolved O_2 control and automatic methanol addition to provide a 10% v/v starter culture is used for the next stage. A 300-L fermenter is employed for the final fermentation production stage. This fermentation is fitted with provisions for automatic methanol addition, pH control by automatic addition of ammonium hydroxide and dissolved oxygen control by proportional addition of oxygen to the sparge air by a mass flow controller. This is run for 20 h and yields a 300 L total volume of high saline and high cell density product subjected to spray dryer processing.

Other Fermenters

1. **Tower fermenters**: As the name suggests, these are vessels characterized by a high height to diameter ratio, between 6:1 and 15:1. They are aerated by gas sparging via a simple sparger usually located near the fermenter base.
2. **Airlift fermenters**: In this, the mixing system (motor, driveshaft and impellors) is replaced by a constant flow of gas introduced into a riser tube. The airlift vessel may be baffled to improve mixing. These vessels provide very gentle mixing, and so are particularly suited to cells that are too shear sensitive to be mixed by an impeller.
3. **Hollow fibre chambers**: It is used to grow anchorage-dependent cells. The system consists of a bundle of fibres and the cells grow within the extracapillary spaces (ECS) of a cartridge. Medium and gas perfuse through the capillary lumea to the ECS, where they are available to the cells. The size of the lumea can be selected such that any product is retained in the ECS or passes through the lumea such that the system acts as a perfusion bioreactor.

FERMENTER DESIGN AND CONTROL

The precise control of a number of parameters is required for proper incubation. The parameter of temperature, pH, dissolved O_2 or redox, agitation, pressure, foam control, auxiliary feed or a combination of these controllers are important for success of fermenter (Fig. 12.2).

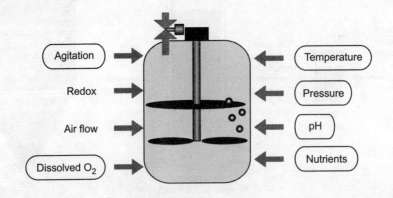

Figure 12.2 Fermenter.

The laboratory-scale vessels could have a capacity of just 10 L or less, whereas production vessels may be as large as several thousand litres. The smallest units may incorporate an electrical heater. The feedstocks (e.g. nutrient and pH control agents) are fed from flasks via peristaltic pumps. The larger vessels have an integral jacket for controlling temperature with hot or cold water. The indirect sterilization is allowed using injected steam. Where larger quantities of feedstock are required, they may be held in separate pressurized tanks and fed via a 'thrust pump' arrangement of valves.

The actual fermentation process is known as the *incubation phase* and is just part of the batch cycle. A complete fermentation cycle can typically include the following steps (depending on vessel design):

1. Empty (blank) sterilization of vessel and pipework using direct steam
2. Injection
3. Charging with base medium
4. Indirect sterilization via steam injected into the vessel jacket
5. Cooling and jacket drain
6. Preinoculation—vessel environment under control
7. Inoculation—injection of a small sample of the monoculture
8. Incubation—fermentation process itself
9. Harvesting—product removed ready for extraction processes

A control system should provide flexibility whereby accurate and repeatable control of the fermentation environment is achieved. It should include the following features:

1. Precise loop control with setpoint profile programming
2. Sequential control for vessel sterilization and more complex control strategies
3. Recipe management system for easy parameterization
4. Secure collection of online data from the fermenter system for analysis and evidence
5. Local operator display with clear graphics and controlled access to parameters

ISOLATION OF MUTANTS

Mutant selection has been the most successful approach for strain improvement. A majority of desirable mutants, especially the minor gene mutants capable of showing increased production, are isolated by screening a large number of clones surviving the mutagen treatment. This is referred to as *secondary screening*. The efforts are therefore, focused on developing techniques for the isolation of particular classes of mutants.

1. Isolation of auxotrophic mutants is the basis for commercial production of amino acid from *Corynebacterium glutamicus* in Japan. The phe-mutants of *C. glutamicus* accumulate tyrosine that is commercially exploitable.
2. Many analogue-resistant mutants have feedback-insensitive enzymes of the biosynthetic pathway. Such mutants tend to overproduce the end product of the concerned pathway.
3. Reversion mutants of appropriate auxotrophs may often be high producers.

4. Sometimes, mutants with altered cell membrane permeability show high production of selective metabolites.
5. Sometimes, revertants from nonproducing mutants of a strain are high producers. For example, the mutant of *Streptomyces viridifaciens* shows over six-fold increase in chlortetracycline production over the original strain from which the nonproducing mutant is obtained.
6. Mutants have been selected to produce altered metabolites, especially in case of aminogycoside antibiotics. For example, *Pseudomonas aureofaciens* produces the antibiotic pyrrolnitrin. A mutant of this fungus yields 4′-fluoropyrrolnitrin.

FACTORS INFLUENCING RATE OF MUTATION

Mutation rate is the frequency with which a gene changes from the wild type to a mutant. It is expressed as the number of mutations per biological unit, which may mean per cell division, per gamete or per round of replication.

Mutation frequency is defined as the incidence of a specific type of mutation within a group of individual organisms, which is expressed as a percentage of the total population being studied.

The factors influencing mutation rate are given below.

Factor 1 (frequency of primary changes in DNA): The mutation rate depends on the frequency of primary changes in DNA. These primary changes may arise from spontaneous molecular changes in the DNA, or be induced by chemical or physical agents in the environment.

Factor 2 (probability of repair): When a change in DNA takes place, it shall be repaired. Most cells posses a number of mechanisms to repair changes in DNA. Most alterations are repaired before they are replicated. If these repair systems are effective, mutation rates will be low. If they are faulty, mutation rates will be increased.

There are mutations that cause increase in the overall mutation rate for other genes. Such mutations usually occur in genes that encode components of the repair mechanisms or repair enzymes.

Factor 3 (probability of recognition): When DNA is sequenced, all mutations are potentially detectable. In practice, however, sequencing is still quite expensive, so most of the mutations are detected by their phenotypic effects. Some mutations may appear more likely to take place simply because they are easier to detect.

Some other factors are listed below:
1. A wide variation in the mutation rates is seen from one gene to another within an organism and between organisms. Generally the rates are low in lower organisms like bacteria as their multiplication rate is very high, whereas rates are higher in higher organisms.
2. Spontaneous mutation rates are low for all organisms (1^{-100} in 10 billion cells for bacteria and viruses, and 1^{-10} per million gametes for eukaryotes).
3. Within each major class of organisms, mutation rates vary considerably. This perhaps is a consequence of biological differences or differences in repair mechanisms or due to different exposure to mutagens.

4. Within a single species, mutation rates can vary per gene. Some regions of DNA seem to be more susceptible to mutations than others. Within single organism, the mutation rate of two genes can differ by a thousand-fold or more, so within species, some mutations may be very rare and others quite common. Exposures to very high doses of very potent mutagens can increase the mutation rate per generation by more than a hundred-fold.
5. It is difficult to measure the mutation rate. The indication of the mutation rate cannot be achieved directly by measuring the frequency of existing mutations in a population as a single mutation may be passed on to many offspring. There is often an increase or decrease in the frequency of a mutation in a population due to selective pressures.
6. Both the nature of the gene and its environment can influence the mutation rate. The size of the gene, its base composition, its position in the genome, and whether or not it is being actively transcribed influence its mutation rate.
7. A rare feature about human mutations is that they mostly arise in the father rather than the mother. The reason behind this is that sperms are produced late in a male's development, compared to eggs and also there is a difference in the number of cell divisions required to produce a sperm versus an egg. Children of older parents are known to have more number of mutations within a gene.
8. Mutations can also change a mutated gene back to the normal, wild-type form of the gene. Such back mutations are typically much rarer than forward mutations.
9. The repair capacity of the organism also decides how many mutations ultimately remain in a genome. In the first case of Bloom syndrome, the ability to repair DNA damage is decreased leading to an elevated mutation rate causing cancer in human beings.

ISOLATION OF FERMENTATION PRODUCTS

Production of Penicillin

The mould from which Alexander Fleming isolated penicillin was later identified as *P. notatum*. A variety of moulds belonging to other species and genera were later found to yield greater amounts of the antibiotic and a series of closely related penicillins. The naturally occurring penicillins differ from each other in the side chain (R group).

Penicillin was produced by a surface culture method early in World War II. Submerged culture methods were introduced in 1943 and are now almost exclusively employed. The production of penicillin needs strict aseptic conditions. The contamination by other microorganisms reduces the yield of penicillin. The widespread occurrence of penicillinase-producing bacteria inactivates the antibiotic. The penicillin production also needs lot of air. In all methods, deep tanks with a capacity of several thousand gallons are filled with a nutrient culture medium. The medium is made up of corn steep liquor, glucose, lactose, nutrient salts, phenyl acetic acid or a derivative and calcium carbonate as a buffer. The medium is inoculated with a suspension of conidia of *P. chrysogenum*. The medium is constantly aerated and agitated, and the mould grows throughout as pellets. After about 7 days, the growth is complete, the pH rises to 8.0 or above, and penicillin production ceases.

When the fermentation is complete, the masses of mould growth are separated from the culture medium by centrifugation and filtration.

The process of extracting the penicillin from the clear fluid is complex. The method involves various extractions with organic solvents and recrystallization. Penicillin is assayed to determine its potency before being bottled and sold.

The potency of a batch of penicillin is determined by a biologic assay in which the unknown is compared with a standard preparation of crystalline sodium penicillin G. Cylinder-plate method is used to determine the potency of penicillin. The stainless steel cylinders open at both the ends are placed on the agar nutrient medium. The test consists of adding nutrient, agar, previously inoculated with a specified strain of *Staphylococcus*, to a sterile Petri dish.

The cylinders are filled with suitable dilutions of the working standards of penicillin and of the unknown sample. The plates are incubated at 37°C for 16–18 h. The diameters of the zones of inhibition of the bacterial growth are measured. The antibiotic activity of the unknown sample is determined by comparing its zones of inhibition with those of the standard penicillin.

Bioproduction of Streptomycin

The strains of *S. griseus* are used in production of streptomycin and various other antibiotics. The spores of this actinomycete are inoculated into a medium to establish a culture with a high mycelial biomass. This is used for introduction into an inoculum tank. With subsequent use of the mycelial inoculum to initiate the fermentation process in the production tank, the bioproduction process of the antibiotic is set in.

The medium contains soybean meal (N-source), glucose (C-source) and sodium chloride. The process is carried out at 28°C, and the maximum production is achieved in the pH range of 7.6–8.0. High agitation and aeration are prerequisites. The process of fermentation lasts for about 10 days.

The fermentation process basically involves three phases. In the first phase of bioproduction, there is rapid growth of the microbe with production of mycelial biomass. The proteolytic activity of the microbe releases NH_3 to the medium from the soybean meal causing a rise in pH. During this initial fermentation phase, there is little production of streptomycin. During the second phase, the production of mycelium is limited. However, the secondary metabolite, streptomycin accumulates in the medium. The glucose and NH_3 released are consumed during this phase. The pH remains fairly constant between 7.6 and 8.0. In the third and final phase, when carbohydrates become depleted, streptomycin production ceases and the microbial cells begin to lyse. The pH increases and bioproduction process normally ceases by this time.

After the process is complete, mycelium is separated from the broth by filtration, and the antibiotic is recovered. The streptomycin is adsorbed onto activated charcoal and eluted with acid alcohol. It is then precipitated with acetone and further purified by use of column chromatography.

Microbial Production of Vitamin B_{12}

Vitamin B_{12} was known to the scientific world in the early 1920s, when two American physicians, Minot and Murphy, demonstrated to the world that they were able to cure pernicious anaemia, a disorder first described in 1835, with a diet containing whole liver. Their initial discovery led to further investigations into identifying the so-called extrinsic factor, giving rise to a whole new scientific research field that has culminated into a number of Nobel Prizes, awarded most notably to Minot and Murphy (together with Whipple) in 1934 for their discoveries concerning liver therapy in cases of anaemia, and to Hodgkin in 1964 for her work on X-ray techniques for structure elucidation of important biochemical substances.

After several years of intensive research, the full chemical synthesis of vitamin B_{12} was achieved by Woodward and Eschenmoser in 1973. This highly complicated synthesis—involving about 70 synthetic steps—makes any industrial production of vitamin B_{12} by chemical methods far too technically challenging and expensive.

Today, vitamin B_{12} is exclusively produced by biosynthetic fermentation processes, using selected and genetically optimized microorganisms. The species capable of producing vitamin B_{12} belong to the genera of *Aerobacter, Agrobacterium, Alcaligenes, Azotobacter, Bacillus, Clostridium, Corynebacterium, Flavobacterium, Micromonospora, Mycobacterium, Nocardia, Propionibacterium, Protaminobacter, Proteus, Pseudomonas, Rhizobium, Salmonella, Serratia, Streptomyces, Streptococcus* and *Xanthomonas*. For the industrial production of cobalamin, it is a common strategy to use random mutagenesis in order to generate strains that produce vitamin B_{12} in high yields. This is achieved by treating the appropriate microorganisms with mutagenic agents like UV light, ethylene amine, nitrosomethylurethane or N-methyl-N'-nitro-N-nitrosoguanidine and selecting the strains with practical advantages, such as productivity, genetic stability, reasonable growth rates and resistance to high concentrations of toxic intermediates present in the medium.

Bioproduction

Vitamin B_{12} is recovered as a by-product of streptomycin and aureomycin antibiotic fermentations. A soluble cobalt salt is added to the fermentation reaction as a precursor to yield vitamin B_{12}. Relatively high amounts of this vitamin B_{12} accumulate in the fermentation medium at concentrations that are not toxic to *Streptomyces* species.

Vitamin B_{12} is also produced on large scale by direct fermentation processes. The bacterial strains of *Propionibacterium shermanii* or *Pseudomonas denitrificans* are used, nowadays, for fermentation processes. *P. shermanii* is grown in anaerobic culture for 3 days at 30°C and in aerobic culture for 4 days. The fermentation medium (growth medium) used for production contains glucose, corn steep liquor (a waste product of starch manufacture), ammonia and cobalt chloride. Ammonia is used in the form of ammonium hydroxide that maintains the pH of the medium at 7. *P. denitrificans* is grown for 2 days in aerobic culture medium containing sucrose, betaine, glutamic acid, cobalt chloride, 5,6-dimethylbenzimidazole and salts.

The vitamin resulted in the fermentation is retained within the cells. The cells are, therefore, collected by high-speed centrifugation, and vitamin B_{12} is recovered by releasing the vitamin from the cells by treating with acid, heating and cyanide treatment. The vitamin thus released from the cell is absorbed on ion-exchange resin IRC-50 or charcoal. It is then purified by phenol and water. The vitamin is finally crystallized from aqueous acetone solutions. The normal yield is 23 mg/L when *Propionibacterium* is used.

Tetracycline Production with Sweet Potato Residue by SSF

Tetracyclines are broad-spectrum antibiotics, and the hydrocarbon derivatives of octahydronaphthalene with four annelated six-membered rings. Submerged culture was usually used for tetracycline production, and around 5500 different types of antibiotics were reported by 1982. However, most of them had no economic value due to toxicity or ineffectiveness in clinical use. The bioproduction of therapeutically important antibiotic was successfully carried out in Taiwan using sweet potato as raw material. Sweet potatoes and its residue are abundantly available in Taiwan. In 1985, the cultivation area of sweet potato was 23,239 hectares and annual production was 3,69,461 tonnes. Starchy materials have a high productivity per hectare and an excellent rate of fermentation by a great number of fast growing microorganisms.

Materials and Methods

1. Sweet potato residue: Sweet potato residue purchased from the local market in Taiwan was screened with 4–16 mesh to remove the dust and large aggregates. It contains 14.0–16.1% moisture, 2.32% crude protein, 3.6% ash, 18.1% crude fibre and 65.4% carbohydrate.
2. Tested organisms: *S. viridifaciens* ATCC 11989 is used for tetracycline and chlortetracycline production, and *B. subtilis* ATCC 6633 is used for bioassay of antimicrobial activity.
3. Culture media and culture conditions: The tested organism was cultivated in a slant of inorganic salt starch agar at 26°C. The medium is made up of the soluble starch 10 g, ammonium sulphate 2.0 g, calcium carbonate 2.0 g, sodium chloride 1 g, potassium hydrogen phosphate 1.0 g, magnesium sulphate 1.0 g, trace element solution 1 mL (contained ferrous sulphate 1.0 g, copper sulphate 0.5 g, zinc sulphate 1.0 g and manganese sulphate 0.1 g/L) and agar 20 g/L. The basal solid medium contains sweet potato residue 100 g, nitrogen source 2.4 g, rice bran or wheat bran or soybean meal 20 g, $CaCO_3$ 1.0 g and NaCl 0.2 g. The medium is mixed thoroughly with conidia and distilled water, and incubated statically in flask (the thickness of medium is about 2 cm) at 26°C for 5–7 days by sting once a day.
4. Extraction of antibiotics: After fermentation, the antibiotic is extracted by shaking with 5 times volume of distilled water at room temperature for 5 min.
5. Observation under scanning electron microscope: Morphogenesis of tested organism on the solid substrate with different moisture contents or different incubation periods is observed under Hitachi S-550 scanning electron microscope (Hitachi Co., Japan) at 20 KV with gold metal shadowing.
6. Moisture content: The samples were dried at 60°C under vacuum for 8–12 h, until their weight remained constant.

7. pH: Initial pH of substrate is determined directly with pH meter, whereas final pH is determined after diluting with 5 times the volume of distilled water.
8. Water activity: Samples with different moisture contents are placed in a sealed container at 25°C, and water activity is determined by a hygrometer or modified Conway method.
9. Bulk density: The dry weight or wet weight of samples per unit volume of 1 mL is the bulk density in dry weight or wet weight, respectively.
10. Extraction of tetracycline: The solid substrate is extracted with 5 times the volume of distilled water. It is noted that 5 min of shaking was enough for tetracycline extraction. Prolonged shaking treatment could not improve the efficiency of extraction, but made the filtration or centrifugation troublesome. The effect of extraction volume on the recovery of tetracycline has been exclusively studied. It is reported that 4–5 times the volume of distilled water is efficient for tetracycline recovery.

The optimum conditions for tetracycline production from sweet potato residue by SSF using *Streptomyces viridifaciens* are moisture content 68–72%, initial pH 5.8–6.0, supplement with $(NH_4)_2SO_4$ 0.5%, $CaCO_3$ 1.8%, NaCl 0.6%, K_2HPO_4 0.4%, $MgCl_2$ 0.8%, methionine 0.2%, sodium glutamate 0.4% and incubation at 26°C for 5 days. Each gram of dry substrate produce 2129 pg of tetracycline.

In case of peanut meal as the nitrogen source, the optimal conditions for tetracycline production are sweet potato residue supplement with peanut meal 20%, $(NH_4)_2SO_4$ 0.5%, $CaCO_3$ 1.8%, NaCl 0.6%, K_2HPO_4 0.4%, $MgCl_2$ 0.8%, soluble starch 10%, methionine 0.2%, histidine 0.8%, sodium glutamate 1.6%. Each gram of dry substrate produced 4720 pg of tetracycline.

QUESTIONS

1. Describe screening of soil for organisms producing antibiotics.
2. Draw a diagram of typical fementers.
3. Write a short note on bioreactors, vitamin B_{12} antibiotics and antimicrobial spectrum.
4. How will you carry out the screening of soil for organisms producing antibiotics?
5. Write about the production of penicillin.
6. Write about the screening of soil for organisms producing antibiotics.
7. What do you know about antimicrobials spectrum?
8. What are antibiotics? Explain about the various methods of their standardization in brief.
9. Name the organisms and fermentations used in the production of streptomycin.
10. Write about the production of tetracyclines.
11. What is a fermenter? Write the factors involved in the designing of a fermenter.
12. Write about a typical batch fermenter.
13. Write a note on primary screening methods of soil for antibiotic-producing organisms.
14. Given an account of historical developments in discovery of antibiotics.

CHAPTER 13 Genetic Recombination

CHAPTER OUTLINE

- Introduction
- Methods of Gene Transfer in Plants
- Horizontal Gene Transfer Technology
- Protoplast Fusion
- Gene Cloning
- Development of Hybridoma by Monoclonal Antibodies
- Some Important Medicines Produced by Biotechnology
- Gene Transfer in Humans
- Applications of Gene Transfer
- Questions

INTRODUCTION

Since early 1990s, scientists have attempted to establish the transfer of genes to human somatic cells as a valid therapy. The importance of gene transfer was highlighted by three trials, involving participants with haemophilia B and two types of severe combined immunodeficiency X-linked and adenosine deaminase-deficient participants. The most successful trials of gene transfer have been unfortunately questioned about their prospects. In the haemophilia B trial, the detection of vector (the agent that carries genes to cells in participants' semen) raised concerns about modifications of the germline. The X-linked severe combined immunodeficiency trial received setback by unexpected vector-induced leukaemia in two participants. Despite these developments, number of gene transfer trials approved worldwide has increased since 1989, of which 77% have been conducted in the United States and UK. Most trials have used virus-based vectors (70%). Several hazards associated with gene transfer have been verified by clinical experience, and others are predicted on theoretical grounds. It is, therefore, worth considering whether the risks observed in studies related to human gene transfer pose any unusual ethical and social challenges, and, if so, what is required to be done to tackle the same.

Gene transfer or DNA uptake refers to the process that moves specific pieces of deoxyribonucleic acid (DNA) into cells. DNA carries within its structure the hereditary information that determines the structures of protein, which is a prime molecule of life. The most important contribution to the fields of biotechnology and molecular biology has come from genetic engineering. Exogenous DNA into organisms and their expression have been achieved successfully. Transgenic animals and plants can be obtained by the introduction of exogenous DNA into targeted animals and plants respectively accompanied by the stable expression.

The techniques for the transfer of DNA into organisms differ from organism to organism. Generally, there are two approaches for DNA transfer. In the first case, the transfer of DNA takes place by natural method, and in the second case the transfer is by artificial method. In a gene transfer, therapeutic DNA is combined with a vector (often of viral origin). Vectors can be injected into recipient's tissue directly or used to modify cells ex vivo for transplantation to the recipient. The choice of methods of DNA transfer depends on the target cells in which transformation is to be performed.

It also depends on the objectives of gene manipulation. The transfection may be either stable or transient. Although, choice of DNA transfer method is very important, the other important steps involved are: (1) selection of gene, (2) its isolation, (3) preparation of recombinant DNA and (4) selection of transformed cells. The regeneration of organism with new characteristics is equally important.

METHODS OF GENE TRANSFER IN PLANTS

Gene transfer is a standard technique in biological research, and it has recently gained clinical significance in the development of human gene therapy and gene marking protocol. The development of recombinant DNA technology and techniques for transferring recombinant genes into plants have revolutionized the field of plant science.

Some important features of gene transfer are as follows:

1. Gene transfer technology has facilitated the molecular cloning by tagging genomic sequence of important genes whose gene products control the normal pattern of growth and differentiation of plants.
2. It has allowed investigations using reporter genes in the transcription regulation of plant genes.
3. Gene transfer technology has applications in the improvement of plant agriculture productivity.
4. The success story of gene transfer demonstrates the potential use of plant as biotechnological and chemical factories.

Gene transfer of plants is basically carried out by two methods (Fig. 13.1).

Vector-Mediated Gene Transfer

This technique uses two types of methods: (1) using *Agrobacterium rhizogenes* and (2) virus-mediated.

Vector-Free DNA-Mediated (Direct) Gene Transfer

Physical Method

These are the techniques of electroporation, microinjection, macroinjections, biolistics, microprojectile bombardment, liposome-mediated gene transfers and ultrasonication employed for effective gene transfer in living cells.

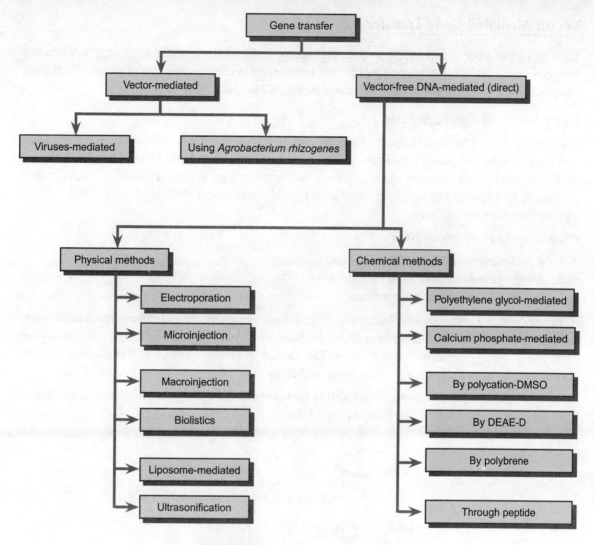

Figure 13.1 Gene transfer in plants.

Chemical Method

This method employs the following:

1. Using polyethylene glycol (PEG)
2. Calcium phosphate-mediated DNA transfer
3. Transfer of DNA by polycation-dimethyl sulphoxide (DMSO)
4. DNA transfer by DEAE-dextran method
5. DNA transfer by polybrene
6. Gene transfer through peptide

Vector-Mediated Gene Transfer

The term *plant gene vector* refers to potential vector both for the transfer of genetic information between plants and the transfer of genetic information from other organisms to plants. Plant gene vectors being used for transfer of genes are plasmid of *Agrobacterium* viruses.

Using Vector of Agrobacterium

For many species of family of Leguminosae, the combination of in vitro techniques for transformation and regeneration has proven difficult. The transgenic plants have been obtained with a selected group of species and varieties. The *A. rhizogenes* approach generating composite plants with a transgenic root system and untransformed wild-type shoot is especially useful for the study of root nodule formation on legumes.

Classification of *Agrobacterium*

1. *A. tumefaciens*-induced (Ti) crown gall disease
2. *A. rhizogenes*-induced hairy root disease
3. *A. radiobacter*—an avirulent strain

A. tumefaciens is a gram-negative bacterium. It commonly occurs in soil and causes crown gall disease in higher plants. It enters the plant through the mechanically wounded tissue site. This infection causes local proliferation of undifferentiated tissue. The transfer of DNA or xDNA of this plasmid encodes substance that acts as plant hormone, causing proliferation of tissue seen in the infected plant.

A. tumefaciens has an extraordinary ability to transfer its DNA segment into the plant genome. This DNA segment is known as *T-DNA* (Fig 13.2).

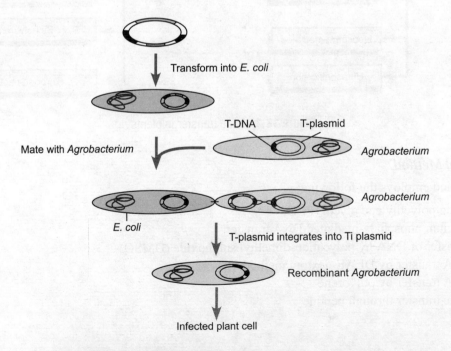

Figure 13.2 Mechanism of gene transfer by using *A. tumefaciens*.

Ti plasmid has four regions as follows:
1. Region A: comprises of T-DNA and responsible for tumour induction
2. Region B: responsible for replication
3. Region C: responsible for conjugation
4. Region D: responsible for virulence, mutation in the region

When *A. tumefaciens* is mixed with cells, it inserts the DNA containing the new gene into the plant chromosomes. Genetically engineered plant cells with the outline resistance gene are able to grow on antibiotic medium. Whole plants are generated from the single cell. Regenerated whole plant and its progeny carry the antibiotic resistance trail.

Vector-Free DNA-Mediated (Direct) Gene Transfer

The species and genotype-independent transformation techniques—wherein no natural vector is involved, but which are linked to the direct delivery of naked DNA to the plant cell—are covered under this category. This is also referred to as *DNA-mediated gene transfer*. A general scheme for the production of transgenic organism using direct DNA delivery method is explained below.

Physical Methods
Electroporation

It is a new technique that transfers genes into plants or animals by using electrical pulses to produce transient pores in the plasma membrane. The microscopic pores induced in biological membrane by the application of electric field are known as *electropores*. These electropores allow the molecules, ions and water to pass from one side of the membrane to the other. The formation of electropores depends on the type of cells used and the amplitude and duration of the electric pulse applied to them.

The electric currents may lead to dramatic heating of the cells that can result in the death of cells. The heating effects are minimized by using relatively high amplitude for a short duration pulse or by using two very short-duration pulses. The membranes reseal after a short-period. If properly treated, the cells can then regenerate cell walls, divide to form a callus and finally regenerate new plants. In mammalian transgenesis, electroporation is an effective technique of introducing exogenous DNA into embryonic stem (ES) cells.

Advantages
1. It can be applied for a number of cell types.
2. Simultaneously a large number of cells can be treated.
3. The technique is fast and less.
5. High percentage of stable transformants may be produced.

Applications
1. It increases efficiency of transformation or transfection of bacterial cells.
2. It is useful in introduction of exogenous DNA into animal cell lines, plant protoplast, yeast protoplast and bacterial protoplast.
3. It is possible to study the transient expression of molecular constructs.

4. Genes encoding selectable marker may be used to introduce genes using electroporation.
5. Electroporation of early embryo may result in the production of transgenic animals.
6. Wheat, maize, tobacco and rice have been stably transformed with frequency up to 1% by this method.
7. Naked DNA may be used for gene therapy by applying electroporation device on animal cells.
8. Hepatocytes, haematopoietic stem cells, epidermal cells, fibroblast, mouse T and B lymphocytes can be transformed by this technique.

Microinjection

In microinjection, DNA can be introduced into cells or protoplast with the help of very fine needles or glass micropipettes with a diameter of 0.5–10 m. Some of the DNA-injected materials may be taken up by the nucleus. The computerized handling of operations, such as holding pipette, needle, microscope stage and video technology have improved the efficiency of this technique.

Advantages

1. Mere precise integration of recombinant gene in limited copy number can be obtained.
2. The method is effective in transforming primary cells, as well as, cells in established cultures.
3. The DNA injected is subjected to less-extensive modifications.
4. Frequency of stable integration of DNA is far better as compared to other methods.

Applications

1. The technique is ideally useful for producing transgenic animal quickly.
2. The process is applicable for plant cells, as well as animal cells but more common for animal cells.
3. This can be applied to inject DNA into plant nuclei.
4. The procedure is important for gene transfer to embryonic cells.
5. The method has been successfully used with cells and protoplast of tobacco, alfalfa, etc.

Limitations

1. Skilled personnel are required.
2. Embryonic cells are preferred for manipulation.
3. This method is useful for protoplasts and not for the walled cells.
4. Knowledge of mating timing and acolyte recovery is essential.
5. The process causes random integration.
6. Rearrangement or deletion of host DNA adjacent to site of integration is common.
7. It is more useful for animal cells.
8. It is a costly technique.

Macroinjection

This method is tried for artificial DNA transfer to cereal plants that show inability to regenerate and develop into whole plants from cultured cells. Needles used for injecting DNA have their diameter greater than cell diameter. DNA is injected with conventional syringe into region of plant that develops into floral tillers. Around 0.3 mL of DNA solution is injected at a point above tiller node until several drops of solution come out from top of young inflorescence.

Advantages

1. Instrument is simple and cheap.
2. This technique does not require protoplast.
3. The method may prove useful for gene transfer into cereals, which do not regenerate from cultured cell easily.
4. It is a simple technology to handle.

Limitations

1. It is less specific.
2. The frequency of transformation is very low.
3. It is less efficient.

Biolistics or microprojectiles for DNA transfer

Biolistic or particle bombardment is a physical technique that uses accelerated microprojectiles to deliver DNA or other molecules into intact tissues and cells. The method was developed initially to transfer genes into plants by Sanford. The technique of biolistics transformation is relatively new, and it is a novel method amongst the physical methods for artificial transfer of exogenous DNA. This method is better in efficiency and does not need protoplast. This technique can be used for any plant cells, root sections, seed embryos and pollen grains. The gene gun is a device that literally fires DNA into target cells. The DNA to be transformed into the cells is coated on to microscopic beads made up of either gold or tungsten. The beads are carefully coated with DNA and are then attached to the end of the plastic bullet and loaded into the firing chamber of the gene gun. An explosive force fires the bullet down the barrel of the gun towards the target cells that lie just beyond the end of the barrel. When the bullet reaches the end of the barrel, it is caught and stopped, but the DNA-coated beads continue towards the target cells. Some of the beads pass through the cell wall into the cytoplasm of the target cells. Here, the bead and the DNA dissociate and the cells become transformed. Once inside the target cells, the DNA is solubilized and may be expressed.

Applications

1. Genomes of subcellular organelles have been accessible to genetic manipulation by biolistic method.
2. It is used successfully to transform cotton, soybean, spruce, sugarcane, corn, papaya, sunflower, maize, rice, tobacco, wheat, etc.
3. Method can be applied to filamentous fungi and yeast.
4. Mitochondria of plants and chloroplast of *Chlamydomonas* are transformed.
5. The particle gun is used with pollen, early stage embryoids, meristems and somatic embryos.

Advantages

1. Walled intact cells can be penetrated.

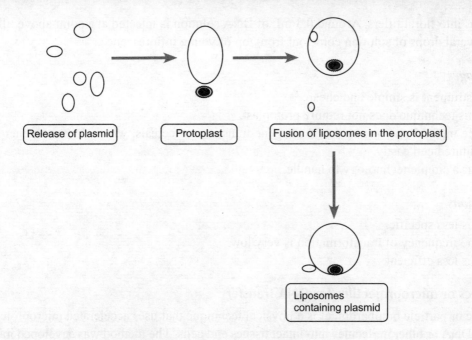

Figure 13.3 Liposome-mediated gene transfer.

2. Requirement of protoplast is avoided.
3. Manipulation of genome of subcellular organelles can be achieved.

Liposome-mediated gene transfer

Introduction of DNA into cell via liposomes is known as *lipoinfection*. Liposomes are small lipid vesicles, in which large numbers of plasmids are enclosed. They can be induced to fuse with protoplast vesicle device, such as PEG. Therefore, they are used for gene transfer. This method can be used for both transient and stable transfection. It can be used for adherent cells, primary cell lines, and suspension cultures (Fig 13.3). This method can also be used in the transfer for the production of transgenic animal.

Some other techniques used for liposomes

For the following protocol, Lipofectamine reagent from Invitrogen Corporation is used. Lipofectamine reagent is a 3:1 (w/w) liposome formulation of the polycationic lipid 2,3-dioleyloxy-N-[2(sperminecardox-ido)ethyl]-N,N-dimethyl-1-propanaminium trifluoroacetate (DOSPA) and neutral lipid dioleoylphosphatidylethanolamine (DOPE) in membrane-filtered water (Catalogue Number 18324; Invitrogen Corporation, Carlsbad, California, US) (Hawley-Nelson, 1993; Shih, 1997).

1. Around 40,000 cells per well of a 24-well plate are placed in 0.5 mL of the appropriate complete growth medium (10% serum is added, if needed).
2. The cells are incubated at 37°C in a CO_2 incubator until the cells are 50–80% confluent. It takes around 20 h.

3. About 3 µg DNA is diluted into 25 µL medium without serum for each well and mixed thoroughly.
4. About 3 µL Lipofectamine reagent is diluted with 25 µL medium without serum for each well and mixed thoroughly.
5. The diluted DNA (*step 3*) and Lipofectamine reagent (*step 4*) are combined together and incubated at room temperature for 30 min. During incubation, the DNA–liposome complexes are formed.
6. The medium in the cells is replaced with 0.2 mL transfection medium without serum.
7. About 0.15 mL medium without serum is added to the tube containing the complexes for each well.
8. The cells are then incubated with the complexes for about 10 h at 37°C in a CO_2 incubator. The incubating time shall be flexible and depends on used cell type.
9. About 0.4 mL growth medium containing double the 2× normal concentration of the serum is added without removing the transfection mixture.
10. Later, the medium is replaced with fresh and complete medium.
11. The cell extracts for transient gene expression are assayed at 24–72 h after transfection, depending on the cell type and promoter activity.
12. To obtain stable transfectants, the cells are passaged 1:10 into the selective medium after 72 h of transfection for the reporter gene transfected.

For use as gene transfer vehicles, liposomes are prepared by adding an appropriate mix of bilayer constituents to an aqueous solution of DNA molecules. In this aqueous environment, phospholipid hydrophilic heads associate with water, whereas hydrophobic tails self-associate to exclude water from within the lipid bilayer. This self-organizing process creates discrete spheres of continuous lipid bilayer membrane enveloping a small quantity of DNA solution. The liposomes are then ready to be added to target cells. Germline transgenesis is possible with liposome-mediated gene transfer and ES cells have been successfully transfected by liposomes.

Advantages

1. Long-term stability
2. Protection of nucleic acid from degradation
3. Low toxicity
4. Simplicity of technique

Chemical Methods

PEG-mediated transfection

PEG stimulates endocytosis, and therefore, DNA uptake occurs. This method is utilized only for protoplast. PEG-based vehicles are less toxic and more resistant to nonspecific protein adsorption making them attractive for nonviral gene delivery.

The protoplasts are kept in solution containing PEG. The molecular weight of PEG used is 8000 Da with final concentration of 15%. Calcium chloride is added to it along with sucrose and glucose that act as osmotic buffering agents. To reduce the effects of nuclease present, the carrier

DNA from salmon or herring sperm may also be added. After exposure of the protoplast to exogenous DNA in presence of PEG and other chemicals, PEG is allowed to be removed. Intact surviving protoplasts are then cultured to form cells with walls and colonies in turn. After several passages in selectable medium, frequency of transformation is calculated. PEG-PBLG nanoparticle-mediated *HSV-TK/GCV* gene therapy is also reported for oral squamous cell carcinoma.

Calcium phosphate-mediated DNA transfer

The process of transfection involves the admixture of isolated DNA (10–100 mg) with solution of calcium chloride and potassium phosphate under conditions that permit precipitation of calcium phosphate. The technique is used for introducing DNA into mammalian cells. The cells are then incubated with precipitated DNA either in solution or in tissue culture dish. A fraction of cells shall take the calcium phosphate DNA precipitate by endocytosis. The transfection efficiencies using calcium phosphate can be quite low (in the range of 1–2%). It can be increased, if very high purity DNA is used, and the precipitate is allowed to form slowly. The technique has been developed whereby cell taking up exogenous DNA could be increased to 20%.

Limitations

1. Integrated genes undergo substantial modification.
2. Integration with host cell chromosome is random.
3. Many cells do not like solid precipitate adhering to them and the surface of their culture vessel.
4. Frequency is very low.

Due to the above limitations, transfection applied to somatic gene therapy is limited. The calcium phosphate method of DNA transfer is reproducible and efficient, but there is a narrow range of optimum conditions.

Transfer of DNA by polycation-DMSO

For DNA transfer by polycation, polybrene is used to increase the adsorption of DNA to the cell surface followed, by a brief treatment with 25–30% DMSO to increase membrane permeability and enhance uptake of DNA. In this technique, carrier DNA is not required and stable transformants are produced. This method works with mouse fibroblast and chick embryo.

DNA transfer by DEAE-D

The DNA can also be transferred with the help of DEAE-dextran. DEAE-dextran may be used in the transfection medium containing DNA. This is polycationic, high molecular weight substance, convenient for transient assays in *cos* cells. When diethylaminoethyl-dextran treatment is coupled with DMSO, up to 80% transformed cells can express the transferred gene. It is established that serum inhibits this transfection. Therefore, cells are washed nicely to make them serum free. It does not appear to be efficient for the production of stable transfectants. The treatment with chloroquinine increases transient expression of DNA.

Advantages

1. It is a simple method.

2. It can be used for transient cells, which can not survive even short exposure to calcium phosphate.
3. It is economically viable and cheep technique.

Different steps involved in these techniques are as follows:

1. The growing cells are harvested exponentially by trypsinization and transferred into 60-mm tissue culture dishes at a density of 105 cells per dish.
2. About 5 mL complete growth medium is added to it.
3. It is then incubated for 24 h at 37°C with 5% CO_2.
4. DNA/DEAE-dextran/TBS-D (Tris-buffered saline with dextrose) solution is prepared by mixing 2 mg of supercoiled plasmid DNA into 1 µg/mL DEAE-dextran in TBS-D.
5. The medium is removed and washed three times with phosphate buffered saline (PBS) and twice with TBS-D. DNA/DEAE-dextran/TBS-D solution (250 µL) is added to it.
6. It is then incubated at 37°C for 60 min with 5% CO_2.
7. DNA/DEAE-dextran/TBS-D solution is removed and washed with TBS-D thrice and twice with PBS.
8. About 5 mL of medium supplemented with serum and chloroquine (0.1 mM) is added to it.
9. It is incubated for 4 h at 37°C with 5% CO_2.
10. The medium is removed and washed with serum-free medium three times.
11. About 5 mL of medium supplemented with serum is added to it and incubated at 37°C for 48 h with 5% CO_2.
12. The cells are harvested after 48 h of incubation.
13. The RNA or DNA is analysed by hybridization or the expressed protein is analysed by radioimmunoassay, immunoblotting or immunoprecipitation.

Use of polybrene

Several polycations, including polybrene (1,5-dimethyl-1,5-diazaundecamethylene poly-methobromide) (Chaney, 1986) and poly-L-ornithine (Nead, 1995) have been used in gene transfection with the DMSO enhancement.

The steps involved are as follows:

1. The cells are harvested exponential by trypsinzationin and replanted at a density of 5000 cells/ mm^2 in 10 mL MEMα containing 10% fetal calf serum.
2. It is incubated at 37°C 24 h in 5% CO_2.
3. The medium is replaced with 3 mL prewarmed medium containing serum, 10 µg DNA and 30 µg polybrene (37°C). The medium is mixed thoroughly before adding polybrene.
4. It is the then incubated for 12 h with agent and shaken each hour.
5. The medium is removed and to it added 5 mL of 30% DMSO in serum-containing medium.
6. After 4 min of incubation, the DMSO solution is aspirated, and the cells are washed twice with warm (37°C) serum-free medium. A total of 10 mL of complete medium containing 10% fetal calf serum is added to it.

Gene transfer through peptide

A series of peptide sequences are able to bind to, and condense, DNA to make it more amenable for entry into cells. For example, the tetrapeptide 'serine-proline-lysine-lysine' (present on the *C* terminal of the histone H1 protein) helps in DNA transfer. Lysine is a positively charged amino acid. Its side chains help to counteract the negatively charged phosphate DNA backbone and allow the DNA molecules to pack closely to each other. Tyrosine-lysine-alanine-(lysine)^8-tryptophan-lysine is a peptide that is very effective to form complexes with DNA. DNA-binding peptides that can be coupled to cell-specific ligands can also be synthesized. It allows receptor-mediated targeting of the peptide/DNA complexes to specific cell types. Rational design of peptide sequences has been used to develop synthetic DNA-binding peptide.

HORIZONTAL GENE TRANSFER TECHNOLOGY

Horizontal gene transfer (HGT) is a process in which a prokaryotic cell can acquire genes from other microbes of the same generation, which, in some cases, can be a different species, or even a different genus than the donor. The technology of HGT was first described in Japan in 1959. The related publication has demonstrated the transfer of antibiotic resistance between different species of bacteria.

There are three types of HGT as follows:

1. Transformation
2. Transduction (bacterial)
3. Conjugation (bacterial)

Transformation

In molecular biology, transformation is the genetic alteration of a cell resulting from the direct uptake, incorporation and expression of exogenous genetic material (exogenous DNA) from its surrounding and taken up through the cell membrane(s). Transformation occurs most commonly in bacteria. It can also be affected by artificial means. Bacteria that are capable of being transformed, whether naturally or artificially, are called *competent*. Transformation is one of three processes by which exogenous genetic material may be introduced into bacterial cell, the other two being conjugation (transfer of genetic material between two bacterial cells in direct contact) and transduction (injection of foreign DNA by a bacteriophage into the host). Transformation is also used to describe the insertion of new genetic material into nonbacterial cells, including animal and plant cells. However, because transformation has a special meaning in relation to animal cells indicating progression to a cancerous state, the term should be avoided for animal cells while describing introduction of exogenous genetic material. Introduction of foreign DNA into eukaryotic cells is usually called *transfection*.

Mechanism of Transformation

1. **Bacteria**: The bacterial transformation may be referred to as a stable genetic change brought about by the uptake of naked DNA (DNA without associated cells or proteins). Competence

refers to the state of being able to take up exogenous DNA from the environment. Two forms of competence exist: natural and artificial.

2. **Natural competence**: About 1% of bacterial species are capable of naturally taking up DNA under laboratory conditions. Many more are able to take it up in their natural environments. Such bacteria carry sets of genes that provide the protein machinery to bring DNA across the cell membrane(s).

3. **Artificial competence**: The artificial competence is induced by laboratory procedures. It involves making the cell passively permeable to DNA by exposing it to conditions that do not normally occur in nature. The efficiency with which a competent culture can take up exogenous DNA and express its genes is known as *transformation efficiency*.

    The calcium chloride transformation is an ideal technique for this purpose. The chilling cells in the presence of divalent cations such as Ca^{2+} (in $CaCl_2$) prepare the cell membrane to become permeable to plasmid DNA. The cells are incubated on ice with the DNA and then briefly heat shocked (at 42°C for 30–120 s) allowing the DNA to enter the cells. This method works very well for circular plasmid DNA. An excellent preparation of competent cells gives $\sim 10^8$ colonies per microgram of plasmid. The noncommercial and good preparations give around 10^5–10^6 transformants per microgram of plasmid. Interestingly, the cells that are naturally competent are usually transformed more efficiently with linear DNA than with plasmid DNA.

4. **Plasmid transformation**: The transformation usually produces a mixture of relatively few transformed cells and an abundance of nontransformed cells. A method is therefore needed to identify the cells that have acquired the plasmid. In order to be stably maintained in the cell, a plasmid DNA molecule must contain an origin of replication. This allows it to be replicated in the cell independent of the replication of the cell's own chromosome. The method usually consists of using a plasmid that contains a gene that gives the bacterial cells resistance to an antibiotic that they are naturally sensitive to. The mixtures of cells are then plated on medium that contains the antibiotic. Only the transformed cells are able to grow, and the cells that do not take up the plasmid are killed in the medium.

5. **Induction in plants**: A number of mechanisms are available to transfer DNA into plant cells.

    a. *Agrobacterium*-mediated transformation is the easiest and most simple plant transformation. Plant tissue (often leaves) are cut into small pieces (10 × 10 mm) and soaked for 10 min in a fluid containing suspended *Agrobacterium*. Some cells along the cut will be transformed by the bacterium that inserts its DNA into the cell. It is then placed on selectable rooting and shooting media to allow plants to regrow. Some plant species can be transformed just by dipping the flowers into suspension of *Agrobacterium*.

    b. Particle combardment: This method also allows transformation of plant plastids. The transformation efficiency is lower than in *Agrobacterium*-mediated transformation, but most plants can be transformed with this method. The particles of gold or tungsten are coated with DNA and then shot into young plant cells or plant embryos. Some genetic materials shall stay in the cells and transform them.

    c. Electroporation: The transient holes in cell membranes are made using electric shock. This treatment allows DNA to enter as described above for bacteria.

Transduction

The technique of transduction was discovered by Norton Zinder and Joshua Lederberg at the University of Wisconsin, Madison in 1951. This is a common tool used by molecular biologists to stably introduce a foreign gene into a host cell's genome. In this process, DNA is transferred from one bacterium to another by a virus. It is also the process whereby foreign DNA is introduced into another cell via a viral vector. The desired genetic material is packed into a suitable plant virus, and this modified virus is allowed to infect the plant. If the genetic material is DNA, it can recombine with the chromosomes to produce transformant cells. However, genomes of plant viruses usually consist of single-stranded RNA that replicates in the cytoplasm of the infected cell. For such genomes, this technique is a form of transfection and not a real transformation, because the inserted genes never reach the nucleus of the cell and do not integrate into the host genome. The progeny of such infected plants is virus free and also free of the inserted gene.

Generalized transduction may occur in two main ways, namely, recombination and headful packaging. If bacteriophages undertake the lytic cycle of infection upon entering a bacterium, the virus shall control the machinery of the cell for use in replicating its own viral DNA. If by chance, bacterial chromosomal DNA is inserted into the viral capsid used to encapsulate the viral DNA, this may lead to generalized transduction. If the virus replicates using headful packaging, it attempts to fill the nucleocapsid with genetic material.

The new virus capsule now loaded with part bacterial DNA continues to infect another bacterial cell. This bacterial material may become recombined into another bacterium upon infection.

When the new DNA is inserted into this recipient cell it can fall to one of the following three fates:
1. The DNA shall be absorbed by the cell and be recycled for spare parts.
2. If the DNA is originally a plasmid, it recircularizes inside the new cell and becomes a plasmid again.
3. If the new DNA matches with a homologous region of the chromosome of recipient cell, it shall exchange DNA material similar to the actions in conjugation.

Conjugation

The process of conjugation involves the transfer of genetic material between bacterial cells by direct cell–cell contact or by a bridge-like connection between two cells. Discovered in 1946 by Joshua Lederberg and Edward Tatum, conjugation is a mechanism of HGT as in case of transformation and transduction; although these two other mechanisms do not involve cell–cell contact.

The bacterial conjugation is often incorrectly regarded as the bacterial equivalent of sexual reproduction or mating, since it involves the exchange of genetic materials. During conjugation, the donor cell provides a conjugative or mobilizable genetic element that is most often a plasmid or transposon. Most conjugative plasmids have systems ensuring that the recipient cell does not already contain a similar element.

The genetic information transferred is often beneficial to the recipient. The benefits derived are antibiotic resistance, and xenobiotics tolerance or the ability to use new metabolites. Such beneficial plasmids may be considered bacterial endosymbionts.

Mechanism of Conjugation

The prototypical conjugative plasmid is the F-plasmid or F-factor. It is an episome with a length of about 100 kb. An episome is a plasmid that can integrate itself into the bacterial chromosome by homologous recombination. It carries its own origin of replication, the *oriV* and an origin of transfer or *oriT*. There can only be one copy of the F-plasmid in a given bacterium, either free or integrated, and bacterium that possesses a copy is called *F-positive* or *F-plus* (denoted F^+). The cells that lack of F-plasmids are called *F-negative* or *F-minus* (F^-) and as such, can function as recipient cells.

The F-plasmid carries a *tra* and a *trb* locus, which together are about 33-kb long and consist of about 40 genes. The *tra* locus includes the *pilin* gene and regulatory genes, which together form pili on the cell surface. The locus also includes the genes for the proteins that attach themselves to the surface of F^- bacteria and initiate conjugation. The pili are not exactly the structures through which DNA exchange occurs. Several proteins coded for the *tra* or *trb* locus seem to open a channel between the bacteria and it is presumed that the *traD* enzyme, located at the base of the pilus, initiates membrane fusion (Fig. 13.4).

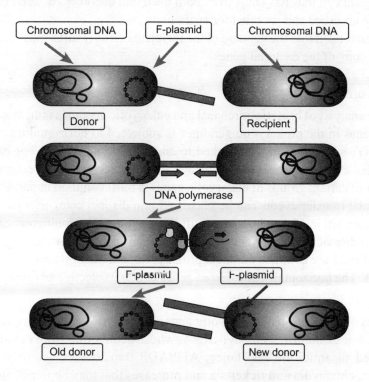

Figure 13.4 Conjugation of bacteria.

When conjugation is initiated by a signal, the *relaxase* enzyme creates a nick in one of the strands of the conjugative plasmid at the *oriT*. The enzyme *relaxase* may work alone or in a combination of over a dozen proteins known as *relaxosome*. In the F-plasmid system, the relaxase enzyme is called *TraI*, and the relaxosome consists of *TraI*, *TraY*, *TraM* and the integration host factor (IHF). The nicked strand or *T-strand*, is then unwound from the unbroken strand and transferred to the recipient cell in a 5'-terminus to 3'-terminus direction. The remaining strand is replicated either independent of conjugative action (vegetative replication beginning at the *oriV*) or in concert with conjugation (conjugative replication similar to the rolling circle replication of lambda phage). Conjugative replication may require a second nick before successful transfer can occur. It has been recently claimed that conjugation is inhibited with chemicals that mimic an intermediate step of this second nicking event.

Specialized Transduction

The specialized transduction occurs as a result of mistakes in the transition from a lysogenic to lytic cycle of a virus. If a virus incorrectly removes itself from the bacterial chromosome, bacterial DNA from either end of the phage DNA may be packaged into the viral capsid. The process of specialized transduction leads to three possible outcomes as follows:

1. The bacterial DNA can match up with a homologous DNA in the recipient cell and exchange it. The recipient cell thus has DNA from both itself and the other bacterial cell.
2. DNA can be absorbed and recycled for spare parts.
3. DNA can insert itself into the genome of the recipient cell as if still acting like a virus resulting in a double copy of the bacterial genes.

HGT in Prokaryotes

The comparative analysis of bacterial, archaeal and eukaryotic genomes indicates that a significant fraction of the genes in the prokaryotic genomes is subjected to horizontal transfer. The amount and source of HGT sometimes may be linked to an organism's lifestyle. For example, bacterial hyperthermophiles exchange genes with archaea to a greater extent as compared to bacteria, whereas, transfer of certain groups of eukaryotic genes is most common in parasitic and symbiotic bacteria. Horizontal transfer events can be classified into distinct categories of (1) acquisition of new genes, (2) acquisition of paralogs of existing genes and (3) xenologous gene displacement whereby, a gene is displaced by a horizontally transferred ortholog from another lineage (xenolog). All these types of HGT are common amongst prokaryotes, but their relative contributions differ in different lineages. The horizontally transferred genes confer a selective advantage on the recipient organism.

Several cases of acquisition of eukaryotic genes by bacteria seem to reveal the evolutionary forces involved. For examples, isoleucyl-tRNA, whose acquisition from eukaryotes by several bacteria is linked to antibiotic resistance, ATP/ADP translocases acquired by intracellular parasitic bacteria, chlamydia and rickettsia and proteases that may be implicated in chlamydial pathogenesis.

PROTOPLAST FUSION

Protoplast fusion is a physical phenomenon wherein two or more protoplasts come in contact and adhere with one another either spontaneously, or in presence of fusion inducing agents. The protoplast is a cell of which cell wall is removed. The cytoplasmic membrane is the outermost layer in such a cell. Specific lytic enzymes are used to remove cell wall and obtain protoplast. By protoplast fusion, it is possible to transfer some useful genes, such as genes for disease resistance, frost hardiness, nitrogen fixation, rapid growth rate, more product formation rate, drought resistance, protein quality, herbicide resistance and heat and cold resistance from one species to another. Protoplast fusion is an important tool in strain improvement for bringing genetic recombinations and developing hybrid strains in filamentous fungi.

Protoplast fusion is used to combine genes from different organisms to create strains with desired properties. This technique is more useful in engineering of microbial strains for desirable industrial properties.

Enzymes Used for Breaking of Cell Walls

For protoplast fusion, it is mandatory that the cell wall of plant and microorganisms is degraded. Various enzymes are used for this process. The enzymes cellulase and pectinase or macerozyme acting on plant cell wall. Bacterial cell walls are degraded by the action of lysozyme. Fungal wall is degraded by Novozyme 234 that includes glucanase and chitinase. *Streptomyces* cell wall degraded by action of lysozyme and achromopeptidase is another example.

Methods of Protoplast Fusion

Different techniques of protoplast fusion are in vogue.
1. Spontaneous fusion: It is the phenomena wherein the protoplast during isolation often fuses spontaneously. During the enzyme treatment, protoplasts from adjoining cells fuse through their plasmodesmata to form multinucleate protoplasts.
2. Induced fusion: Fusion of freely isolated protoplasts from different sources with the help of fusion inducing chemical agents is known as *induced fusion*. Normally, isolated protoplasts do not fuse with each other, because the surface of isolated protoplast carries negative charges (–10 mV to –30 mV) around the plasma membrane. There is a strong tendency witnessed in the protoplasts to repel each other due to same charges. This type of fusion needs a fusion-inducing chemical that actually reduces the electronegativity of the isolated protoplast and allows it to fuse with each others. The isolated protoplast can be then induced to fuse by adopting any one technique.
3. Mechanical fusion: In this process, the isolated protoplast is brought into intimate physical contact mechanically under microscope using micromanipulator or perfusion micropipette.
4. Chemofusion: It is a nonspecific and inexpensive technique. It can cause massive fusion product. It can be cytotoxic and nonselective and possesses less fusion frequency. Several

chemicals are used to induce protoplast fusion, such as PEG, sodium nitrate and calcium, calcium ions (Ca^{++}).

5. Electrofusion: Mild electric stimulation is used to fuse protoplast. In this two glass capillary microelectrodes are placed in contact with the protoplast. An electric field of low strength (10 kvm^{-1}) gives rise to dielectrophoretic dipole generation within the protoplast suspension that leads to pearl chain arrangement of protoplasts. Subsequent application of high strength of electric fields (100 kvm^{-1}) for some microseconds results in electric breakdown of membrane and subsequent fusion.

Protoplast Fusion in Plants

The somatic fusion is achieved in the following steps:

1. The removal of the cell wall of one cell of each type of plant using cellulase enzyme to produce a somatic cell called *protoplast*.
2. The cells are then fused using both an electric shock (electrofusion) to join the cells and the nuclei fused together or by chemical treatment. The resulting fused nucleus is called *heterokaryon*.
3. The somatic hybrid cell has its cell wall induced to form using hormones.
4. The cells are then grown into calluses that are then further grown to plantlets and finally to a full plant known as a *somatic hybrid*.

Protoplast Fusion in Fungi

Production and regeneration of protoplasts is a useful technique for fungal transformations. Commercial preparation of enzymes that contain mixture of products to digest fungal cell wall is used for protoplast fusion. Novozyme 234 includes enzyme mixture of glucanases and chitinaze. It is added to rapidly growing fungal tissue suspended in an osmotic buffer (e.g. 0.6 mol^{-1} KCl, 1.2 mol^{-1} sorbitol or 1.2 mol^{-1} MgSO$_4$). The protoplasts and DNA are mixed in the presence of 15% (w/v) PEG 6000 and pH buffer (TRIS-HCl). Around 10 mL^{-1} PEG causes clump formation in protoplasts. The mycelium is grown on cellophane at 37°C and placed on agar overnight. It is then incubated with enzyme at 30°C for 1.5 h in empty Petri dish containing KCl. The protoplasts are then filtered and washed in KCl, centrifuged and resuspended.

Protoplast Technology for *Streptomyces* spp.

For obtaining protoplasts from *Streptomyces*, lysozyme is used, which breaks glycan portion of peptidoglycan wall. The cultures from spore suspension (2 days in shaker at 30°C) are harvested by centrifugation, resuspended in 0.03 mol^{-1} sucrose, washed and finally reharvested and resuspended in lysozyme solution in protoplasting medium (30 min; 2 h at 30°C).

Protoplast Fusion in Bacteria

In Gram-positive bacteria, protoplast fusion can be carried out at low frequency. For Gram-negative bacteria, it is possible to obtain protoplasts, but regeneration is difficult.

GENE CLONING

Cloning

The word *clone* is used in many different contexts in biological research. In its most simple and strict sense, it refers to a precise genetic copy of a molecule, cell, plant, animal or human being. Cloning refers to established technologies that have been part of agricultural practice for a very long time and currently form an important part of the foundations of modern biological research.

The genetically identical copies of whole organisms are common. In the plant breeding world they are commonly referred to as 'varieties' rather than clones. Many valuable horticultural or agricultural strains are maintained solely by vegetative propagation from an original plant, reflecting the ease with which it is possible to regenerate a complete plant from a small cutting. The developmental process in animals does not usually permit cloning, as easily as, in plants. Many simpler invertebrate species, such as certain kinds of worms, are capable of regenerating a whole organism from a small piece, *even* though this is not necessarily their usual mode of reproduction. The vertebrates have lost this ability entirely, although regeneration of certain limbs, organs or tissues can occur to varying degrees in some animals.

A single adult vertebrate cannot generate another whole organism. The cloning of vertebrates does occur in nature, in a limited way. The twins occur by chance, in humans and other mammals with the separation of a single embryo into halves at an early stage of development. The resulting offsprings are genetically identical, having been derived from one zygote that resulted from the fertilization of one egg by one sperm.

It is a process that gives rise to genetically identical organisms. Cloning of higher organisms leads to duplicity or creation of identical twins. It is an asexual reproduction (replication) of mammals. In mammalian cloning, DNA of adult cell is injected into oocyte or unfertilized egg cell whose DNA is removed. So, in this process, DNAs from two different sources are not combined. Instead, DNA from one source combines with DNA (cell) regulating substances of other cell. The special regulatory (chemical) substances present in oocyte reprogrammed DNA, so that whole organism can develop from the DNA. Thus, the egg cell develops into embryo instead of adult cell. Later, these embryos are transferred to foster mothers. Finally, foster mother delivers baby lamb identical to donor or adult cell.

Gene Targeting

It is a modified form of the cloning technique. It involves modification or manipulation of somatic cell DNA before it is injected into enucleated oocyte. It involves homologous recombination of target gene and host DNA. Somatic cell is first transformed, transfected with gene (DNA) of interest, which is meant for gene targeting. The rest of the technique is identical to that of cloning gene targeting.

1. Gene knockout: It involves targeted inactivation of gene. The mutations are introduced into somatic cell DNA to make gene inactive. This DNA is later expressed in oocyte. By this, the clones produced lacks gene that is inactivated.
2. Gene knockin: It involves targeted insertion of gene. Gene is expressed in clones.

A gene of interest is inserted into somatic DNA before it is injected into enucleated oocyte.

Applications

1. Gene targeting is used to study genetics of human diseases and gene therapy. In gene therapy, inactive copy of gene is replaced with functional gene.
2. This can be applied in any animal species that allows selective breeding of animals for generation.
3. This method is successfully done in mice, sheep and flies.
4. This is used to establish function of gene or gene product, such as enzyme or hormone, etc.
5. In pharmaceutical industry, gene targeting is used to overexpress proteins of therapeutic value in milch animals.

Stem Cells

Isolation of human stem cells by using two different approaches was reported in 1998. These cells are able to differentiate into all types of cells or tissues. Further, they can be maintained as undifferentiated cells in culture and are able to reproduce themselves throughout the lifespan of organisms.

The stems cells exist in growing and adult humans. Stem cells are isolated from embryo, cord blood, bone marrow, liver, brain, etc. Only ES cells are able to differentiate into any cell type. Others produce only narrow range of cells. Generally the developmental potential of a stem cell is restricted to differentiate cells of the tissue in which they are present.

Applications

Stem cells have many potential uses in medicine and molecular biology.

1. They are used for replacement of lost or degenerated tissue.
2. They are used in gene therapy, e.g. neural stem cells are used for gene therapy of tumour growth suppression.
3. They are useful in delivering therapeutic gene products directly into the tissue.
4. Genetically engineered stem cells express therapeutic genes in the tissue of choice.
5. They are useful for exploring normal process of tissue development.

DEVELOPMENT OF HYBRIDOMA BY MONOCLONAL ANTIBODIES

Hybridoma Technology

Hybridoma technology involves the preparation of hybridoma (hybrid) cells to produce monoclonal antibodies. Monoclonal antibodies are antibodies produced by one cell line (clone). They are directed against one specific antigen. The lymphocytes in the body are polyclonal (multiple cell lines), and they can produce many types of antibodies (polyclonal) against antigens. Separation of single cell line that produces only one antibody from the mixture of polyclonal cells is a hard task.

Hybridomas

A simple method of producing single cell line of polyclonal cells involves fusion of two cell populations. Fusion generates hybrid cells with desired properties. These hybridoma cells produce monoclonal antibodies.

Different steps involved in hybridoma technology are summarized below.
1. A mouse is immunized with an antigen.
2. The spleen of the mouse is removed after several weeks. During this period, antibody-producing cells are formed in sufficient amounts in the spleen.
3. The B lymphocytes are removed from spleen and fused with myeloma cells in the presence of PEG. Myeloma cells are used for fusion because lymphocytes obtained after immunization are incapable of continuous growth in culture. The fusion of lymphocytes with myeloma cells leads to immortalization (infinite lifespan) of cells. Moreover, the myeloma cells lack HGPRTase that is used as genetic marker to identify fused cells.
4. The fused cells are allowed to grow in medium containing hypoxanthine-aminopterin-thymidine (HAT).
5. The spleen cells die in the medium because they are not immortalized, and myeloma cells also die because they cannot survive in HAT medium due to lack of HGPRTase.
6. Only hybridoma cells formed by the fusion of lymphocytes with myeloma cells proliferate, because they are immortal and contain HGPRTase of lymphocytes. Thus, the hybridoma cells are obtained. Further, the DNA of hybridoma cells is derived from myeloma cells and lymphocytes.

The hybridoma cells are screened for monoclonal cells by using antibodies (Fig. 13.5).

Hybridization Techniques

These techniques are based on the tendency of two DNA strands to form duplex. Interstrand hydrogen bonding between complementary sequences favours duplex formation. A DNA strand and RNA strand containing complementary sequences can also form duplex. Usually, two DNA strands involved in duplex formation come from two different sources. In case of DNA–RNA duplex, DNA and RNA may have different origins. Thus, duplex formed are hybrids, and the process of their formation is hybridization technique. Some of these hybridization techniques are described below.

Southern Blot

It is a technique useful in the detection of specific DNA fragment from mixture of DNA fragments. A cDNA probe of gene of interest is employed in this technique. This probe hybridizes with the gene of interest, thus leading to its identification. Different steps involved in southern blot techniques are given below.
1. The chromosomal DNA is purified and cleaved by restriction endonuclease. Mixtures of DNA fragments are obtained.
2. The fragments are separated by agarose gel electrophoresis. The gel is prepared from agarose in a suitable buffer. The DNA fragments are loaded into a well at one end of the gel and

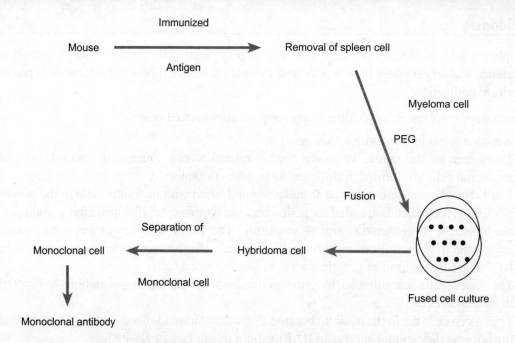

Figure 13.5 Production of monoclonal antibodies.

subjected to electrophoresis. Under conditions of experiment, DNA fragments carry net negative charge, and hence they move towards anode. Since the gel acts as molecular sieve, movements of DNA fragments towards anode depends on their sizes. Small fragments move faster, whereas big fragments get retarded. The DNA fragments are separated according to their sizes.

3. The DNA in the separated fragments is denatured and made single stranded by soaking gel first in HCl followed by NaOH solution.
4. The DNA is transferred to nitrocellulose by placing nitrocellulose sheet on the gel. The buffer in the gel is removed with the help of several blotting papers.
5. The DNA sticks to nitrocellulose, and only the buffer passes through nitrocellulose and absorbed by a blotting paper. Thus, a perfect nitrocellulose print of the gel is obtained.
6. The nitrocellulose sheet is incubated with buffer containing cDNA probe. cDNA probe is P32 radiolabeled. It hybridizes with gene of interest and excess probe is washed off.
7. Nitrocellulose sheet is dried and placed next to X-ray film. The fragments to which cDNA probe is bound appears as bands on film when it is developed.

Northern Blot

It is conceptually similar to southern blot. It is mainly used to detect specific RNA fragment from mixture of RNA fragments.

1. Mixture of RNA fragments are obtained from RNA by action of ribonuclease (RNase).
2. Fragments are separated by agar gel electrophoresis.
3. Nitrocellulose print of fragments is obtained.

4. cDNA probe is undertaken and excess probe is washed.
5. Visualization of hybrid on X-Ray film as band.

Western Blot

It is also similar to southern blot in many ways, except in the nature of probe used. In this blotting technique, radiolabeled (polyclonal or monoclonal) material is used as probe. This technique is used to identify a specific protein from mixture of proteins. The steps involved in the technique of western blot are given below.

1. The protein mixture containing desired protein is subjected to electrophoresis to separate proteins.
2. The nitrocellulose print is obtained.
3. Antibody probe (radiolabeled) is added.
4. Visualization of antigen–antibody complex is done on X-ray film.

SOME IMPORTANT MEDICINES PRODUCED BY BIOTECHNOLOGY

Humulin

Humulin is the brand name for a group of biosynthetic human insulin products, originally developed by Genentech in 1978. The generic names are regular insulin, insulin isophane and the Lente series consisting of Lente (generically known as insulin zinc suspension, a mixture of Ultralente and Semilente) acquired by Eli Lilly and Company, who arguably facilitated the product's approval with the US Food and Drug Administration. The Lente series was discontinued by Eli Lilly and Company in 2005.

Humulin is synthesized in a laboratory strain of *Escherichia coli* bacteria that has been genetically altered with recombinant DNA to produce biosynthetic human insulin. Humulin R consists of zinc–insulin crystals dissolved in a clear fluid. The synthesized insulin is then combined with other compounds or types of insulin that affect its shelf life and absorption. For example, Humulin N is combined with protamine to extend the time-activity profile of Humulin R for an extended period of time.

1. Humulin R regular U-500 (concentrated) insulin human injection, USP (rDNA origin) is a stronger concentration (500 units/mL) of Humulin R.
2. Humulin R (regular human insulin injection [rDNA origin]) is short-acting insulin that has a relatively short duration of activity as compared with other insulins.
3. Humulin 70/30 (70% human insulin isophane suspension, 30% human insulin injection [rDNA origin]) is a mixture of different types. It is an intermediate-acting insulin combined with the onset of action of Humulin.
4. Humulin N (human NPH insulin injection [rDNA origin]) is intermediate-acting insulin with a slower onset of action and a longer duration of activity than Humulin R.
5. Humulin 50/50 (50% human insulin isophane suspension, 50% human insulin injection [rDNA origin]) is a mixture insulin. It is intermediate-acting insulin combined with the onset of action of Humulin R.

Activase

Activase is known as *tissue plasminogen activator (tPA)*, and can be used to treat patients with acute ischaemic stroke. For certain patients, activase may improve the chances of recovery from stroke with little or no disability. The patients can receive activase only if they begin treatment within 3 h after their stroke symptoms start and only after they have had a scan to rule out bleeding in the brain.

Activase is approved for the treatment of acute myocardial infarction (heart attack) and acute ischemic stroke (blood clots in the brain) within 3 h of the onset of symptoms. It is also useful in treatment of acute massive pulmonary embolism (blood clots in the lungs).

Humatrope

Humatrope (somatotropin or somatropin) is a polypeptide hormone of rDNA origin manufactured by Eli Lilly and Company. It is used to stimulate linear growth in paediatric patients, who lack adequate normal human growth hormone. It has 191 amino acid residues and a molecular weight of 22,125 Da.

Its amino acid sequence is identical to that of human growth hormone of pituitary origin (anterior lobe). Humatrope is synthesized in a strain of *E. coli* modified by the addition of a gene for human growth hormone.

Hepatitis B Vaccine (Recombinant)

Hepatitis B vaccine contains the purified surface antigen (HBsAg) of the hepatitis B virus (HBV). It is produced by fermenting genetically engineered yeast cells containing a plasmid that carries the HBsAg. Energix-B (hepatitis B vaccine [recombinant]) is indicated for immunization against infection caused by known subtypes of HBV. Hepatitis D (caused by the delta virus) should also be prevented by Engerix-B, since delta virus replicates only in presence of HBV infection. The vaccination against HBV has been shown to reduce the overall incidences of HBV infection. It is also useful in reducing the incidences of the complications of HBV infection, such as chronic active hepatitis, cirrhosis and primary hepatocellular carcinoma.

Engerix-B does not prevent hepatitis caused by other agents, such as hepatitis A virus, non-A/non-B hepatitis viruses or other pathogens known to infect the liver.

Vaccination is recommended for persons of all ages or weight and for persons at risk of exposure to HBV, such as health care personnel, selected patients and patient contacts including adult haemodialysis patients and patients requiring frequent and/or large volume transfusions of blood or clotting concentrates. The vaccination is also given to residents and staff of institutions for the mentally handicapped; subpopulations with a known high incidence of HBV infections, such as immigrants from Southeast Asia; persons who have travel to areas where HBV infection is endemic and receive blood from or have close contact with the native populations; infants born to mothers who are carriers of HBsAg; persons at increased risk of HBV infection due to their sexual practices; prisoners; military personnel identified as being at increased risk, and users of illicit injectable drugs.

 ## GENE TRANSFER IN HUMANS

The application of gene transfer technology to human gene therapy has been intensively investigated during the last one decade. Immunotherapies with interleukin-2 and tumour infiltrating lymphocytes (TILs) have previously been shown to induce regression of metastatic melanoma in approximately half the patients.

The first experiment in humans was approved by the National Institutes of Health, United States (January 1989), and permission was granted for the infusion into patients with advanced melanoma of antilogous TIL that had been subjected to retrovirus-mediated gene transfer.

Techniques of Gene Transfer in Human Cells

Different approaches are considered to deliver a given DNA sequence known as *transgene* to the nucleolus of a cell, where this genetic material has to be integrated and potentially be transmitted to the following generation. One can thereby alter a group of cells in order to produce a protein that may be directed towards the deficient organ, such as genetically controlled microfactory.

The retroviruses seem to be the most eligible means for efficient gene transfer, but only in cells undergoing at least one mitotic cycle to obtain integration of the viral genome.

Herpes Virus–Derived Vector

These vectors are defective herpes viruses derived from thermosensitive mutants, incapable of replicating at temperature above 31°C and processing a natural tropism for the cells of the nervous system.

Adenoviruses

The recombinant adenoviruses offer first choice alternative to retroviruses when addressing the transfer of gene into differentiated cells is incapable of proliferating, such as neuron muscular cells, or when the target is expandable such as pulmonary epithelial cells. Importantly, the recombinant adenoviruses due to their very high infectious nature, and their ubiquity can be administered in vivo intravenously

 ## APPLICATIONS OF GENE TRANSFER

Role of Gene Transfer in Human Diseases

The gene transfer has an effective role to play in treatment of different human ailments.
  1. **Arthritis**: The gene therapy has application in a nonlethal disease, rheumatoid arthritis (RA). Intra-articular transfer of IL-1 receptor antagonist (IL-1Ra) cDNA reduces disease in animal

models of RA. For establishing safety and feasibility in humans, a phase I clinical study was carried out involving nine postmenopausal women with advanced RA, who required unilateral silastic implant arthroplasty of the 2nd–5th metacarpophalangeal (MCP) joints. It is possible to transfer a potentially therapeutic gene safely to human rheumatoid joints and to obtain intra-articular transgene expression. This justifies additional efficacy studies and encourages further development of genetic approaches to the treatment of arthritis and related disorders.

2. **Pulmonary hypertension**: The primary pulmonary hypertension associated with progressive pulmonary hypertension and right ventricular failure, is a life-threatening disease. The transfection of the *PGIS* gene to the liver resulted in a drug delivery system of prostacyclin for the lung, and therefore this may be an effective treatment for pulmonary hypertension. The intrinsic prostacyclin drug delivery using gene therapy may be useful.

3. **Cardiovascular disease**: Despite the substantial advances made during the last two decades in the prevention and treatment of cardiovascular events, cardiovascular disease (CVD) is still the main cause of deaths in the world. Application of gene therapy in the discipline of cardiovascular disorders has been the subject of intensive research work during recent times. The human clinical trials carried out in recent years have shown that the injection of naked DNA encoding vascular endothelial growth factor promotes collateral vessel development in patients with critical limb ischaemia or chronic myocardial ischaemia. Application of gene transfer to other CVDs shall require the development of integrated technologies, as well as a better definition of cellular and gene targets.

4. **Parkinson's disease**: After nearly twenty years of preclinical experimentation with various gene delivery approaches in animal models of Parkinson's disease (PD), clinical trials are finally underway. The risk/benefit ratio for these procedures is now considered acceptable under approved protocols by regulating agencies. The recombinant adeno-associated viral vector, a nonpathogenic and non–self-amplifying, is an ideal candidate for gene delivery to human brain. The candidate genes tested in PD patients encode glutamic acid decarboxylase, which is injected into the subthalamic nucleus to catalyse biosynthesis of the inhibitory neurotransmitter aminobutyric acid. It essentially mimics deep brain stimulation of the nucleus. It also facilitates encoding of aromatic L-amino acid decarboxylase that converts L-dopa to dopamine.

5. **Cancer**: A retroviral vector containing the wild-type *p53* gene under control of a β-actin promoter was produced to mediate transfer of wild-type *p53* into human non–small-cell lung cancers by direct injection. No clinically significant vector-related toxic effects were observed in nine patients whose conventional treatments failed, for a period of five months after treatment.

    In situ hybridization and DNA polymerase chain reaction demonstrated vector *p53* sequences in post treatment biopsies. Apoptosis (programmed cell death) was more frequent in posttreatment biopsies than in pretreatment biopsies. While tumour regression was observed in three patients, tumour growth stabilized in other three patients.

6. **Growth factor in neuropathic pain**: Peripheral nerve injury is a commonly occurring phenomenon that sometimes results in chronic neuropathic pain with hyperalgesia and allodynia. The treatment of neuropathic pain is a major clinical challenge because of our poor

understanding of the underlying mechanisms. Analgesics, especially opioids, are relatively effective, but their use is restricted and limited by patient tolerance. Gene transfer is a novel means of expressing identified transgenes, including those for γ-aminobutyric acid, glutamic acid decarboxylase, μ-opioid receptor and neurotrophic factors. The gene delivered into the nervous system by retrograde axonal transport, followed by its repeated intramuscular transfer using liposomes containing the haemagglutinating virus of Japan (HVJ) has given positive results of therapy.

7. **Gene transfer in cellular immunotherapy**: Adoptive cellular therapy is a potentially powerful method of eradicating established tumours. As established in animal models and subsequent clinical studies, the T cells have been found to be particularly potent effector cells. It is apparent that the stimulation of certain subpopulations of T cells that are reactive to tumour antigens can lead to more therapeutic T cells. The use of gene transfer techniques has facilitated development of more effective and specific methods to generate these tumour-specific T cells. Another area of tremendous interest is in the adoptive transfer of DCs manipulated to present tumour antigen to resting naive T cells. Gene transfer techniques offer more optimal ways to generate therapeutic dendritic cells (DCs). Adoptive immunotherapy may ultimately have its greatest use in patients undergoing cellular rescue after ablative chemotherapy. The infusion of immunocompetent T cells, genetically modified stem cells or the programmed DCs may offer a possibility to direct a patient's immune response to eliminate residual microscopic disease.

8. **Genetic disease treatment**: It was well before the advent of recombinant DNA technology that the interest in gene therapy was shown in the middle of 1960. At that time, the first speculation about the possible treatment of genetic disorders by introducing functional genes via virus-mediated gene transfer was made. In 1990, the first gene therapy clinical trial for the treatment of patients with melanoma was conducted. The results of this study indicated that retrovirus-mediated gene transfer in patients was safe.

Gene Transfer Technology in Other Areas

In grapevine, current lack of high-throughput genetic techniques (e.g. induced mutant collections) and the difficulties associated with genetic mapping (allele diversity, chimerism, and generation time) highlight the critical role of transgenic technology for characterizing gene function. The presence of original features in grapevine, including perennial status, vegetative architecture, inflorescence/tendril, flower organization (corolla) and fleshy fruit of considerable acidity with various flavonoids compounds, makes functional genomics an essential approach to link a gene to a trait.

The gene transfer techniques are important in the production of enzyme reference materials, which could be of great therapeutic and biochemical significance. The gene transfer techniques are also used in the management of marine resources. Tetrodotoxin obtained from frog and pseudopterosin extracted from Caribbean coral are used as CVS stimulant and anti-inflammatory agent, respectively. The technique of gene transfer can systematize management of important marine sources.

QUESTIONS

1. Define genetic recombination and describe various methods of genetic recombination.
2. Describe the applications of genetic engineering for production as pharmaceuticals.
3. Differentiate amongst transformation, conjugation and transduction.
4. Define protoplast fusion. Write a note on its applications.
5. Write short notes on screening of soil, historical development of antibiotics and Humatrope.
6. What is gene cloning? Give its applications.
7. Explain the different methods of genetic recombination.
8. Define the term 'cloning' and give its applications.
9. What is genetic recombination? Name three types of genetic recombination.
10. Write in brief about protoplast fusion.
11. What are gene transduction and transformation?

CHAPTER 14: Medicine and Edible Vaccines

CHAPTER OUTLINE

- Introduction
- Edible Vaccines
- Questions

INTRODUCTION

Medicine evolves the science and art of healing humans. It consists of a variety of health care practices aimed at maintaining wellness by the prevention and treatment of illness. Before introduction of medicine, arts of healing were based on ritual practices that were developed out of religious and cultural traditions.

Besides medical practitioners, many highly trained health professionals are involved in the delivery of modern health care. These include nurses, emergency medical technicians and paramedics, radiographers, pharmacists, dieticians, physiotherapists, respiratory therapists, speech therapists, occupational therapists and bioengineers. All these segments of health care professionals contribute to sound public health.

Public Health

WHO defines *public health* as the art and science of preventing disease, prolonging life and promoting health and efficiency through organized community efforts. The spectrum of activities in public health covers the following:

1. Education of the individuals in the principles of hygiene
2. Organization of medical and nursing services for the early diagnosis and preventive treatment of diseases
3. Sanitational issues for effective control of communicable infections. It also involves total development of social machinery that ensures a standard of living adequate for the maintenance of health to every individual in the community.

The major objectives of the public health are identified as follows:

1. Promotion of health and efficiency
2. Prevention of disease and prolongation of life

3. Elevation of the standard of living adequate for health maintenance
4. Provision of the right to health and longevity

Social medicine: It is defined as the science and art of preventing disease, prolonging life and promoting health and efficiency of population by intercepting social factors that have direct or indirect relationship with the disease process.

Community medicine: It is a branch of medical practice that is concerned with promoting, maintaining and when necessary, restoring the health of human communities, rather than with clinical use of individual patients. It takes into account the following:

1. Defined consumer population or community
2. Comprehensive and integrated service
3. Defined delivery system of health care
4. An epidemiological understanding of community health problem and
5. Management-oriented approach for the solution of these problems.

Basic Sciences of Medicine

The subject of medicine has gained its strength from various disciplines of biosciences. They include the following:

1. **Anatomy**: It is the study of the physical structure of organisms.
2. **Biochemistry**: It is an interdisciplinary science, involving study of the chemistry taking place in living organisms, especially the structure and functions of their chemical components.
3. **Biomechanics**: The study of the structure and function of biological systems by means of the methods of mechanics.
4. **Biostatistics**: Knowledge of statistics as applied to biological studies is essential in the planning, evaluation and interpretation of medical research. It is also essential to understand epidemiology and evidence-based medicines.
5. **Biophysics**: It is an interdisciplinary science that uses the methods of physics and physical chemistry to study biological systems.
6. **Cytology**: It is the microscopic study of individual cells.
7. **Embryology**: It refers to the study of early development of organisms.
8. **Epidemiology**: It covers the study of the demographics of disease processes and includes, but is not limited to, the study of epidemics.
9. **Genetics**: It is the study of genes and their role in biological inheritance.
10. **Histology**: It is the branch of science dealing with the study of structures of biological tissues by light microscopy, electron microscopy and immunohistochemistry.
11. **Immunology**: It is the study of immune system, which includes the innate and adaptive immune system in humans.
12. **Microbiology**: The study of microorganisms, including protozoa, bacteria, fungi and virus.
13. **Molecular biology**: It is the study of molecular underpinnings of the process of replication, transcription and translation of the genetic material.

14. **Neuroscience**: It includes those disciplines of science that are related to study of the nervous system. A main focus of neuroscience is on biology and physiology of the human brain and spinal cord.
15. **Pathology**: It is the science dealing with study of disease, the causes, course, progression and resolution thereof.
16. **Pharmacology**: It is that branch of bioscience that deals with the study of drugs and their action on living organism.
17. **Photobiology**: It is the study of interactions between nonionizing radiation and living organisms.
18. **Physiology**: It is the study of normal functioning of the body and the underlying regulatory mechanisms.
19. **Radiobiology**: It encompasses study of the interactions between ionizing radiation and living organisms.
20. **Toxicology**: It is the study of hazardous effects of drugs and poison on human beings.

EDIBLE VACCINES

Vaccines play an important role in the health care system of any country. Many diseases are eradicated, and the spread of many ailments are effectively controlled by the use of live or attenuated causative agents as vaccines. However, high cost of production, maintaining a chain for vaccine distribution, cultivation of pathogens, poor reliability of the attenuation process, dangers due to contamination attached to the production and use of many free or attenuated vaccines are the main hurdles in the production of vaccines.

Plants may serve as useful production units for vaccines, since large amounts of antigens could be produced at a relatively low cost using agricultural technique, instead of expensive and sophisticated cell culture-based expression systems. Antigens expressed in a plant can be administered orally as edible parts of the plant or by parenteral routes (IM or IV) after isolation and purification from the plant tissue. The edible part of the plant to be used as vaccine is usually fed in raw condition in order to prevent possible denaturation that takes place during cooking. It also enables to avoid cumbersome purification protocols.

For production of edible vaccines, it is always desirable to select a plant whose products are consumed raw to avoid degradation during cooking (Table 14.1).

Table 14.1 Some antigens produced by edible plants

Antigen	Plant
CT-B	Potato
Foot and mouth virus	*Arabidopsis*
Herpes virus B surface antigen	Tobacco
Human cytomegalovirus glycoprotein B	Tobacco
Rabies glycoprotein	Tomato

Abbreviation: CT-B, cholera toxin B subunit.

Table 14.2 Production of antigen vaccines from transgenic plants

Protein	Plant
Rabies virus glycoproteins	Tomato
CT-B protein	Potato, tobacco
HBsAg	Tobacco
Norwalk virus capsid protein	Tobacco
Protein of foot-and-mouth disease virus	Tobacco
Mouse glutamate decarboxylase	*Arabidopsis*
Glycoproteins of swine transmissible	Potato
Gastroenteritis coronavirus protein	*Arabidopsis*

Abbreviations: CT-B, cholera toxin B subunit; HBsAg, hepatitis B surface antigen.

The production of edible vaccines was first reported in 1990 in tobacco (at 0.02% of total leaf protein level). The production of vaccines in potatoes was reported in the bacterium, *Agrobacterium tumefaciens*, commonly used to deliver the DNA for bacterial or viral antigens. A plasmid carrying the antigen gene and an antibiotic resistance gene are incorporated into the bacterial cells. The cut piece of potato leaves are exposed to an antibiotic that can kill the cells lacking the new genes. The surviving cells can multiply and form a callus. This callus is allowed to sprout shoots and roots that are grown in soil to generate plants. In about three weeks time, the plants bear potatoes with antigen vaccines. The examples of antigen vaccines in plants are given in Table 14.2.

The antigens in transgenic plants are delivered through bioencapsulation. The tough outer wall of plant cells protects them from gastric secretions and finally breaks up in the intestine. The antigens are released, taken up by microfold cells (M cell) in the intestinal lining that overlie Peyer's patches and gut-associated lymphoid tissue (GALT), passed on to macrophages, other antigen-presenting cells and local lymphocyte populations. This generates serum immunoglobulin G (IgG), immunoglobulin E (IgE) responses, local immunoglobulin A (IgA) response and memory cells, which would promptly neutralize the attack by the real infectious agents (Fig. 14.1).

In future, the global immunization program could be possible with edible vaccines. The distinctive advantages with the production of edible vaccines are as follows:

1. Low cost of production
2. Separation and purification is easy
3. Pathogenic contamination can be avoided, safe for use
4. Aseptic conditions are not required for oral immunization
5. No constricted criteria for its storage; it can be stored near the site of use

Important Steps in the Production of Edible Vaccines

For production of edible vaccines or antibodies, it is desirable to select a plant whose products are consumed raw to avoid degradation during cooking. Thus, plants like tomato, banana and cucumber are generally the plants of choice. While the expression of a gene that is stably integrated

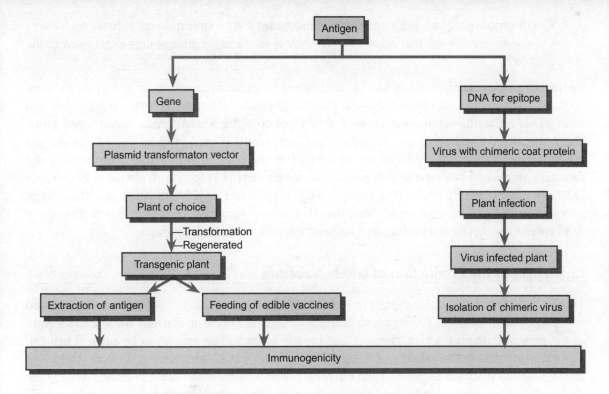

Figure 14.1 Antigens produced in transgenic plants.

into the genome allows maintenance of the material in the form of seeds, some virus-based vectors can also be used to express the gene transiently to develop the products in a short period.

1. **First generation**: Foreign DNA can be introduced into plant's genome by bombarding embryonic suspension cell cultures using gene gun or more commonly through *Agrobacterium tumefaciens*, a naturally occurring soil bacterium, which has the ability to get into plants through some kind of wound, scratch, etc. It possesses a circulartumour-inducing (Ti) plasmid, which enables it to infect plant cells, integrate into their genome and produce a hollow tumour (crown gall tumour), where it can live. This ability can be exploited to insert foreign DNA into plant genome. Prior to this, the plasmid needs to be disarmed by deleting the genes for auxin and cytokinin synthesis, so that it does not produce tumour.

2. **Second generation**: Successful expression of foreign genes in plant cells and/or its edible portions has given a potential to explore further and expand the possibility of developing plants expressing more than one antigenic protein. Multicomponent vaccines are obtained by crossing two plant lines harbouring different antigens. Adjuvant may also be coexpressed along with the antigen in the same plant.

Approaches to mucosal vaccine formulation include the following:

1. Gene fusion technology, creating nontoxic derivatives of mucosal adjuvants
2. Genetically inactivating antigens by deleting an essential gene

3. Coexpression of antigen and a cytokine that modulates and controls mucosal immune response
4. Genetic material itself that allows DNA/RNA uptake and its endogenous expression in the host cell.

Chimeric virus: Certain viruses can be redesigned to express fragments of antigenic proteins on their surface, such as cowpea mosaic virus (CPMV), alfalfa mosaic virus, tobacco mosaic virus (TMV), cauliflower mosaic virus (CaMV), potato virus X and tomato bushy stunt virus. Technologies involved are overcoat and epicoat technology. It may even be possible to present a cocktail of specific HIV epitopes on the surface of the plant virus. CPMV is genetically and thermally stable and can survive in highly acidic condition (pH 1) for 1 h. Wide range of epitopes has been expressed in CPMV. This includes HIV-1 gp41 and gp120, human rhinovirus, canine parvovirus, foot-and-mouth disease virus, fungal epitopes, mammalian epitopes from hormones or from colon cancer cells and protozoan epitopes from *Plasmodium falciparum*.

Challenges in the Production of Edible Vaccines

Three successful human clinical trials have shown that adequate doses of antigen can be achieved with plant-based vaccines. To determine the right dosage, one needs to consider the age and weight of the person, fruit/plant's size, ripeness and protein content. The amount to be eaten is critical, especially in infants, who might spit it, eat a part of eat or eat all and throw it up later. Too low a dose would fail to induce antibodies and too high a dose would instead cause tolerance. Giving the vaccine into a teaspoon of baby food may be more practical than administering it in a whole fruit. The transformed plants can also be processed into pills, puddings, chips, etc. Regulatory concerns would include lot-to-lot consistency, uniformity of dosage and purity. Attempts at boosting the amount of antigens often lead to stunted growth of plants and reduced tuber/fruit formation, as too much mRNA from the transgene causes gene silencing in plant genome.

Some of the techniques to overcome these limitations are as follows:
1. Optimization of coding sequence of bacterial/viral genes for expression as plant nuclear genes
2. Expression in plastids
3. Plant viruses expressing foreign genes
4. Coat-protein fusions
5. Viral-assisted expression in transgenic plants

Nonscientific Challenge
At present, small companies are undertaking most of the research on edible vaccines with markets of developing countries in view. A few international aid organizations and some national governments are rendering support, but the effort remains largely underfunded.

Regulatory Issues
It is still unclear whether the edible vaccines would be regulated under food, drugs or agricultural products. The licensing of vaccine component antigen itself—genetically engineered fruit or transgenic seeds—is not yet clear.

Clinical Trials

Antigen expression in plants has been successfully shown in the past, like heat-labile enterotoxin subunit B (LT-B) of enterotoxigenic *Escherichia coli* (ETEC) in tobacco and potato, rabies virus G protein in tomato, HBsAg in tobacco and potato, Norwalk virus in tobacco and potato and CT-B of *Vibrio cholerae* in potato. Ethical considerations usually preclude clinical trials from directly assessing protection, except in a few cases.

Some Edible Vaccines Produced by Plants

1. **Vaccines of potato**: The transgenic potatoes were created and grown by Charles Arntzen, Hugh S. Mason and their colleagues at the Boyce Thompson Institute for Plant Research, an affiliate of Cornell University. The volunteers ate bite-sized pieces of raw potatoes, which had been genetically engineered to produce part of the toxin secreted by the *E. coli*. An edible vaccine could stimulate an immune response in humans.

2. **Vaccines of tobacco**: In 1990, the first report of the production of edible vaccine in tobacco—at 0.02% of total leaf protein level—was published in the form of a patent application under the International Patent Cooperation Treaty. Since acute watery diarrhoea is caused by enterotoxigenic *E. coli* and *V. cholerae* that colonize the small intestine and produce one or more enterotoxin, an attempt was made towards the production of edible vaccine by expressing LT-B in tobacco.

3. **Vaccines of banana**: For edible vaccines or subunit vaccinations, bananas can be conveniently used. The advantage of bananas as desired vectors is that they can be eaten raw as compared to potatoes or rice that needs to be cooked and banana can also be consumed in a pure form.

4. **Vaccines of tomato**: Tomatoes serve as ideal candidates for HIV antigen, unlike many other transgenic plants that carry the protein. Tomatoes are edible and immune to any thermal process, which helps retain its healing capabilities. The introduction of the artificial protein gene into the tomato germs was accomplished with the help of a needle. The germs were then cultivated on a special nutrient medium. The plants that grew roots were planted into the soil and grown in the hothouse until they reached maturity and produced fruit. The plants were then tested for the protein that could be

found in the leaves and the tomato plant itself.

5. **Vaccines of maize**: More than 2 billion people are infected every year with hepatitis B, and about 350 million of these are at high risk of serious illness and death from liver damage and liver cancer. The Egyptian scientists have genetically engineered maize plants to produce a protein used to make hepatitis B virus vaccine.

6. **Vaccines of rice**: When the predominant T cell epitope peptides are derived from Japanese cedar, pollen allergens are expressed in rice seeds and are delivered to the mucosal immune system. The development of an allergic immune response of the allergen-specific T_h2 cell is suppressed. It is also observed that not only specific IgE production and release of histamine from mast cells are suppressed, but also inflammatory symptoms of pollinosis, such as sneezing, are also found to be suppressed. These results indicate the feasibility of using an oral immunotherapy agent derived from transgenic plants that accumulate T-cell epitope peptides of allergens for allergy treatment.

Advantages and Limitations of Edible Vaccines

Edible vaccines have many advantages over traditional injected vaccines. Edible vaccines are cost effective and can be administered by different routes. They lower the risk of contamination and eliminate the cost of transportation. The transgenic plants can be cost efficient in storage, preparation, production and transportation. These transgenic plants do not require cold chain storages. The various advantages of edible vaccines are enlisted as follows:

1. Production of vaccines with low costs
2. Easy for mass production system by breeding compared to an animal system
3. Eliminate cost of transportation
4. Edible plants are very effective as delivery vehicles for inducing oral immunization
5. Excellent feasibility of oral administration in comparison to injection
6. Easy for separation and purification of vaccines from plant materials
7. Convenience and safety in storing and transporting vaccines
8. Adjuvant for immune response is not necessary
9. Effective prevention of pathogenic contamination from animal cells
10. Effective retention of activity of vaccines by controlling the temperature in plant cultivation
11. Storage near the site of use
12. Heat stable, eliminating the need for refrigeration
13. Antigen protection through bioencapsulation is possible
14. Reduced need for medical personnel and sterile injection condition
15. Subunit vaccine (not attenuated pathogens) means improved safety

16. Reduced need for medical persons and sterile injection conditions

However, it has the following disadvantages:
1. Dosage of vaccines would be variable
2. Not convenient for infants

Edible vaccines offer promising future for providing therapeutic agents in order to make global immunization program very successful, covering large contingent of people in developing and underdeveloped countries.

QUESTIONS

1. What are edible vaccines?
2. What are major objectives of public health?
3. Give an account of basic sciences related to medicine.
4. Write a note on development of edible vaccines.
5. Write a note on challenges in production of edible vaccines.

CHAPTER 15: Broad Applications of Pharmaceutical Biotechnology

CHAPTER OUTLINE

- Introduction
- Applications of Biotechnology in the Production of Biomolecules
- Applications of Biotechnology in the Diagnosis of Diseases
- Applications of Biotechnology in Agriculture
- Applications of Ultrasound to Biotechnology
- Testing of Pharmaceuticals
- Questions

INTRODUCTION

Biotechnology is one of the world's fastest growing and most rapidly changing technology, involving biosciences that deal with genetically manipulating organisms. Biotechnology may be considered as a great technical revolution in the last two decades. This branch of biosciences has revolutionized the concept of production of drugs and therapeutics to save the mankind from a lot of deadly diseases. The major impact of biotechnology is on the introduction of molecular biology into medicine. It has led to powerful techniques known as *genetic engineering*, which may be effectively used for synthesis of newer proteins as therapeutics using microorganisms. Recent advancements in biotechnology have provided a new insight into many diseases and in the developments of newer tools and techniques, which have brought revolutionary changes in diagnosing, treating and preventing life-threatening diseases. Some of the important biotechnological products that have revolutionized the field of medicine are as follows:

1. **Therapeutics**: Antibiotics, monoclonal antibodies (mAbs), insulin, growth hormones, interferons, cytokines and monokines, colony-stimulating factors (CSFs), hormone-releasing factor, immunosuppressive factor, clotting factor, clot-dissolving factor, neuroactive peptides, anticoagulant factor, lipoproteins, vitamins and enzymes.
2. **Vaccines**: Hepatitis vaccine, herpes vaccine, rabies vaccine, polymyositis vaccine, influenza vaccine, cholera vaccine, filariasis vaccine, AIDS vaccine and leishmaniasis vaccine.
3. **Delivery system**: Liposomes, biopolymers, plasmids and mAbs.

APPLICATIONS OF BIOTECHNOLOGY IN THE PRODUCTION OF BIOMOLECULES

Production of Antibiotics

Antibiotics are generally defined as low molecular mass microbial secondary metabolites, which at low concentration inhibit the growth of other microorganisms. Around 10,000 antibiotic substances have been isolated and characterized so far. Several fungal genera are known to produce antibiotics. Penicillin, the first antibiotic to be used medically, was extracted from *Penicillium notatum*.

Today, around 124 different antibiotics are available in the market, belonging to the different classes, such as β-lactams, tetracyclines, aminoglycosides, macrolides, ansamycin, peptides and glycopeptides.

Production of Vaccines

With the development of genetics, molecular biology and plant biotechnology in recent years, production of vaccines (living vector vaccine, nucleic acid vaccine, etc.) has received special impetus. In particular, the technology of transgenic plants to produce human or animal therapeutic vaccines received increasing attention. Expressing vaccine candidates in vegetables and fruits open up a new avenue for producing oral/edible vaccines. Transgenic plant vaccines are very promising as they are low-cost products and are easy to store and convenient for immune inoculation. They can be consumed directly without isolation and purification. Till date, many transgenic plant vaccines have been produced.

Numerous vaccines are known today to protect mankind. Once inside the body, the vaccines, in general, create a state that alarms the host in the same way, as in the case of a particular pathogen attacking the body. Recent advances in immunology have led to the development of new and promising strategies for vaccine production (Table 15.1).

New Generation of Vaccines

1. **Hepatitis B vaccines**: Numerous hepatitis B virus (HBV) vaccines are available in the market. They can be broadly classified into hepatitis B surface antigen and recombinant HBV.

Table 15.1 Use of vaccines in different diseases and their strategic development

Disease	Type of vaccine
Tuberculosis	Attenuated
Cholera	Inactivated
Plague	Inactivated
Polio	Inactivated attenuated
Yellow fever	Inactivated exotoxin
Diphtheria/tetanus	Inactivated exotoxin

2. **Liposomal vaccines**: HBsAg particle contains three peptides (small, middle and long peptide) and lipids. The particle on encapsulation into liposomes serves as adjuvant carrier.
3. **Contraceptive vaccines**: The application of immunological approach for control of fertility is a reality today. Birth control vaccines for human use have reached the stage of clinical trials both in India and abroad. Today, vaccines against hormones such as luteinizing-hormone-releasing hormone, follicle-stimulating hormone and human chorionic gonadotropin hormone are in the process of development.
4. **AIDS vaccines**: Since HIV was identified as the causative agent of AIDS, enormous efforts have been made to develop safe and effective vaccines.
5. **Malaria vaccines**: Malaria is a serious protozoan disease responsible for nearly one million lives every year. A total of 85% of deaths occur in children below 5 years of age. Malaria is caused by various species of the genus *Plasmodium* of which *P. falciparum* is the most virulent and prevent. Current vaccine strategies are aimed at producing synthetic subunit vaccines, consisting of epitomes that can be recognized by T cell and B cell.
6. **DNA vaccines**: These can serve as an alternative to immunization with purified protein and are effective in generating antibody, helper and cytotoxic response and protective immunity.
7. **Transgenic plants as edible immunogenic concept**: A novel and practical approach for development of vaccines could be through introduction of genes and encoding microbial antigens in plants. An alternative approach to deliver the gene is by infecting the plant with *Clavibacter michiganensis*, tobacco mosaic virus or by using gene gun.

Human Therapeutics from Recombinant DNA Technology

One of the greatest advantages of the recombinant DNA technology has been the production of human therapeutics, such as hormones, growth factors and antibodies that are not only scarcely available, but also are very costly for human use.

1. **Recombinant hormones**: Insulin (and its analogues), growth hormone, follicle-stimulating hormone and salmon calcitonin
2. **Blood products**: Albumin, thrombolytics, fibrinolytics and clotting factors (factor VII, factor IX, tissue plasminogen activator, recombinant hirudin)
3. **Cytokines and growth factors**: Interferons, interleukins and CSFs (interferons α, β and γ), erythropoietin, interleukin-2, granulocyte-macrophage CSF (GM-CSF) and granulocyte-CSF (G-CSF)
4. **Monoclonal antibodies and related products**: Mouse, chimeric or humanized; whole molecule or fragment; single chain or bispecific and conjugated (rituximab, trastuzumab, infliximab and bevacizumab)
5. **Recombinant vaccines**: Recombinant protein or peptides, DNA plasmid and anti-idiotype, hepatitis B surface antigen (HBsAg) vaccine and human papillomavirus (HPV) vaccine
6. **Recombinant enzymes**: Dornase alpha (Pulmozyme), alglucosidase alfa (Myozyme), alpha-L-iduronidase (Aldurazyme), urate oxidase, etc.
7. **Miscellaneous products**: Bone morphogenic protein, conjugated antibody, pegylated recombinant proteins and antagonists

Production in Lactic Acid

Lactic acid can be produced by either fermentation technology or chemical synthesis. Due to environmental concern and the limited nature of petrochemical feedstock, the route of fermentation technology has received considerable interest recently.

Lactic acid is widely used in food, cosmetics, pharmaceuticals and chemical industries as indicated below.

1. Food preservative, flavour, pH regulator and mineral fortification
2. Cosmetic moisturizer, skin-lightening agent and antiacne agent
3. Pharmaceutical and chemical industries—parenteral/IV solution, dialysis solution, tableting and surgical sutures

Plant Tissue Culture for Production of Plant Secondary Metabolites

Recent advances in the molecular biology, enzymology and fermentation technology of plant cell cultures suggest that these systems may become a valuable source of important secondary metabolites.

Plant tissue culture refers to growing and multiplication of cells, tissue and organs of plants on defined solid or in liquid nutrient media under aseptic and controlled conditions. Some important examples of secondary metabolite production from plant tissue culture are given in Table 15.2.

Table 15.2 Recently produced bioactive secondary metabolites from plant tissue culture

Plant	Active ingredient	Type of culture
Allium sativum	Proteolytic enzymes	Callus and tissue culture
Cassia acutifolia	Anthraquinone	Suspension culture
Catharanthus roseus	Indole alkaloid	Suspension culture
Cephaelis ipecacuanha	Cephaeline	Callus culture
Cephaelis ipecacuanha	Emetine	Callus culture
Cinchona ledgeriana	Quinoline alkaloids	Globular cell suspension culture
Dioscorea deltoidea	Diosgenin	Callus culture
Mentha arvensis	Terpenoid	Shoot culture
Papaver somniferum	Sanguinarine	Cell suspension culture
Panax ginseng	Ginsenoside	Root culture
Podophyllum hexandrum	Podophyllotoxin	Suspension culture
Salvia fruticosa	Rosmarinic acid	Callus and suspension culture
Solanum elaeagnifolium	Solasodine	Callus culture
Taxus species	Taxol	Suspension culture
Withania somnifera	Withaferin	Shoot culture

APPLICATIONS OF BIOTECHNOLOGY IN THE DIAGNOSIS OF DISEASES

The maximum benefits of biotechnology have been utilized in the production of potential drugs (biotechnology-derived proteins and polypeptides). Currently, there are about 35 biotechnology-derived therapeutics and vaccines approved by the USFDA alone for medicinal use. Additionally, more than 500 drugs and vaccines are in the process of reaching the global market.

In medicine, modern biotechnology finds promising applications in the areas of drug production, pharmacogenomics, gene therapy, genetic testing techniques in molecular biology and detection of genetic diseases. Amniocentesis and chorionic villus sampling can be used for testing the developing fetus for Down's syndrome.

APPLICATIONS OF BIOTECHNOLOGY IN AGRICULTURE

The techniques of modern biotechnology may be used for development of a high-yield crop variety with a new endogenic character. However, while increase in crop yield is the most obvious application of modern biotechnology in agriculture, it is also the most difficult one. Current genetic engineering techniques work best for effects that are controlled by a single gene. Many of the genetic characteristics associated with yield (e.g. enhanced growth) are controlled by a large number of genes, each of which has a minimal effect on the overall yield.

Reduced Vulnerability of Crops to Environmental Stresses

The biotechnological techniques may be successfully used for development of crops containing genes that enable them to withstand biotic and abiotic stresses. The discovery of genes that enable plants to overcome the limiting factors of drought and excessive salty soil has been made possible. One of the latest developments is the identification of an antistress plant gene, *At-DBF2*, from a tiny weed *Arabidopsis thaliana*. This is because it is very easy to grow, and its genetic code is well mapped out. When this gene was inserted into tomato and tobacco cells, the cells were able to withstand environmental stresses like drought, cold and heat, far more than ordinary cells.

Reduced Dependence on Fertilizers, Pesticides and Other Agrochemicals

The important application of modern biotechnology in the field of agriculture is on reducing the dependence of farmers on agrochemicals. For example, *Bacillus thuringiensis* (Bt) is a soil bacterium that produces a protein with insecticidal qualities. An insecticidal spray from this bacterium is produced using fermentation technology. In this form, the Bt toxin occurs as an inactive protoxin, which requires digestion by an insect to be effective. The crops may also be genetically engineered to acquire tolerance to broad-spectrum herbicides. The lack of herbicides with broad-spectrum activity and no crop injury is a consistent limitation in crop weed management. Multiple applications of numerous herbicides are routinely used to control a wide range of weed species detrimental to agronomic crops.

Production of Novel Substances in Crop Plants

The techniques of biotechnology are also being applied for novel uses in plant; for example, oilseed can be genetically modified to produce fatty acids for detergents, substitute fuels and petrochemicals. Potato, lettuce, safflower, tomato, rice, tobacco and other plants are genetically engineered to produce insulin and certain vaccines. If future clinical trials prove successful, the advantages of edible vaccines would be enormous, especially for developing countries. The transgenic plants may be grown locally and with economic viability.

APPLICATIONS OF ULTRASOUND TO BIOTECHNOLOGY

The research related to application of ultrasound to biotechnology is of recent origin. Several processes that take place in the presence of cells or enzymes are activated by high-intensity ultrasonic waves, which can break the cells and denaturize the enzymes. Low-intensity ultrasonic waves are capable of modifying cellular metabolism or improving mass transfer of reagents and products through the boundary layer or cellular wall and membrane. In case of enzymes, the increase in the mass transfer rate of reagents to the active site seems to be the most important factor. As compared to native enzymes, immobilized enzymes are more resistant to thermal deactivation produced by ultrasound waves.

TESTING OF PHARMACEUTICALS

By using biotechnological tools, the raw materials and their finished products of natural origin, such as starch, gum, gelatin, talc, etc., can be tested for detection of various microorganisms. Biotechnology has applications in evaluation of disinfectants and cleaning agents, for testing biological activity of an antibiotic preparation and testing of endotoxin (bacterial endotoxin test).

The application and promise of biotechnology raised many ethical, economic, social and environmental concerns across the globe. There has been resistance by certain groups of people for the commercial use of genetically modified food or agricultural produce. Their main concerns are related to adverse effects of this technology on health, nutrition and also on economic environment prevailing in developing countries. If certain genuine concerns related to their misutilization are taken care of, the science of biotechnology can offer very promising future to mankind.

QUESTIONS

1. What are the pharmaceutical applications of biotechnology?
2. Write about different scopes of biotechnology.
3. Give an account of applications of biotechnology in the production of biomolecules.
4. Give an account of applications of biotechnology in diagnosis of disease.

CHAPTER 16: Biotechnology in Drug Discovery

CHAPTER OUTLINE

- Introduction
- Current Trends in Modern Pharmaceutical Analysis for Drug Discovery
- Recent Pharmacokinetic Advances in Drug Discovery and Development
- Natural Product Drug Discovery
- Questions

INTRODUCTION

Drug discovery is the process by which drugs are discovered and/or designed in the disciplines of medicine, biotechnology and pharmacology. In the past, most of the drugs were discovered either by identifying the active ingredient from medicinal plants or by serendipitous discovery. The quest to discover healing drugs has always been influenced by prevailing social and cultural factor. A new approach has been developed based on understanding how diseases and infections are controlled at the molecular and physiological levels. The process of drug discovery involves identification of candidates, synthesis, characterization, screening and assays for therapeutic efficacy. Once a compound has exhibited its efficacy in these tests, the process of drug development begins prior to clinical trials. Despite advances in technology and understanding of biological systems, drug discovery is still a lengthy, difficult and expensive process with low rate of success.

In ancient times, accumulated knowledge and wisdom could only be passed on by word of mouth from generation to generation. Ancient carvings and writing on the walls of caves held clues to the early use of drugs. However, it was only the written accounts that led our ancestors to seek out remedies for treating disease, such as those complied by the inhabitants of ancient civilizations of India, Egypt, Mesopotamia (particularly Greece) and China.

The beginning of the 21st century has witnessed paradigm shift in the fields of biotechnology and drug discovery. The research pertaining to discovery and the development of new drug entities using computational methodology is undertaken in a big way. The success of modern drug research depends on the rapid and universal publication of scientific results.

A drug may also interact elsewhere to produce an undesired effect. Plants contain a complex and variable mixture of chemicals, and the risk of unwanted side effects occurring is always high.

The simplest way to minimize this is to isolate the compound that produces the desired response. While this may be obvious today, it was not the reason behind the successful isolation of active principles in the second decade of the 19th century (Figs. 16.1 and 16.2).

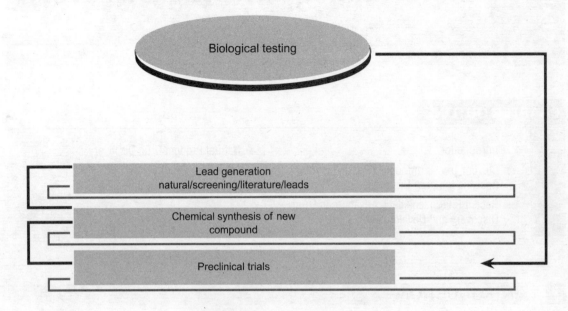

Figure 16.1 Drug discovery cycle.

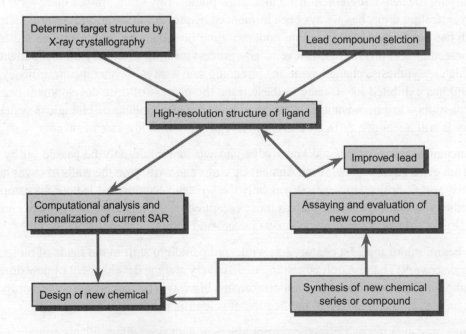

Figure 16.2 Structure-based drug design cycle.

CURRENT TRENDS IN MODERN PHARMACEUTICAL ANALYSIS FOR DRUG DISCOVERY

Successful drug discovery and development depends on close interactions between various disciplines with inputs from biotechnology, biomedical engineering, proteomics and genomics inter alia. Currently, drug discovery efforts are being revolutionized by high-input technologies, combinatorial chemistry, informatics, genomics, proteomics and miniaturization. Today, pharmaceutical analysis is the backbone of the whole drug discovery and development process. It is used to provide accurate and precise data, not only supporting drug discovery and development, but also postmarket surveillance, high-throughput analysis, structural analysis, purity analysis and quantitative estimation.

The rational approach to drug design has been adopted in developing specific inhibitors of the human cellular enzyme and purine nucleoside phosphorylase (PNP). The free purine released can then be used in the biosynthesis of new nucleic acids within the cell.

$$\text{Purine base} \longleftrightarrow \text{Sugar + Pi-Free Purine + Sugar P}$$

RECENT PHARMACOKINETIC ADVANCES IN DRUG DISCOVERY AND DEVELOPMENT

Pharmacokinetics (PK) is one of the major crucial areas to evaluate whether promising clinical candidates can be developed into marketable drugs. This is also one of the important considerations that accounts for the costly and high late-stage failure rates in drug development. So far, poor PK profiles have accounted for around 40% attrition in late stage of drug development, which in turn had substantial effect on R&D costs. Therefore, the working culture in the pharmaceutical industry today has shifted to conducting PK investigations at a much earlier stage in drug discovery. In addition to the application of chemometrics in the early stages of drug development in silico technology has emerged as a new and promising tool for early ADME–Tox investigations. Commercial ADME–Tox software is now available to screen compounds in the early stage of drug development. However, more work needs to be performed to refine current in silico models, because these predictive models are based on limited PK and toxicity data sets. Another exciting novel innovation for ADME studies in early clinical drug development is human microdosing, using two ultrasensitive analytical techniques, accelerator mass spectrometry (AMS) and positron emission tomography (PET).

NATURAL PRODUCT DRUG DISCOVERY

Natural products are important sources of new structures, leading to drugs in all major disease areas. Finding new habitats of fungi, bacteria or plants and identifying new living organisms have been a continuous challenge. Traditionally, drug discovery programmers within the pharmaceutical industry relied on screening various biological specimens for potential drugs. The examples include digoxin and digitoxin, as well as morphine, codeine, taxol, etc. Plants are the sources of drugs as they produce a vast array of novel bioactive molecules, many of which probably serve as chemical defences against infection. If an interesting activity is described, larger quantities of the plant material are collected, from which chemists purify and characterize the active principles. Microorganisms, mainly bacteria and fungi, have also proved to be rich sources of bioactive molecules, such as antibiotics, anticancer and anti-infective agents.

Testing and identifying biologically active natural products from plant extracts or microbes in today's high-throughput screening (HTS) environment pose a challenge to sample preparation and single component identification. An enhancement of productivity levels in the drug discovery process has been achieved with the implementation of library chemistry and HTS technologies. Despite these efforts, the numbers of new chemical entities reaching the market have not increased. Only one drug originated from a de novo combinatorial chemistry approach. However, the natural products remain an important source of structures contributing to mostly semisynthetic or synthetic drugs used in wide spectrum of human ailments.

Natural products have been the single most productive sources of leads for the development of drugs. Natural products are important sources of unsurpassed structural diversity and functional density to identify screening hits. The therapeutic effects of natural product-derived drugs are predominantly achieved in antibiotic therapies, immunoregulation and oncology. It is less likely to identify potent natural products against molecular targets of human diseases in indications, such as cardiovascular or nervous systems. Those targets are generally not related to the biological environment of the producing organism. However, higher hit rates are generally obtained for natural product libraries in HTS campaigns in comparison to medchem or combichem libraries.

Around one hundred new natural products are in clinical development, particularly as anticancer agents immunomodulators, hepatoprotectives and anti-infectives. Application of molecular biological technique is increasing the availability of novel compounds that can be conveniently produced in bacteria or yeasts. The combinatorial chemistry approaches are being based on natural product scaffolds to create screening libraries that closely resemble drug-like compounds. Various screening techniques have been developed to improve the ease with which natural products can be used in drug discovery campaigns and data mining. The virtual screening techniques are also being successfully applied to databases of natural products.

Screening of Natural Products

Ethnobotany is the study of the use of plants in the society, and ethnopharmacology is an area inside ethnobotany, which is focused on medicinal uses of plants. Both can be used in selecting starting materials for future drugs. A collection of plant, animal and microbial samples from rich ecosystems may give rise to novel biological activities. One example of a successful use of this strategy is the screening for antitumour agents by the National Cancer Institute, United States, started in the 1960s.

Two main approaches exist for the finding of new bioactive chemical entities from natural sources: (1) random collection and screening of material and (2) exploitation of ethnopharmacological knowledge in the selection. The former approach is based on the fact that only a small part of biodiversity of the planet has ever been tested for pharmaceutical activity, and organisms living in a species-rich environment need to evolve defensive and competitive mechanisms to survive.

Current research in drug discovery from medicinal plants involves a multifaceted approach combining botanical, phytochemical, biological and molecular techniques. Medicinal plant or herb drug discovery continues to provide new and important leads against various pharmacological targets, including cancer, HIV/AIDS, Alzheimer's disease, malaria and pain. Several natural product drugs of plant origin have either recently been introduced into the US market, including arteether, galantamine, nitisinone and tiotropium, or are currently involved in late-phase clinical trials. As part of National Cooperative Drug Discovery Group (NCDDG) research project, numerous compounds from tropical rainforest plant species with potential anticancer activity have been identified.

Artemisinin, an antimalarial agent from sweet wormtree *Artemisia annua*, used in Chinese medicine since 200 BC is one drug used as a part of combination therapy for multiresistant *Plasmodium falciparum*. Several compounds mainly from plants used as dietary supplements have been extracted for their use as chemoprotective agents.

Structural Elucidation

The elucidation of the chemical structure of a natural product is critical to avoid the rediscovery of a chemical agent that is already known for its structure and chemical activity. The UV-visible infrared spectroscopy, high-performance liquid chromatography (HPLC) and elemental analysis are useful techniques for establishing quality profiles of chemical moieties. Mass spectrometry (MS), often used to determining structure is an important technique by which individual compounds are identified based on their mass/charge ratio, after ionization. The chemical compounds usually exist in nature as mixtures. Hence, the combinations of liquid databases of mass spectra for known compounds are available. Nuclear magnetic resonance (NMR) spectroscopy is another important technique for determining chemical structures of natural products. NMR yields information about individual hydrogen and carbon atoms in the structure, allowing detailed reconstruction of the molecule's architecture.

QUESTIONS

1. Write a note on the design of drug discovery system for biotechnological production.
2. Write a note on current trends in modern pharmaceutical analysis for drug discovery.
3. Explain about natural product drug discovery.

CHAPTER 17: Immobilization of Enzymes

CHAPTER OUTLINE

- Introduction
- History of Enzymes
- Enzyme Structure
- Enzyme Kinetic
- Bioproduction of Enzymes
- Enzyme Production
- Enzyme Immobilization
- Immobilization of Plant Cells
- Enzyme Engineering
- Applications of Enzymes
- Biosensors
- Questions

INTRODUCTION

Enzyme is a biocatalyst that accelerates biochemical reactions. The concept of biocatalysts is very wide, and it includes pure enzyme, crude cell extract, viable plant and animal cells, microbial cells and intact nonviable microbial cells. The sources of enzymes used in commerce are microorganisms, higher plants and animals. Most important plant enzymes are papain, proteases, amylases and soybean lipoxygenase, whereas commonly used animal enzymes are lipase, trypsin, rennet, etc. The enzyme papain, extracted from papaya fruit, is a meat tenderizer, and pancreatic protease is used in leather softening and manufacture of detergents.

The enzyme-containing detergents have been known since 1913, but their use was limited because of their instability in detergent formulations. In 1965, a new stable enzyme (e.g. protease) was introduced for application in detergent production. The first commercial process was developed for production of fructose from glucose in the 1970s through the isomerization of glucose. Malt is used as the source of enzymes in brewing industry. The production of primary and secondary metabolites by microorganisms is possible only due to involvement of various enzymes. They are of two types: (1) extracellular (secreted outside the cell) and (2) intracellular enzymes (secreted within cell). There is a wide range of extracellular enzymes produced by pathogenic and saprophytic microorganisms, such as cellulase, polygalacturonase, polymethyl galacturonase, pectin methyl esterase, etc. These enzymes help in establishment in host tissues or decomposition of organic substrates. The intracellular enzymes such as asparaginase, invertase and uric oxidase are of high economic value and difficult to extract as they are produced inside the cell. They can be obtained by breaking the cells by means of a homogenizer and extracting them through the biochemical processes. The process of enzyme purification is difficult, as the cell debris and nucleic acid are not easily removed.

Microbial enzymes have the following two distinct advantages over the animal and plant enzymes:

1. They can be produced on large scale within limited space, and their production is economical with respect to time. The amount produced depends on type of microbial strain used, size of fermenter and growth conditions. It can be easily extracted and purified.
2. The technical advantages associated in production of enzymes using microorganism are given below.
    a. They can grow in a wide range of environmental conditions.
    b. They are capable of producing a wide variety of enzymes.
    c. They show genetic flexibility, and they can be genetically manipulated to the yield of enzymes.
    d. They have short generation times.

At present, more than 2000 enzymes have been isolated and characterized, of which about 1000 enzymes are recommended for various applications. Amongst them, around 50 microbial enzymes have industrial applications.

In recent years, industrial production of enzymes has considerably increased. The total global production of enzymes was estimated around 65,000 tonnes with a value of about US$ 4×10^8 around a decade ago. Now the cost is expected to be doubled.

HISTORY OF ENZYMES

As early as the late 18th and early 19th centuries, the digestion of meat by stomach secretions and the conversion of starch to sugar by plant extracts and saliva were known. However, the mechanism by which this occurred had not been identified.

In the 19th century, while studying the fermentation of sugar to alcohol by yeast, Louis Pasteur came to the conclusion that this fermentation was catalysed by a vital force contained within the yeast cells called *ferments* that were thought to function only within living organisms.

In 1877, German physiologist Wilhelm Kühne first used the term *enzyme*, which comes from Greek word *enzymos*. The word *enzyme* was used later to refer to nonliving substances such as pepsin, and the word *ferment* was used to refer to chemical activity produced by living organisms.

In 1897, Eduard Buchner submitted his first paper on the ability of yeast extracts that lacked any living yeast cells to ferment sugar. He named the enzyme that brought about the fermentation of sucrose *zymase*. In 1907, he received the Nobel Prize in chemistry for his biochemical research and his discovery of cell-free fermentation. Following Buchner's example, enzymes are usually named according to the reaction they carry out. Typically, to generate the name of an enzyme, the suffix -*ase* is added to the name of its substrate (e.g. lactase is the enzyme that cleaves lactose) or the type of reaction (e.g. DNA polymerase for DNA polymers).

Many early workers noted that enzymatic activity was associated with proteins, but several scientists (such as Nobel laureate Richard Willstätter) argued that proteins were merely carriers for the true enzymes and that proteins per se were incapable of catalysis. However, in 1926, James B. Sumner showed that the enzyme urease was a pure protein and crystallized it. Sumner did likewise for the enzyme catalase in 1937. The conclusion that pure proteins can be enzymes was definitively proved by Northrop and Stanley (1930), who worked on the digestive enzymes pepsin, trypsin and chymotrypsin. These three scientists were awarded the 1946 Nobel Prize in chemistry.

This discovery that enzymes could be crystallized eventually allowed their structures to be solved by X-ray crystallography. This was first done for lysozyme, an enzyme found in tears, saliva and egg whites that digest the coating of some bacteria. The structure was solved by a group led by David Chilton Phillips in 1965. This high-resolution structure of lysozyme marked the beginning of the field of structural biology and the effort to understand how enzymes work at an atomic level of detail.

ENZYME STRUCTURE

The enzymes are proteins, and their function is determined by their complex structure. The reaction takes place in a small part of the enzyme called the *active site*, while the rest of the protein acts as scaffolding. This is shown in this diagram of a molecule of the enzyme trypsin, with a short length of protein being digested in its active site. The amino acids around the active site attach to the substrate molecule and hold it in position while the reaction takes place. This makes the enzyme specific for one reaction only, as other molecules would not fit into the active site (Fig. 17.1).

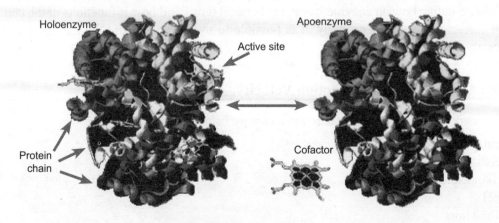

Figure 17.1 Structure of enzyme.

Many enzymes need cofactors or coenzymes to work properly. These can be metal ions (such as Fe^{2+}, Mg^{2+}, Cu^{2+}) or organic molecules (such as haem, biotin, FAD, NAD or coenzyme A). Many of these are derived from dietary vitamins. The complete active enzyme with its cofactor is called a *holoenzyme*, while just the protein part without its cofactor is called the *apoenzyme*.

ENZYME KINETIC

Kinetics is that branch of enzymology that deals with the factors that affect the rate of enzyme-catalysed reactions. *Enzyme kinetics* is the study of the chemical reactions that are catalysed by enzymes. In enzyme kinetics, the reaction rate is measured and the effects of varying the conditions of the reaction investigated. Studying an enzyme's kinetics in this way can reveal the catalytic mechanism of this enzyme, its role in metabolism, how its activity is controlled, and how a drug or a poison might inhibit the enzyme.

The kinetic studies on enzymes that only bind one substrate, such as triose-phosphate isomerase, aim to measure the affinity with which the enzyme binds this substrate and the turnover rate. When enzymes bind multiple substrates, such as dihydrofolate reductase enzyme kinetics can also show the sequence in which these substrates bind and the sequence in which products are released. Examples of enzymes that bind a single substrate and release multiple products are proteases, which cleave one protein substrate into two polypeptide products.

Importance of Enzyme Kinetics

1. It is essential for detailed study of an enzyme.
2. Influence of various factors on enzyme limits can be reflected in enzyme kinetics.
3. Catalytic function can be studied by measurement of the rate of the catalysed reaction.
4. It helps in understanding various biological phenomenon.
5. It helps define how an enzyme works in chemical terms and how it functions in the cell.
6. If factors are analysed properly, it is possible to totally understand the nature of enzyme-catalysed reaction.

Factors Affecting Enzyme Reaction Velocity

The factors influencing enzyme reaction velocity are listed below.

1. Enzyme concentration
2. Substrate concentration
3. Temperature
4. pH
5. Inhibitors
6. Activators

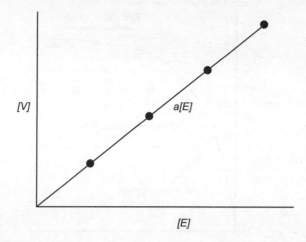

Figure 17.2 Effect of enzyme concentration on rate of enzyme reaction.

Effect of enzyme concentration on rate of enzyme reaction (Fig. 17.2): The rate of enzyme (E)-catalysed reaction is proportional to the enzyme concentration (provided S is saturating E).
$$v\,[E];\ v = k\,[E]$$
As E increases rate of reaction increases in a linear manner. However, some deviations occur, which are as follows:

Upward curve: In the beginning the rate is low, but as E concentration is increased rate increases. This is due to presence of a dissociable activator or coenzyme in the enzyme preparation. Binding of substrate A
$$E + A \rightarrow EA$$
$\uparrow E \rightarrow \uparrow A$ makes E active (e.g. some proteases).

This is due to presence of highly toxic impurity in the reaction mixture (not in E solution). So, when E is in small amount it is inhibited, but as its concentration is increased, it overcomes the toxic impurity and rate increases.

Downward curve: This is more commonly observed. As E concentration is increased beyond a certain point, the rate decreases. This may be due to the following:
1. A limitation in the capacity of the method of estimation. This is not a true decrease, but occurs as the assay method cannot give higher reading (e.g. in spectrophotometer the maximum OD is 2.0).
2. Substrate may be used up.
3. Presence of a reversible inhibitor in the enzyme preparation.
$$E + I \rightleftarrows EI$$
Where E = enzyme and I = inhibitor.

As E concentration increases, I increases and inhibits E.

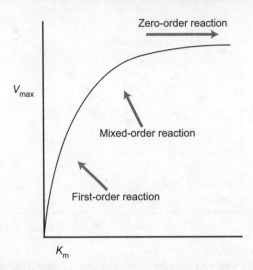

Figure 17.3 Effect of temperature on enzyme-catalysed reactions.

Effect of substrate concentration on the rate of enzyme: It is the most important factor in determining velocity of enzyme reaction.

1. At low substrate concentration [S], rate of reaction is low and a straight line is obtained. As S concentration is increased a mixed-order reaction is obtained and the curve reaction is obtained.
2. As S is increased further, the rate does not change and becomes constant. This is because E active sites are all filled and E is saturated with S. At this point, the velocity is equal to maximum velocity (V_{max}).
3. The S concentration at half V_{max} ($V_{max}/2$) is called *Michaelis constant* (K_m). This is a constant for an E and a specific substrate. It gives the affinity between E and S.
4. High K_m indicates low affinity while low K_m indicates high affinity.
5. However, some E do not obey Michaelis–Menten kinetics and do not give a hyperbolic curve, but give a sigmoidal curve. These are allosteric enzymes. These are regulatory enzymes and have quaternary structure.
6. When [S] is plotted versus V, the saturation curve is sigmoidal. This indicates cooperative binding of S to multiple sites. Binding of one site affects binding at other sigmoidal curve cooperative effect.

Allosteric E has multiple binding sites.

1. Active sites binds S and converts to product [P].
2. Modulatory site binds S and other modulatory molecules and this binding affects the activity of active site.

    Modulators may be +ve modulators → ↑activity, –ve modulators → ↓activity.

Effect of temperature on enzyme-catalysed reactions (Fig. 17.3): For example, human E: ~25–37°C.

DNA polymerase in Taq polymerase is active at a temperature up to 100°C.
1. As temperature is increased, rate of reaction increases.
2. At very low temperature, e.g. 0°C, the rate of reaction may be almost zero.
3. For every 10°C rise of temperature the rate is doubled. This is temperature coefficient (Q10).
4. This occurs as the kinetic energy of the molecules increases.
5. But this occurs only up to a specific temperature which is known as *optimum temperature*. Beyond this temperature, the rate decreases sharply. This occurs as the enzyme is denatured and the catalytic activity is lost.
6. For most E, optimal temperatures are at or slightly above those of the cell in which E occurs. Some E in bacteria, which survive in hot springs has high optimal temperature.

Effect of pH on enzyme-catalysed reactions: When E activities are measured at several pH values, optimal activities are generally observed between pH values of 5 and 9. However, some enzymes, such as pepsin has low pH optimum (2.0), whereas others have high pH optimum (e.g. alkaline phosphatase pH ~ 9.5).

The shape of pH activity curve is determined by the following:
1. Enzyme is denatured at high or low pH.
2. Alteration in the charge state of E or S or both pH can affect activity by changing the structure or by changing the charge on which functional S are binding.

For example, $Enz^{--} + SH = Enz.SH$

where, SH = charged substrate.
At low pH the enzyme is protonated and loses its negative charge,
$$Enz^- + SH = Enz - SH.$$
At high pH, the substrate loses its proton and positive charge
$$SH^+ \rightarrow S^{--} = H + So, Enz^{--} + S^{--} \rightarrow No\ reaction$$

Effect of inhibitors on rate of enzyme-catalysed reaction
1. Inhibitors (I) are substances that combine with E and decrease its activity.
2. Presence of I decreases the rate of E-catalysed reaction.
3. Inhibitors may be of different types as follows:

   a. *Irreversible inhibitor*
   $$E + I \rightarrow EI$$
   The inhibition increases with time. This inhibitor cannot be removed by dialysis or other means.
   1. Inhibition of succinate dehydrogenase by malonate. Inhibition of methanol dehydrogenase by ethanol.
   2. E may also undergo changes to conformation when pH is changed and this shall affect the activity of the E.

The examples of irreversible inhibitors are CN inhibits xanthine oxidase, nerve gas inhibits cholinesterase, iodoacetamide, heavy metal ions (Hg^{++}), oxidizing agents.

b. *Reversible inhibitors*

$$E + I \rightleftharpoons EI$$

The reaction is reversible and the inhibitor (I) can be removed by dialysis or other means. These are of three types: competitive, noncompetitive and uncompetitive.

Examples of reversible inhibitors:
1. Inhibition of methanol dehydrogenase by ethanol.
2. Inhibition of succinate dehydrogenase by malonate.
3. E may also undergo changes to conformation when pH is changed and this shall affect the activity of the E.

Activators
1. Some E requires activators to increase the rate of reaction.
2. Activators cause activation of E-catalysed reaction by either altering the velocity of the reaction or the equilibrium reached or both.
3. Essential activators: Essential for the reaction to proceed. These are recognized as substrate that is not change in the reaction, e.g. metal ion such as Mg^{++} for kinases.
4. Some E requires activators to increase the rate of reaction. Activators cause activation of E-catalysed reaction by either altering the velocity of the reaction or the equilibrium reached or both.
5. Nonessential activators: Activator may act to promote a reaction, which is capable of proceeding at an appreciable rate in the absence of activator.

BIOPRODUCTION OF ENZYMES

Microorganisms

There are number of enzyme producing microorganisms, which differ with respect to substrates (Table 17.1).

Table 17.1 Enzymes from various microorganisms

Enzyme	Microorganism
	Bacteria
Penicillinase	*Bacillus cereus*
α-Amylase	*Bacillus coagulans*
α-Amylase, protease	*Bacillus licheniformis*
Penicillin acylase	*Bacillus megaterium*
L-Asparaginase	*Citrobacter* spp.
Penicillin acylase	
	Escherichia coli
Pullulanase	*Klebsiella pneumonia*

	Actinomycetes
Glucose isomerase	*Actinoplanes* sp.
	Fungi
Urate oxidase	*Aspergillus flavus*
Amylase, protease, pectinase	*Niger*
Amylases, lipase, protease	*Aspergillus oryzae*
Esterase, invertase	*Aureobasidium pullulans*
Lipase	*Yarrowia lipolytica*
Tyrosinase	*Neurospora crassa*
Dextranase	*Penicillium funiculosum*
Glucose oxidase	*Penicillium notatum*
Lipase	*Rhizopus* sp.
Invertase	*Saccharomyces fragilis*
Cellulase	*Trichoderma reesei*
Cellulase	*Trichoderma viride*

Enzyme Properties

1. **Specificity of species**: The macromolecules, including proteins, are species specific. The enzyme types (protease, α-amylase, lactase), which are found in many species, shall have properties that vary as much as the other properties of the organisms. For example, protease of two closely related species differs in several ways in spite of some similarities. The variation in these molecules, it is believed, causes phylogenetic development involving microbial variation.

2. **Activity and ability variation**: The extracellular enzymes commonly employed are influenced externally by temperature, pH, etc. Unlike extracellular enzymes, the intracellular enzymes are little influenced by external environmental factors.

    The optimum stability and activity of extracellular enzymes are very much close to optimum conditions for microbial growth. The activity and stability of enzymes also differ. Xylose isomerase is stable in the pH range of 4.0–8.5, but shows optimum activity at pH between 5.5 and 7.0. Similarly, temperature also influences enzyme activity. The enzyme activity gradually increases with increase in temperature, but at certain stage, temperature inactivates the rate of reaction and at high temperature the enzyme is denatured, as it is proteinaceous in nature. Thermal stability in the target enzyme may be a useful attribute during production of enzyme itself, as heat may be used to destroy contaminant enzyme activity.

    The stability of enzymes is also influenced by following factors:
    a. High concentration of respective enzymes
    b. Presence of its substrate and/or product (e.g. α-amylase shows more stability in presence of starch)
    c. Presence of ions (e.g. α-amylase is denatured within 4 h in absence of Ca^{++})

d. Reduced amount of water content in reaction mixture, e.g. β-glactosidase results in production of glucose and galactose from hydrolysis of lactose in normal conditions. But the same enzyme produces some glucose and galactose and mixture of trisaccharides in abnormal conditions.
3. **Specificity of substrate**: Organic matter in a complex matrix contains various constituents such as cellulose, hemicellulose, lignin, etc. In nature, these are decomposed and mineralized by a variety of microorganisms. However, it is not possible for a single microbe to decompose all the constituents. It is also possible that a particular microbe secretes an enzyme in higher amount and utilizes the substrate more rapidly than others. This inherent capacity makes the microbes capable to compete in microbial competition for substrate utilization. Due to this high enzyme-producing ability, commercial exploitation of microorganisms is possible, e.g. the fungus, *T. reesei* secretes cellulase in high amount, which may be used for commercial production of cellulose.
4. **Inhibition and activation**: Some of the enzymes obtained from different sources show variation in responses to a given activator or inhibitor. For example, cobalt activates β-galactosidase isolated from bacteria and inhibits the same when it is obtained from fungi. Other examples of some activators of enzymes used commercially are proteins (for proteases), starch (for α-amylase), cellulose (for cellulase) and pectin (for pectinase).

ENZYME PRODUCTION

The following steps are involved in production of enzymes:
1. **Isolation, strain development and preparation**: The overall objective of developing suitable strain of microorganism is to (1) ensure production of enzyme in high amount and other metabolites in low amount, (2) to complete fermentation process in short time and (3) ensure utilization of low cost culture medium by microorganism. A microorganism is isolated on suitable culture medium following microbial techniques. Its enzyme producing ability is optimized by improving strains and formulating culture medium (pH and temperature). The strains of microorganisms are developed by using mutagens (mutagenic chemicals and ultraviolet light). Inoculum of enzyme producing strains developed after treatment of mutagens is prepared by multiplying its spores and mycelia in liquid broth.
2. **Formulation and preparation of medium**: The formulation of culture medium is undertaken in such a way that it should provide all nutrients required for enzyme production in high amount, but not for good microbial growth. An ideal nutrient medium should have a cheap source of carbon, nitrogen, amino acids, trace elements, growth promoters and little amount of salts. Enough care must be taken to maintain desired pH during fermentation. The conditions of pH, temperature and ingredients of culture medium are optimized for a specific microbe prior to inoculation. It is observed that enzyme production is accelerated with enriched culture medium with following typical constituents used for enzyme fermentation:
    a. *Carbohydrates:* Molasses, barley, corn, wheat and starch hydrolysate.
    b. *Proteins:* Meals of soybean, peanut, cotton seed, corn steep liquor and yeast hydrolysate. Central Food Technological Research Institute (CFTRI), Mysore, has developed

technology for conversion of tapioca starch to glucose using fungal enzymes. Enzymes have also been isolated from bacterial cultures which convert glucose to fructose in starch hydrolysate.

3. **Sterilization of medium, maintenance of culture and fluid**: The culture medium is usually sterilized batchwise in a large-size fermenter using continuous sterilization method. The sterilized medium is then inoculated with sufficient amount of inoculum to initiate fermentation process, which is identical to production of antibiotic. Traditionally, the surface culture technique where inoculum remains on upper surface of broth is the method of enzyme production. This technique is in use for production of some of the fungal enzymes, for example, amylase (from *Aspergillus* sp.), protease (from *Mucor* sp. and *Aspergillus* sp.) and pectinase (from *Penicillium* sp. and *Aspergillus* sp.).

Nowadays submerged culture method is most widely practised because of possibility for more yield of enzymes and least chance of contamination. The parameters such as pH, temperature and oxygen are maintained in fermenter at optimum level. The requirements of these parameters differ from microbe to microbe and even in the same species of a microbe. A little amount of oil is added to fermenter to control foaming during fermentation. The extracellular enzymes are produced by the inoculated microbe in culture medium after 30–150 h of incubation. Most of the enzymes are produced on completion of exponential phase of growth, but in a few cases, they are produced during exponential phase. After the fermentation is over, the broth is kept at 5°C to avoid contamination. The recovery of enzymes from the fermented broth (fluid) of bacterium is more difficult than that of filamentous fungus. The broth is directly filtered or centrifuged after pH adjustment. The bacterial broth is usually treated with calcium salts to precipitate calcium phosphate, which helps in separation of bacterial cells and colloids. The liquid is then filtered and centrifuged to remove cell debris.

The extracellular enzymes and other metabolites (10–15%) are also produced in the fermented broths, which are removed after enzyme purification.

4. **Enzyme purification**: It is a complex process and it involves following steps:
   a. Preparation of concentrated solution under vacuum at low temperature or by ultrafiltration
   b. Clarification of concentrated enzyme by a polishing filtration to remove other microbe
   c. Addition of preservatives or stabilizers such as calcium salts, proteins, starch, sugar, alcohols, sodium chloride (18–20%), sodium benzoate, etc.
   d. Precipitation of enzymes with acetone, alcohol or organic salt (ammonium sulphate or sodium sulphate)
   e. Drying the precipitate by freeze drying, vacuum drying or spray drying
   f. Packaging for commercial supply

ENZYME IMMOBILIZATION

Many enzymes are produced by microorganisms on a large scale and their production is cost effective even if they are used only once. However, more enzymes are such that they affect the cost of their production and may not be economical, if not reused. This necessity of reuse of enzymes led to the development of immobilization techniques, which basically involve process conversion of water-soluble enzyme protein into a solid form of catalyst using several methods. It is only possible to immobilize microbial cells by similar techniques. Immobilization is the imprisonment or arresting of an enzyme in a distinct phase that allows exchange with, but is separated from the bulk phase in which the substrate, inhibitor or effector molecules are dispersed and monitored. The first commercial application of immobilized enzyme technology was reported from Japan in 1969 with the use of *A. oryzae* amino acylase for the industrial production of L-amino acids. Consequently, pilot plant processes were reported for production of 6-aminopenicillanic acid (6-APA) from penicillin G and for conversion of glucose to fructose by immobilized glucose isomerase.

Advantages of Immobilized Enzymes

The advantages of immobilized enzymes are as follows:
1. They can be reused.
2. The production is less labour intensive.
3. The continuous use is possible.
4. There are less chances of contamination with their use.
5. The reaction time required is minimized.
6. The saving in capital cost is possible.
7. High enzyme: substrate ratio can be achieved.
8. They are more stable as compared to conventional means.
9. Improved process control is a reality.

The first immobilized enzymes to be scaled up for industrial production are immobilized amino acid acylase, penicillin G acylase and glucose isomerase. Some other industrially important enzymes produced by this technique are aspartase, esterase and nitrilase.

Techniques of Enzyme Immobilization (Fig. 17.4)

The different techniques of enzyme immobilization are (1) physical immobilization, (2) chemical immobilization, (3) adsorption, (4) covalent bonding, (5) entrapment, (6) copolymerization or crosslinking and (7) encapsulation.

Physical Immobilization
It forms no covalent bonds between the enzymes and the supporting matrix. Earlier approaches include the adsorption of the enzyme onto animal charcoal or alumina, but current advancements make use of ionic adsorption technique onto ion-exchange resins, especially those of the sephadex-type and controlled-pore glass.

Advantages
1. Simplicity, general applicability and high yield.
2. It confers ability to replace the immobilized enzyme when its catalytic activity is decreased below an acceptable level.

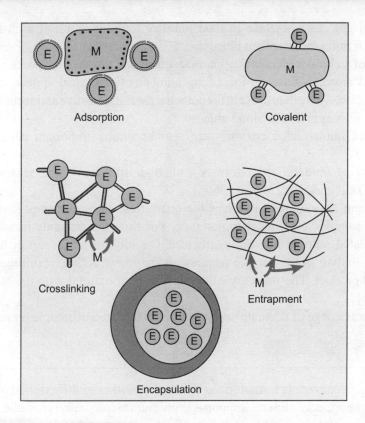

Figure 17.4 Immobilization of enzymes.

Limitations

1. It needs to control the working conditions for the use of immobilized enzyme to prevent its desorption.
2. Products of enzyme entrapment in liposomes (artificially produced concentric spheres of phospholipids bilayers), polyacrylamide and agarose suffer from poor flow properties, inefficiency and progressive leaching of the enzyme.
3. Desorption of the protein resulting from changes in temperature, ionic strength and hydrogen ion concentration.

Chemical Immobilization

The covalent bond formation between the enzyme and the matrix procedures are identical to those used in affinity chromatography. A loss of activity would be observed if attachment involves amino acid residues at the active site of the enzyme. The matrix may be polysaccharide or polymers (nylon or inorganic carriers, such as glass and titanium dioxide).

Applications

The first reported application is the production of L-amino acids from immobilized aminoacylase. The other applications are as follows:

1. Immobilized glucose isomerase in food industry for production of high fructose syrup as fructose is 16 times sweeter than glucose.
2. Production of a chiral molecule for pharmaceutical and research purposes.
3. Immobilized enzymes have an increasing number of analytical applications, especially in clinical situation, where they differ in the potential for fast, sensitive and accurate determinations of analytes, such as urea and blood glucose.
4. A number of immobilized enzymes are also becoming important parts of a variety of biosensors.
5. Combination of immobilized enzymes with high specificity and sensitivity, and use in electroanalytical chemistry is possible.
6. Another recent development involves the quantification of enzymes, substrates and other analytes in samples of clinical importance. For this, the reagents including the enzyme are impregnated onto an immobile structure. On addition of a drop of test sample to the matrix, the analyte in the sample triggers an enzyme reaction, resulting in the formation of coloured product. The intensity of colour can be measured with the help of reflectance spectroscopy.
7. It offers an advantage of accurate analysis without the need of sample preparation.

Adsorption

In this technique, an enzyme is immobilized by bonding either to the external or internal surface of a carrier or support, e.g. mineral support (aluminium oxide, clay), organic support (starch), modified sapharose and ion-exchange resins. In this process, bonds of low energy, such as ionic interactions, hydrogen bonds, van der Waals forces, etc., are involved. If the enzyme is immobilized externally, the carrier particle size must be very small (500 Å to 1 mm) in order to achieve an appreciable surface of bonding. In this technique of external immobilization, no pore diffusion limitations are encountered. The enzyme immobilized on an internal surface is protected from abrasion, inhibitory bulk solutions and microbial attack, and a more stable and active enzyme system can be achieved. The pore diameters of carriers may be optimized for internal surface immobilization.

Immobilization by adsorption is achieved by the following procedures:
1. **Static process**: The solution containing the enzyme is allowed to contact the carrier without stirring for immobilizing the enzyme.
2. **Dynamic batch process**: In this process, the carrier is placed into the enzyme solution and mixed by stirring or agitated continuously in a shaker.
3. **Reactor loading process**: The carrier is placed into the reactor, then the enzyme solution is transferred to the reactor and the carrier is loaded in a dynamic environment by agitating the carrier and enzyme solution together.
4. **Electroposition process**: The carrier is placed proximal to one of the electrodes in an enzyme bath, and the current is put on. The enzyme migrates to the carrier, and it is deposited on the surface and immobilization takes place.

5. **Covalent bonding**: It is formed between the chemical groups of enzyme and the surface of carrier. Covalent bonding is utilized under a broad range of pH, ionic strength and other variable conditions. Immobilization steps are attachment of coupling agent followed by an activation process or attachment of a functional group, and finally attachment of the enzyme. Different types of carriers are used in this technique of immobilization, such as carbohydrates, proteins, inorganic carriers, amine-bearing carriers, etc. Covalent attachment may be directed to a specific group (e.g. amine hydroxyl, tyrosyl, etc.) on the surface of the enzyme. Hydroxyl and amino groups are the main groups of the enzyme with which bonds are formed. Different methods of covalent bonding are as follows:

   a. *Diazotization:* It refers to bonding between the amino group of the support (e.g. aminobensyl cellulose, aminosilanized porous glass, amino derivatives) and a tyrosyl or histidyl group of the enzyme.
   b. *Formation of peptide bond:* It refers to formation of bond between the amino or carboxyl group of the support and amino or carboxy group of the enzyme.
   c. *Group activation:* Cyanogen bromide is used for bonding to a support containing glycol group, i.e. cellulose, syphadex, sepharose, etc.
   d. *Polyfunctional reagents:* It involves use of a bifunctional or multifunctional reagent, such as glutaraldehyde, that forms bonding between the amino group of the support and amino group of the enzyme.

   The major problem encountered with covalent bonding is that the enzyme may be inactivated due to changes in conformation while undergoing reaction at active site. However, this problem may be overcome through immobilization in presence of enzyme's substrate or a competitive inhibitor or protease. The most commonly used activated polymers are celluloses or polyacrylamides onto which diazo, carbodiimide or azide groups are incorporated.

6. **Entrapment**: The enzymes can be physically entrapped inside a matrix (support) of water soluble polymer (e.g., polyacrylamide-type gels) and naturally derived gels (e.g. cellulose triacetate, agar, gelatin, carrageenan, alginate, etc.). The pore size of matrix is adjusted to prevent loss of enzyme from the matrix due to excessive diffusion. There is possibility of leakage of low molecular weight enzymes from the gel. Agar and carrageenan have large pore sizes (<10 μm).

   Different methods for enzyme entrapment used are as follows:
   a. *Gel entrapment:* In this process, the enzyme is entrapped in gels.
   b. *Fibre entrapment:* The enzyme is entrapped in fibre format.
   c. *Microcapsule entrapment:* The enzymes are entrapped in microcapsules of monomer mixtures, such as polyamine and polybasic chloride, polyphenol and polyisocyanate. The entrapment of enzymes has been widely used for sensing application, but not much success has been achieved with industrial process.

7. **Crosslinking or copolymerization**: It is widely used in commercial applications. It is characterized by covalent bonding between various molecules of an enzyme via

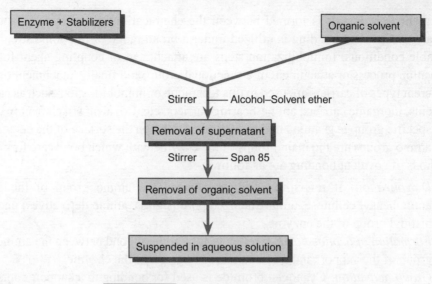

Figure 17.5 Encapsulation of enzyme.

a polyfunctional reagent, such as glutaraldehyde, diazonium salt, hexamethylene disocyanate, and N-N' ethylene bismaleimide. The polyfunctional reagent is capable of denaturing the enzyme. This technique is simple and economical, but not often used with pure proteins, because it produces very little of immobilized enzyme that has very high intrinsic activity.

8. **Encapsulation**: Chent has described the method of encapsulation in 1977. It is the process of enclosing a droplet of solution of enzyme in a semipermeable membrane capsule made up of cellulose nitrate and nylon. The method of encapsulation is simple and economical. Although, the catalyst is very effectively retained within the capsule, the effectiveness of this technique largely depends on stability of enzyme. It has limited applications to medical sciences. In this method, a large quantity of enzyme is immobilized, but main limitation of this technique is that only small substrate molecule is utilized with the intact membrane (Fig. 17.5).

Cell Immobilization

This technique is now well developed and successfully used for industrial-scale production. In cell immobilization technology, the enzymes are active and stable for a long period of time.

The methods of cell immobilization is same as described for enzyme immobilization, involving the processes of adsorption, covalent bonding, cell-to-cell crosslinking, encapsulation and entrapment in polymeric network. Since a long time, adsorption of cells to preformed carrier has been undertaken. The cell attachment to the surface of preformed carrier is done by covalent bonding. The woodchips are used as carrier for *Acetobacter* in production of vinegar since 1823. Preformed carrier of specific choice is used (Table 17.2).

Table 17.2 Whole-cell immobilization

Support material	Cell	Reaction
Gelatin	*Lactobacilli*	Lactose
Glass	*Saccharomyces*	Glucose/ethanol
Cotton	*Zymomonas mobilis*	Glucose/ethanol
Cellulose	*Nocardia erythropolis*	Steroid conversion
Carboxymethyl cellulose	*Bacillus subtilis*	L-Histidine/uronic acid
Diazotized diamines	*Streptomyces*	Glucose/fructose
Glutaraldehyde	*E. coli*	Fumaric acid/L-aspartic acid
Chitosan	*Lactobacillus brevis*	Glucose/ethanol
Aluminium alginate	*Candida tropicalis*	Phenol degradation
Calcium alginate	*Saccharomyces cerevisiae*	Glucose/ethanol
Carrageenan	*E. coli*	Fumaric acid/L-aspartic acid
Polyester	*Streptomyces* sp.	Glucose/fructose
Alginate–polylysine	Hybridoma cells	Monoclonal antibodies

IMMOBILIZATION OF PLANT CELLS

The plant cells are immobilized in alginate, where they have been shown to retain their biological activity. Such systems can be utilized for bioconversions.

The substance utilized for the production of pharmaceuticals and food additives are mostly secondary metabolites. They are often present in very low concentration in plant tissues. Large amounts of plant materials are required for the isolation of given substance in substantial amount. Nevertheless, a great many pharmaceuticals are at present produced from higher plants, such as steroids, alkaloids, glycosides, isoprenoids and tannins.

Method of Immobilization

The secondary metabolites of viable plant cells are produced with the cells immobilized in a porous inorganic support. Immobilization includes the following steps:

1. Preparing a support comprising a substantially uniform and porous matrix of inorganic material having a tensile strength of at least 500 MPa
2. Introducing a culture of viable plant cells into the pores of said matrix
3. Entrapping the plant cells by coating the matrix with a sol or colloidal suspension not interfering with the cell viability
4. Immobilizing the entrapped cells within the matrix with a reactive gas, including a carrier gas saturated with volatile SiO_2 or organic modified SiO_2 precursors

The tensile strength may be provided by impregnating the matrix with a gelling solution of SiO_2 precursor for increasing stiffness of the matrix. The matrix may be an SiO_2 or inorganic oxide matrices, in which the weight ratio between cell load and inorganic matrix ranges between

1×10^{-4} and 1×10^{-2}. The immobilized cells are not released in solution over a period of 6 months and maintain their viability while producing secondary metabolites.

The biosynthetic capability of one of the immobilized species, *Coronilla vaginalis*, was studied by periodically monitoring the production of umbelliferone and marmesin that constituted the major secondary metabolites produced by in vitro cultured cells of this species. The results were evaluated in order to determine the versatility of the method and its potential for exploitation in continuous industrial-scale production of rare and fine chemicals.

Immobilized Plant Cell Reactors

The use of plant cells for the production of chemicals and therapeutics represents a new area of biotechnological exploration. The techniques envisioned for industrial processes are related to those developed for microorganisms. A strong emphasis is placed on immobilized cell systems. The spectrum of products that are synthesized by higher plants and the immobilization techniques that are suited to entrap plant cells from suspension culture are described from time to time. Different reactor configurations are also explained. Both packed bed reactors with alginate-entrapped cells and hollow fibre cartridges with sequestered cells have utility for the continuous production of biochemicals.

Immobilization of Bacterial Cell

Immobilization of bacteria by entrapment within a gel bead or by attachment on a surface greatly improves bioreactor performance. When compared to their free-cell counterparts, immobilized cell bacterial bioreactors have higher productivity per unit. The bacterial biomass can sustain higher liquid flow rates, occupy less space and are more resistant to contamination by microbes, as well as by exposure to high concentrations of chemical pollutants.

The yeast cells are entrapped in calcium alginate gels by using the similar techniques as in enzyme immobilization. Other cell entrapment media that have been previously attempted include polyacrylamide, gelatin, chitosan and k-carrageenan gels.

The immobilized cell reactor is employed to convert glucose into ethanol anaerobically. The reasons for choosing this system of microorganism and product are many folds. First, the anaerobic condition eliminates the need for aeration that causes many technical problems. Secondly, the lack of oxygen prevents the uncontrolled growth of aerobic contaminants in an unsterilized fermenter. The presence of high levels of ethanol also discourages most microorganisms from taking over the fermenter. To reduce further the chance of contamination by bacteria, the pH of the fermenter is kept low. The pH of 4.0 drastically slows down the growth of most bacteria, but only slightly affects the ethanol-producing capacity of the yeast. The production of ethanol in an immobilized bioreactor is a relatively well-studied process. As high as 95% of the theoretical yield of alcohol based on glucose (8.5% ethanol from 14% glucose) has been reported. A high space velocity, defined as the volume of nutrient feed per hour per gel volume of 0.4–0.5 h^{-1}, is commonly used to maximize the ethanol productivity. An ethanol productivity of 20 g/L per hour can be achieved.

Procedural Details
Immobilized cell preparation

1. About 9 g of sodium alginate is dissolved in 300 mL of growth medium, following the same procedure adopted in enzyme immobilization to avoid clump formation. It is then stirred until all sodium alginate is completely dissolved. The final solution contains 3% alginate by weight.
2. About 250 g of wet cells are suspended in the alginate solution prepared in the previous step. The air bubbles are allowed to escape.
3. The yeast–alginate mixture is dripped from a height of 20 cm into 1000 mL of crosslinking solution. The crosslinking solution is prepared by adding an additional 0.05 M of $CaCl_2$ to the growth medium. The calcium crosslinking solution is agitated on a magnetic stirrer.
4. Gel formation is achieved at room temperature, as soon as the sodium alginate drops come in direct contact with the calcium solution. Relatively, small alginate beads are preferred to minimize the mass transfer resistance. A diameter of 0.5–2 mm can be readily achieved with a syringe and a needle.
5. The beads are allowed to fully harden. It takes around 2 h. The concentration of $CaCl_2$ is about one-fourth of the strength used for enzyme immobilization.
6. The beads are washed with a fresh calcium crosslinking solution and dried.

Immobilized cell reactor construction and operation

The immobilized cell reactor is constructed with a 500 mL Erlenmeyer flask. The hardened beads are placed in the flask, and it is then sealed with a rubber stopper with appropriate hose connections. All necessary connections are made, and the experiment is started by filling the flask with the growth medium (100 g/L glucose) to the working volume of 350 mL.

Following sequence of events is monitored both online and offline. The responsibilities of online data acquisition and offline sample collection along with analysis are shared. A microcomputer is programmed to take data on the glucose concentration and the rate of NH_4OH addition needed to maintain the pH at 4.0. The offline samples are analysed for the optical density (for free-cell concentration), glucose concentration and ethanol concentration. Furthermore, the liquid and gas flow rate are measured with a graduated cylinder as indicated.

The reactor is operated in a batch manner until no more glucose is utilized. This can be detected with the levelling off in the glucose concentration.

Substrate feeding is commenced at the rate of 0.4 per hour. The substrate flow rate is then recorded. The approach to the first steady state during the startup is followed.

Various parameters (nitrogen consumption rate, carbon dioxide evolution rate, ethanol concentration, glucose concentration and free-cell level) at the high steady state are recorded.

The substrate feeding rate is decreased to 0.2 per hour. The substrate flow rate is measured, and the transient approach is followed to the new low steady state. The flow rate is shifted, and more information on steady states is obtained. The bioreactor is operated until noticeable deterioration in the performance is detected due to gel swelling, cell death or severe contamination.

Some Important Enzymes
Amylase

It is an enzyme that breaks starch down into sugar. As diastase, amylase was the first enzyme to be discovered and isolated by Anselme Payen in 1833. It is classified as α-amylase and β-amylase. Amylase is present in human saliva, where it begins the chemical process of digestion. Foods that contain much starch but little sugar, such as rice and potato, taste slightly sweet as they are chewed because amylase turns some of their starch into sugar in the mouth. The pancreas also makes amylase (α-amylase) to hydrolyse dietary starch into disaccharides and trisaccharides that are converted by other enzymes into glucose to supply the body with energy. The plants and some bacteria also produce amylase.

It is used in bread making and to break down complex sugars such as starch (found in flour) into simple sugars. Yeast then feeds on these simple sugars and converts it into the waste products of alcohol and CO_2. This imparts flavour and causes the bread to rise. While amylase enzymes are found naturally in yeast cells, it takes time for the yeast to produce enough of these enzymes to break down significant quantities of starch in the bread.

When used as a food additive, amylase has enzyme number *E1100*, and may be derived from swine pancreas or mould mushroom. Bacillary amylase is also used in clothing and dishwashing detergents to dissolve starches from fabrics and dishes. Workers in factories that work with amylase for any of the above uses are at increased risk of occupational asthma.

Hyaluronidases

Hyaluronidases are a family of enzymes that degrade hyaluronic acid. Hyaluronidases lower the viscosity of hyaluronic acid, thereby increasing tissue permeability. They are therefore used in medicine in conjunction with other drugs to speed their dispersion and delivery. The most common application is in ophthalmic surgery, in combination with local anaesthetics. Brand names of some animal-derived hyaluronidases include Hydase (developed and manufactured by Prima Pharm, Inc., distributed by Akorn, Inc.), which has been FDA approved as a 'thimerosal-free' animal-derived hyaluronidase, Vitrase (ISTA Pharmaceuticals) and Amphadase (Amphastar Pharmaceuticals).

Protease

Protease is the single class of enzymes that occupy a pivotal position due to its wide applications in detergent, pharmaceutical, photography, leather, food and agricultural industries. An important biotechnological application of protease is in bioremediation processes. Among the various proteases, bacterial proteases are the most significant, compared with animal and fungal proteases. *Bacillus* species are specific producers of extracellular proteases. These proteases have wide applications in pharmaceutical, leather, laundry, food and waste-processing industries. Thermostable proteases are advantageous in some applications, because higher processing temperatures can be employed. Global requirements of thermostable biocatalysts are far greater than those of the mesophiles of which proteases contribute two-thirds.

B. subtilis is one of the most widely used bacteria for the production of specific chemicals and industrial enzymes and also a major source of amylase and protease enzymes. Proteases are also used in baking, brewing, meat tenderization, peptide synthesis, medical diagnosis, cheese making, certain medical treatments of inflammation and virulent wounds and in unhairing of sheep skins.

Streptokinase

The substance that causes fibrinolysis, by the growth of hemolytic streptococci, was discovered by Tillett and Garner in 1938. The active principle has been purified and concentrated by Christensen. This substance is an enzyme and has been given the name streptokinase. It is readily separable from the bacterial cells by centrifugation or filtration and is plentiful in cell-free material, from which it may be isolated in a partially purified and highly concentrated form. The enzymatic activity is abundantly present in human group A haemolytic *streptococci*. Streptokinase and streptodornase are derived from the Lancefield Group C, H 46 A strain. This organism is nonpathogenic to human beings.

Penicillinase

Many kinds of penicillin biosensors have been developed so far. In general, enzymes are immobilized to the pH-sensitive material by the use of crosslinking or enzyme entrapment methods. In this method, the enzyme penicillinase is adsorptively immobilized directly onto the tantalum pentoxide surface. For the immobilization, the enzyme solution is prepared by dissolving the enzyme penicillinase in a 200 mM triethanolamine (TEA) buffer, pH 8. 10-µL enzyme solution per sensor is pipetted onto the samples and incubated at room temperature for about 1–2 h. After rinsing and drying at room temperature, the sensors are stored in 0.2 mM multicomponent polymix buffer pH 8, at 4°C.

The main advantages of the adsorptive immobilization procedure are the cheapness of this method without any loss of the enzyme activity, the simplicity of the technique and the possibility of sensor (enzyme) regeneration, i.e. the possibility of a repeated enzyme immobilization in case of loss of the enzyme activity.

ENZYME ENGINEERING

The term *enzyme engineering* is used to denote the modification of structure of enzyme by alteration of genes that code enzymes. An enzyme produced by a modified gene is structurally new, and it has great promise to create a new enzyme. Enzyme engineering covers the knowledge of (1) production of new enzymes, (2) structural features related to stability, (3) increase in stability by changing amino acid composition and (4) production of stable enzymes by genetically engineered microbial cells.

Since genes encode enzymes, the changes in gene certainly bring about alteration in structure of enzyme. The technique of alteration of genes by site directed mutagenesis has become much popular in enzyme engineering. It produces single amino acid substitution in the primary structure of enzymes.

APPLICATIONS OF ENZYMES

Enzymes in general are limited in the number of reactions they have evolved to catalyse and also by their lack of stability in organic solvents and at high temperatures. Consequently, protein engineering is an active area of research and involves attempts to create new enzymes with novel properties, either through rational design or in vitro evolution.

The enzymes and cells have wide spectrum of applications that can be grouped into four broad categories: (1) therapeutic, (2) analytical, (3) manipulative and (4) industrial.

1. **Therapeutic applications**: Some inborn errors of metabolism occur due to missing of enzyme where specific genes are introduced to encode specific missing enzymes. Certain diseases are treated by administering the appropriate enzyme. For example, virilization of a disease developed due to loss of hydroxylase enzyme from adrenal cortex and introduction of hydroxyl group (–OH) on 21-carbon in ring structure of steroid hormone. The missing enzyme synthesizes aldosterone in excess leading to masculinization of female baby and precocious sexual activity in males within 5–7 years. Similarly, treatment of leukaemia (a disease in which leukaemic cells require exogenous asparagines for their growth) is possible by administering asparaginase of bacterial origin.
2. **Analytical applications**: The enzymes are used in kinetic analysis. Endpoint analysis refers to total conversion of substrates into products in a few minutes in the presence of enzymes, whereas kinetic analysis involves the rate of reaction and substrate/product concentration. The analysis of antibodies and immunoglobins, necessary for human use, is also possible with enzymes. The usable enzymes are alkaline phosphatase, β-galactosidase, lactamase, etc. Another important utility of enzyme is in biosensor. It is a device of biologically active material displaying characteristic specificity with chemical or electronic sensor to convert a biological compound into an electronic signal. A simple carbon electrode, an ion sensitive electrode, oxygen electrode or a photocell may be used as a biosensor.
3. **Manipulative applications**: A variety of enzymes isolated from different sources are used in genetic engineering as one of the biological tools.
4. **Industrial applications**: The industrial uses of enzymes may be broadly categorized as follows:
    a. *Detergent industry:* The stains on cloth can be easily removed by adding proteolytic enzyme to the detergent. The enzyme attacks on peptide bonds and therefore, dissolves protein. The alkaline serine protease obtained from *B. licheniformis* is commonly used in manufacture of detergent. In addition, the serine protease of *Bacillus amyloliquefaciens* that contains α-amylase is also extensively used.
    b. *Starch industry:* Currently, various enzymatic processes are successfully applied for different products. Glucose isomerase is an important enzyme used commercially in conversion of glucose to fructose via isomerization. Fructose is used for fructose syrup

preparation. The enzyme glucose isomerase is widely used in production of fructose syrup.

The reaction mixture at the end contains 42% fructose, 52% glucose and 6% dextrins. The mixture is sweeter than glucose and as sweet as sucrose. The technique has been developed to obtain 55% fructose concentration in syrup.

c. *Rubber industry:* Catalase is used to generate oxygen from peroxide to convert latex into foam rubber.

d. *Photographic industry:* Protease (ficin) dissolves gelatin of scrap film allowing recovery of its silver content.

e. *Dairy industry:* For a long time, calf rennet has been used in dairy industry. In recent years, calf rennets are replaced by microbial rennets (e.g. *Mueor michei*) that are acid aspartate proteases. Lactase produced by *Bacillus stearothermophilus* is used for hydrolysis of lactose in milk. The enzyme lipase is used for flavour development in cheeses.

f. *Brewing industry:* The enzymes commonly used in brewing industry are α-amylase, β-glucanase and protease that are required for malt in substitution of barley. The biological source for these enzymes is *B. amyloliquefaciens*. The neutral protease helps in the inhibition of alkaline protease by an inhibitor, β-glucanase that takes care of filtration problems due to poor quality of malt.

g. *Wine industry:* The pectic enzymes such as pectin *trans*-eliminase (PTE), polymethyl galacturonase (PMG), polygalacturonase (PG), pectinesterase (PE), etc. are used in wine industry for high yields with improved quality. The peptic enzymes give good result when combined with other enzymes, e.g. protease glucoamylase.

h. *Pharmaceutical industry:* The enzyme penicillin ON acylase is widely used in the production of semisynthetic penicillins. All penicillins consist of an active β-lactam ring, i.e. 6-aminopenicillanic acid (6-APA) group combined with different side chains (R group). The enzyme penicillin ON acylase removes ON group resulting in the separation of 6-APA and R groups. Finally, new synthetic side chains are coupled with 6-APA to synthesize new semisynthetic penicillins.

Enzyme reaction is represented as follows:

$$\text{Penicillin G} \xrightarrow{\text{Penicillin G acylase}} \text{6-APA + G side chain}$$

$$\text{Penicillin V} \xrightarrow{\text{Penicillin V acylase}} \text{6-APA + V side chain}$$

$$\text{6-APA} \xrightarrow{\text{Side chain addition}} \text{Semisynthetic penicillin}$$

The *E. coli* strains are the most explored and exploited ones for production of penicillin G acylase. The biosynthesis of penicillin acylase in *E. coli* is controlled by alterations in concentrations of nutrients and culture conditions. Sudhakaran and Berkar (1989) reported the effect of growth substrate, inducers and regulators on formation of enzyme. The strain *E. coli* NCJM-2400 produced penicillin G acylase intracellularly when grown in nutrient broth containing phenylacetic acid (PAA). It was observed that PAA (20 mM) stimulated enzyme synthesis by 8–10 folds. Phosphate and yeast extract were found essential for both the growth and enzyme biosynthesis, whereas glucose, sorbitol, lactose, acetate and lactate (all 0.1%) catabolically repressed the enzyme formation.

Penicillin V acylase occurs in fungal and actinomycetes sources. However, its activity has also been reported in many bacteria, such as *Bacillus sphaericus*, *Erwinia aroideae* and *Pseudomonas acidovorans*. Lowe et al. (1986) have identified a strain of *Fusarium oxysporum* that exhibited intracellular penicillin V acylase activity, which was induced by phenoxyacetic acid in culture. The enzyme was partially purified and concentrated from disrupted cells (cells hydrolysed with 5% penicillin V solution) by fractional precipitation with miscible solvents.

i. *Paper industry:* Amylases, xylanases, cellulases and ligninases degrade starch to lower viscosity, aiding sizing and coating paper. Xylanases reduce bleach required for decolourizing. Cellulases smoothen fibres, enhance water drainage and promote ink removal. Lipases reduce pitch and lignin-degrading enzymes remove lignin to soften paper.

BIOSENSORS

A biosensor is an analytical device consisting of an immobilized layer of biological material (e.g. enzyme, hormone, antibody, nucleic acid, organelle or whole cells) in intimate contact with a transducer, i.e. sensor (a physical component) that analyses the biological signals and converts them into an electrical signal (Gronow, 1984). A sensor can be a single carbon electrode, an ion-sensitive electrode, a photocell, an oxygen electrode or a thermistor.

The principle of the biosensor is simple. The biological material is immobilized on the immobilization support permeable membrane. The substance to be measured passes through the membrane and interacts with the immobilized material to yield the product. A product (i.e. the product-monitored substrate) may be gas (oxygen), heat, hydrogen ions, electrons or ammonium ions. The product passes through another membrane to transducer. The transducer converts product into an electric signal, which is then amplified. The signal-processing equipment converts the amplified signals into a display, most commonly the electric signal, which can be read out and recorded.

The glucose electrode is built up by immobilizing glucose oxidase in polyacrylamide gel around a platinum–oxygen electrode, which is separated by a teflon membrane. Potassium chloride solution surrounds the platinum–oxygen electrode. From upper surface, glucose oxidase is intimately covered by a cellulose acetate membrane. When glucose solution is brought into contact with membrane, glucose and oxygen pass through membrane into the enzyme layer, and as a result of oxidation–reduction reaction, convert into gluconic acid and hydrogen peroxide in presence of water, oxygen and glucose oxidase. Consequently, concentration of oxygen in the gel around electrode is lowered down. Hydrogen peroxide brings about a change in current, i.e. measurable signal. The electrode records the rate of reaction. The rate of diminution of oxygen concentration is proportional to glucose concentration of the sample. It responds linearly to glucose concentrations over a range of 10^{-1}–10^{-5} mol dm^{-3} with a response time of 1 min. Yellow Springs Instruments Co., USA in 1987 developed the first biosensor for diagnostic purposes useful in measuring glucose in blood plasma. It is a machine that measures six components of blood plasma (glucose, urea, sodium, nitrogen, potassium and chloride).

An indigenous glucose sensor capable of giving electrical signal for a glucose concentration as low as 0–15 mM has been developed by the scientists at Central Electrochemical Research Institute (CECRI), Karaikudi.

Types of Biosensors

Biosensors are of different types based on the use of biological material and sensor devices; a few of them are discussed below.

1. **Electrochemical biosensor**: It is developed using electronic devices, such as field effect transmitters or light-emitting diode. The former measures charge accumulation on its surface, whereas the latter measures photo response generated in silica-based chip as an alternating current. Hence, the field effect transmitter measures a biochemical reaction at the surface and induces into current (Gronow et al., 1988). Moreover, the field effect transmitters can be modified to ion-sensitive, enzyme-sensitive or antibody-sensitive ones by using selective ions, enzymes or antibodies, respectively.

2. **Enzyme electrode**: It is a new type of biosensor designed for the amperometric or potentiometric assay of substrates, such as glucose, urea, alcohol, amino acid and lactic acid. The electrode consists of an electrochemical sensor in close contact with a thin permeable enzyme membrane capable of reacting with the given substrate. The enzyme is embedded in the membrane and produces O_2, H^+, NH_4^+, CO_2 or other small molecules depending on enzymatic reaction, which is then detected by the specific sensor. The magnitude of the response determines the concentration of substrates (Table 17.3).

3. **Amperometric biosensor**: This biosensor measures the reaction of analyte with enzyme and generates electrons directly or through a mediator. The amperometric biosensor contains either enzyme electrode or chemically modified electrodes. The oxygen and peroxide-based

Table 17.3 Enzyme-based biosensors

Substance	Enzyme	Response time	Range
Amines	Monoamine oxidase	4 min	50–200 µ mol dm^{-3}
Cholesterol	Cholesterol oxidase	2 min	10^{-2} to 3×10^{-5} mol dm^{-3}
Carbon monoxide	CO: acceptor	15 s	0–65 µ mol dm^{-3}
Glucose	Glucose oxidase	20 s	2×10^{-3} to 3×10^{-6} mol dm^{-3}
Penicillin	Penicillinases	25 s	1–10 m mol dm^{-3}
Sucrose	Invertase	6 min	10^{-2} to 10^{-3} mol dm^{-3}
Uric acid	Uricase	30 min	5×10^{-3} to 5×10^{-5} mol dm^{-3}

biosensor and others are enzyme electrode biosensors. This biosensor is suitably modified using a mediator. The other advanced types of biosensors are the direct electron transfer systems.

The principle of a mediated biosensor is explained in a redox reaction catalysed by an enzyme directly coupled to an electrode. The electrons are transferred from the substrate to the electrode via enzyme and redox mediator. The oxidase replaces the oxygen requirement of the enzymes.

4. **Thermistor containing biosensor**: Thermistors have been developed by immobilizing enzymes, such as cholesterol oxidase, glucose oxidase, invertase, tyrosinase, etc. They are used to record even a small temperature change (between 0.1 and 0.001°C) during biochemical reaction. In case of thermometric enzyme-linked immunoabsorbent assay (ELISA), thermistors are employed for the study of antigen–antibody reaction with very high sensitivity (10^{-13} mol dm^{-3}).

5. **Bioaffinity sensor**: It measures the concentration of the determinant, i.e. substrate based on equilibrium binding with a high degree of selectivity. Bioaffinity sensors are of recent origin, and they are of diverse nature because of the use of radiolabelled, enzyme-labelled or fluorescene-labelled substance. In this biosensor, a receptor is radiolabelled and allowed to bind with determinant analogue immobilized onto the surface of a transducer. When concentration of a determinant is increased, the labelled receptor forms an intimately bound complex with determinant.

The radiolabelled receptor–determinant complex is removed from the immobilized determinant analogue, resulting in the increased concentration of labelled receptor. It is then measured by a reduction in signal of the labelled receptor. The uses of lectin receptors for saccharide estimation, drug receptors for drug, hormone receptors for hormone, antibodies receptors for antigens and nucleic acid (as gene probe) for inherited diseases and finger printing have been reported by Gronow et al. (1988).

6. **Microbial biosensor**: In these biosensors, either immobilized whole cell of microorganisms or organelles are used. These react with a large number of substrates and show generally slow response. Immobilized *Azotobacter vinelandii* coupled with ammonia electrode shows sensitivity range between 10^{-5} and 8×10 mol dm^{-3}. It measures the concentration of nitrate within 5–10 min. The commonly used microbial biosensors containing oxygen electrodes are given in Table 17.4.

Table 17.4 Microbial biosensors

Microorganism	Sensor	Response time (min)	Range
Brevibacterium lactofermentum	Assimilase sugars	1–10	Above 1 m mol dm^{-3}
B. subtilis	Mutagen screening	90–100	1–6 µg cm^{-3}
Methylomonas flagellate	Methane	1	Up to 6.6 m mol dm^{-3}
Trichosporon brassicae	Acetate	6–10	Up to 22.5 mg dm^{-3}
T. brassicae	Ethanol	10	Below 22.5 mg dm^{-3}

7. **Electronic biosensor**: In this type of biosensor, either enzyme or antibody is immobilized on the surface of a membrane. When a substrate is catalysed to yield product, change in colour is measured by using a light emitting diode and a photodiode. For measuring colour, biosensor with enzyme and dye are immobilized to a membrane.

Applications of Biosensors

The biosensors are becoming popular for their applications in different areas due to the small size, low cost, greater sensitivity and selectivity, and rapid and easy handling. The applications of biosensors are broadly categorized as follows:

1. **Biosensors in medical sciences**: Biosensors have wide applications in the field of medical sciences. The first glucose analyser using biomolecule for the detection of blood glucose was commercialized by Yellow Springs Instruments Company, USA, in 1979. A device with minimum quantity of insulin has been manufactured to deliver insulin to diabetes patient based on glucose level of blood. On the basis of the information provided by biosensor, the device delivers accurate amount of insulin required by the patient suffering with diabetes. The mutagenicity of mitomycin, an aflatoxin causing cancer in infants can be detected by using the biosensor. Abnormal toxic substances produced in body due to infectious diseases can also be detected.

2. **Biosensors in environmental sciences**: The biosensors are helpful in environmental monitoring and pollution control, since they can be miniaturized and automated. The biosensors are useful in quality control of drinking water and detection of pesticides in water. The whole-cell biosensor developed by immobilizing *Salmonella typhimurium* and *B. subtilis* in conjugation with oxygen electrode can be used to measure mutagenicity and carcinogenicity of several chemical compounds. A biosensor coupled with oxygen electrode and immobilized *Trichosporon cutaneum* is effectively used for measuring biological oxygen demand (BOD).

3. **Biosensors in food industries**: The biosensors are used for measuring the fermentation products to improve the feedback control, to carry out rapid sampling and rejection of below standard raw materials and to improve the efficiency of workers. Tokyo University Research Centre of Advance Science and Technology has recently developed a ion-sensitive field-effect transistor (ISFET), which is highly sensitive to changes in ion concentrations. With this biosensor, it is possible to measure the odour, freshness and taste of food products. In determining freshness of marine fishes, either ATPase or aminoxidase or putrescine oxidase is used. The ATPase detects the presence of ATP in fresh fish muscle. ATP is absent in stale food products, therefore, signals do not occur and detection becomes easy. The biosensor (with enzyme cholesterol oxidase) that measures cholesterol levels in butter has been developed at Cransfield Institute of Technology, UK.

4. **Biosensors in military services**: The development of biosensors that can detect toxic gases, including chemical warfare agents, could be of great help for military operations. Such biosensors have advantages over the traditional methods of sensing of chemicals employed by military forces worldwide.

Biochips—the Biological Computer

Until the development of silicon microchips, setting up of computers was very costly and space occupying. But with introduction of silicon microchips, computers are available at affordable prices. Further reduction in size of computers and improvement in computing powers is difficult, because there is inherent limit beyond which circuits cannot be squeezed onto a silicon chip. The width of the circuit cannot be shorter than the wavelength of light. Close placing beyond a limit of many electrical circuits on the same microchip results in electron tunnelling, which creates short circuits in the whole system. Moreover, cramming together a large number of circuits cause heat generation, which cause total failure of the system.

Biochips are developed with suitable biotechnological modifications of microchips. One of the important features of macromolecules (e.g. proteins) is their self-shaping into predetermined three-dimensional structure. This property of protein helps in biochip designing, as the circuits can be crammed around three-dimensional protein structure. A semiconducting organic molecule is inserted into a protein framework, and the whole unit is fixed on a protein support. In biochips, the electrical signals can pass through the semiconducting organic molecule in the same way as that of silicon microchip.

Biochips have many advantages over silicon microchip, and they can be briefly summarized as follows:
1. The protein molecule exhibits less electrical resistance, and less heat is generated during the course of production of electrical signals. Moreover, a large number of circuits can be placed together that is not possible in silicon microchips.
2. The width of electrical circuits is not more than that of one protein molecule and is smaller than the silicon microchip.
3. The problem of electron tunnelling is not severe in biochips.

Applications

1. Biochips can be used as a heart–heat regulator, which may solve the problem of users of pacemakers.
2. Biochips can respond to nerve impulses making it more natural when implanted into the artificial limbs.
3. They can be designed to sense light and sound and convert them to electrical signals, and these signals stimulate brain and senses sight and sound.
4. They could be developed in such a way that they are immune to disastrous effects of electromagnetic waves that are generated due to nuclear explosion and therefore used in military operations.

QUESTIONS

1. How is streptokinase produced? What are its uses?
2. Describe the immobilized enzyme reactors in brief.
3. Write short notes on the use of covalent bonding in immobilization enzymes.
4. Write short notes on use of covalent bonding in immobilization of enzymes, penicillinase and streptokinase.
5. What are the immobilized enzymes? Discuss briefly various methods of immobilization of enzymes. Explain briefly the application of immobilized enzymes.
6. Explain the factors that affect the rate of enzyme-catalysed reactions.
7. Explain the biotechnological production and application of amylases and proteases.
8. Give an account of types of enzymes obtained from microorganisms.
9. Write a note on immobilization of cells.
10. Write a short note on the immobilization of bacteria and plant cells.
11. Give the applications of immobilization of plants and bacterial cells.
12. Discuss the various techniques of immobilization of enzymes.
13. Give the industrial application of enzyme proteases.
14. Give the biological source and use of the enzyme penicillinase.
15. Discuss the types and applications of biosensors.
16. Write about the techniques of enzyme immobilization.
17. Write the source, production, mechanism of action and application of streptokinase.

CHAPTER 18
Nanobiotechnology

CHAPTER OUTLINE

- Introduction
- Computational Gene
- Nanolithography
- Nanomedicine and Nanodevices
- Medical Applications of Molecular Nanotechnology
- Questions

INTRODUCTION

Nanotechnology is a collective term that refers to technological developments on the nanometre scale, i.e. 0.1–100 nm (1 nanometre equals one thousandth of a micrometre or one millionth of a millimetre).

Nanobiotechnology is the branch of nanotechnology that deals with biological and biochemical applications or uses of nanotechnology. It often covers existing elements of nature in order to fabricate new devices such as biosensors.

Nanobiotechnology is used to describe the overlapping multidisciplinary activities associated with biosensors, particularly where photonics, chemistry, biology, biophysics, nanomedicine and engineering converge.

The term *bionanotechnology* is used interchangeably with nanobiotechnology. *Nanobiotechnology* refers to the use of nanotechnology to further the goals of biotechnology, whereas *bionanotechnology* refers to any overlap between biology and nanotechnology, including the use of biomolecules as part of or as an inspiration for nanotechnological devices.

Nanobiotechnological research involves nanospheres coated with fluorescent polymers. Researchers design polymers whose fluorescence is quenched when they encounter specific molecules. Different polymers would detect different metabolites. The polymer-coated spheres could then become part of new biological assays, and the technology might someday lead to particles that could be introduced into the human body to track down metabolites associated with tumours and other health problems. Nanobiotechnology is relatively new to medical, consumer and corporate bodies. Another example from a different perspective would be the evaluation and therapy at the nanoscopic level, i.e. the treatment of nanobacteria (25–200 nm sized) as practised by NanoBiotech Pharma.

Antibody–nanoparticle computational modelling: The conjugation of antibodies and nanoparticles with high affinity and specificity through receptor–ligand recognition modes is of paramount significance in the development of vehicles that can be used for diagnosis in the treatment of cancer and various other diseases, and for immunodiagnostic nanobiosensors. The bio–nano complex formed between an artificial nanomaterial (nanoliposomes and nanoparticles) and a biological entity, such as an antibody, is brought about by the formation of covalent bonds based on their specific structural and chemical properties, such as biocompatibility, water solubility and biodegradability. There is a requirement of an understanding of the relationship of the thermodynamic and kinetic aspects of antibody–membrane association, translational rotational mobility of membrane-bound antibodies, interactions with the diverse cell surface, circulating molecules and various artificial nanomolecules, and the conformation. In order to gain in-depth understanding of the detailed interactions of the nanoparticles and the antibody, molecular dynamics simulation is carried out.

Lately, it has been noted that certain natural proteins and antibodies can recognize specific nanoparticles. For the modelling study, the initial coordinates of the antibody can be made available from the Protein Data Bank (PDB). The basic assumptions, as a first approximation during the modelling study, would be that the hydrophilic derivatizations do not play a critical role in the predominantly hydrophobic nanomaterial–antibody interactions, and the electronic structure remains undisturbed during the conjugation. The nanoparticle is docked into a proposed binding site—polar-hydrogen potential function (PARAM19) and a modified TIP3P water-solvent model for the protein.

The simulation involves approximately 300 steps of minimization, using the steepest descent and the Newton–Raphson method. To reduce the necessary simulation time, a highly efficient method for simulating the localized interactions in the active site of a protein, the stochastic boundary molecular dynamics (SBMD) is used. The reference point for partitioning the system in SBMD chosen closer to the centre of the nanomaterial that is assumed to be a uniform sphere. The complex nano–bio system can be separated into spherical reservoir and reaction zones; the latter is further subdivided into a reaction region and a buffer region. The atoms in the reaction region are propagated by molecular dynamics, whereas those in the buffer region involving Langevin dynamics are retained, using harmonic restoring forces.

COMPUTATIONAL GENE

The computational gene is a molecular automaton, consisting of a structural part and a functional part. The structural part is a naturally occurring gene, used as a skeleton to encode the input and the transitions of the automaton. Its design is such that it might work in a cellular environment. The conserved features of a structural gene (e.g. DNA polymerase-binding site, start and stop codons and splicing sites) serve as constants of the computational gene, whereas the coding regions, the number of introns and exons, the position of start and stop codon and the automata theoretical variables (symbols, states and transitions) are the design parameters of computational gene. The constants and the design parameters are linked by several logical and biochemical constraints (e.g. encoded automata theoretic variables must not be recognized as splicing junctions). The inputs of the automaton are molecular markers indicated by single-stranded DNA (ssDNA) molecules. These markers are signalling aberrant (e.g. carcinogenic) molecular phenotype and turn on the self-assembly of the functional gene. If the input is accepted, the output encodes a double-stranded DNA (dsDNA) molecule, a functional gene, which is successfully integrated into the cellular transcription and translation machinery, producing a wild-type protein or an antidrug. Otherwise, a rejected input shall assemble into a partially dsDNA molecule, which cannot be translated.

Application in Therapy of Cancer

Computational genes could be useful in future to correct aberrant mutations in a gene or group of genes that can trigger disease phenotypes. A most prominent example is the tumour suppressor *p53* gene that is present in every cell and acts as a guard to control the growth. Mutations in this gene may abolish its function, permitting a cancerous growth. For instance, a mutation at codon 249 in the *p53* protein is characteristic for hepatocellular cancer. This ailment may be treated by the cytoplasmic domain of band 3 (CDB3) peptide that binds to the *p53* core domain and stabilizes its fold. The computational genes might allow implementation in situ of a therapy as soon as the cell starts developing defective material. The computational genes combine the techniques of gene therapy that allows to replace an aberrant gene in the genome by its healthier counterpart, as well as to silence the gene expression (similar to antisense technology).

Challenges

Although, mechanistically simple and quite robust on molecular level, following issues are needed to be addressed before an in vivo implementation of computational gene is considered. The transfer of DNA or RNA through biological membranes is a key step in the drug delivery. The DNA material must be internalized into the cell, specifically into the nucleus. Some results show that nuclear localization signals can be irreversibly linked to one end of the oligonucleotides, forming an oligonucleotide–peptide conjugate that allows effective internalization of DNA into the nucleus.

The DNA complexes should have low immunogenicity to guarantee their integrity in the cell and their resistance to cellular nucleases. Current strategies to eliminate nuclease sensitivity include modifications of the oligonucleotide backbone, such as methylphosphonate and phosphorothioate oligodeoxynucleotide (S-ODN), Along with their increased stability, modified oligonucleotides often have altered pharmacological properties.

The DNA complexes similar to any drug may cause nonspecific and toxic side effects. In vivo applications of antisense oligonucleotides showed that toxicity is largely due to impurities in the oligonucleotide preparation and lack of specificity of the particular sequence used. Undoubtedly, the progress in antisense biotechnology shall also result in a direct benefit to the model of computational genes.

NANOLITHOGRAPHY

Nanolithography is the branch of nanotechnology related to the study of fabricating nanometre-scale structures and their applications, meaning patterns with at least one lateral dimension between the size of an individual atom and approximately 100 nm. Nanolithography (Fig. 18.1) is a very active area of research in academia and in industry. It is used in the fabrication of leading-edge semiconductor integrated circuits (nanocircuitry) or nanoelectromechanical systems.

Optical Lithography

Optical lithography requires the use of liquid immersion and a host of resolution-enhancement technologies of phase-shift masks (PSM) or optical proximity correction (OPC) at the 32-nm node. Optical lithography is the predominant patterning technique since the advent of the semiconductor age. It is capable of producing sub-100 nm patterns with the use of very short wavelengths (193 nm). However, many experts feel that traditional optical lithography techniques will not be cost effective below 22 nm. For this, it may have to be replaced by a next-generation lithography (NGL) technique.

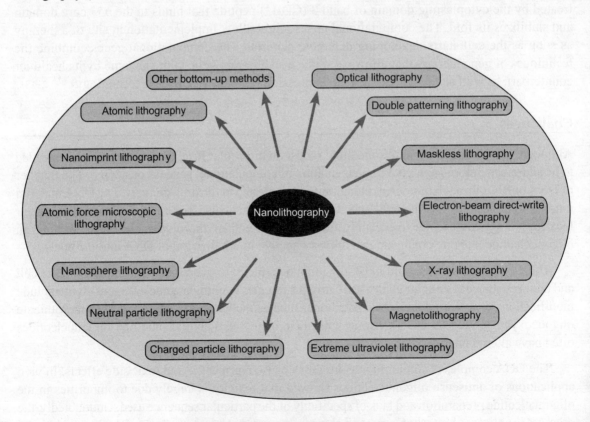

Figure 18.1 Branches of nanolithography.

Other Nanolithography Techniques

1. X-ray lithography can be extended to an optical resolution of 15 nm by using the short wavelength of 1 nm for illumination. This is implemented by the proximity printing approach. This technique requires no lenses. The technique is developed to the extent of batch processing. The extension of the method relies on near-field X-rays in Fresnel diffraction: a clear mask feature is demagnified by proximity to a wafer that is set near to a critical condition. This condition determines the mask-to-wafer gap and depends on size of the clear mask feature and wavelength.
2. Optical maskless lithography is a tool that uses a digital micromirror array to directly manipulate reflected light without the need of an intervening mask. Throughput is inherently low, but the elimination of mask-related production is economically not viable. The system might be more cost effective in case of small production runs of state of the art circuits, such as in a research laboratory, where tool throughput is not a concern.
3. Electron-beam direct-write (EBDW) lithography is the most common nanolithographic technique using a beam of electrons to produce a pattern typically in a polymeric resist material.
4. Extreme ultraviolet (EUV) lithography is a form of optical lithography using ultrashort wavelengths (13.5 nm).
5. Charged particle lithography, ion or electron-projection lithographies are also capable of very high-resolution patterning. Ion-beam lithography uses a focused or broad beam of energetic lightweight ions (like He^+) for transferring pattern to a surface. Using ion-beam proximity lithography (IBL), nanoscale features can be transferred on nonplanar surfaces.
6. Neutral particle lithography (NPL) uses a broad beam of energetic neutral particle for pattern transfer on a surface.
7. Nanoimprint lithography (NIL), and its variants, such as step-and-flash imprint lithography, lithographically induced self-assembly (LISA) and laser-assisted direct imprint (LADI), are promising nanopattern replication technologies. This technique can be combined with contact printing.
8. Scanning probe lithography (SPL) is a promising tool for patterning at the deep nanometre scale. For example, individual atoms may be manipulated using the tip of a scanning tunnelling microscope (STM). Dip pen nanolithography (DPN) is the first commercially available SPL technology based on atomic force microscopy.
9. Atomic force microscopic (AFM) nanolithography is a chemomechanical surface-patterning technique that uses an atomic force microscope.
10. Magnetolithography (ML) is based on applying a magnetic field on the substrate, using paramagnetic metal masks known as *magnetic mask*. Magnetic mask is analogue to photomask, and it defines the spatial distribution and shape of the applied magnetic field. The second component is ferromagnetic nanoparticles (analogue to the photoresist) that are assembled onto the substrate according to the field induced by the magnetic mask.

Bottom-up Methods

Nanosphere lithography uses self-assembled monolayers of spheres (typically made of polystyrene) as evaporation masks. This method has been used to fabricate arrays of gold nanodots with precisely controlled spacings.

It is possible that molecular self-assembly methods shall take over as the primary nanolithography approach, due to ever-increasing complexity of the top-down approaches. The self-assembly of dense lines, less than 20 nm wide, in large prepatterned trenches has been demonstrated. The degrees of dimension and orientation control, as well as prevention of lamella merging, are still to be addressed for this to be an effective patterning technique. The important feature of line edge roughness is also highlighted by this technique.

Another emerging form of bottom-up lithography is self-assembled ripple patterns and dot arrays formed by low-energy ion-beam sputtering. Aligned arrays of plasmonic and magnetic wires and nanoparticles are deposited on these templates via oblique evaporation. The templates are easily produced over large areas with periods down to 25 nm.

NANOMEDICINE AND NANODEVICES

Nanomedicine is the medical application of nanotechnology. It ranges from the medical applications of nanoparticles to nanoelectronic biosensors, and also covers possible future applications of molecular nanotechnology. Current problems for nanomedicine are mainly understanding of the issues related to their toxicity and environmental impact of nanoscale materials.

Nanomedicine research in India has been receiving funding from the US National Institute of Health to set up four nanomedicine centres in the country. The journal *Nature Materials* in April 2006 estimated that 130 nanotech-based drugs and delivery systems were being developed worldwide.

Nanomedicine seeks to deliver a valuable set of research tools and clinically useful devices in the near future. The US National Nanotechnology Initiative expects new commercial applications in the pharmaceutical industry that may include advanced drug delivery systems, new therapies and in vivo imaging. Neuroelectronic interfaces and other nanoelectronic-based sensors are areas of research priorities. The speculative field of molecular nanotechnology believes that cell-repair machines could revolutionize medicine and the medical field.

Nanomedicine is today a large industry, with nanomedicine sales reaching US$10 billion and with over 200 companies having around 60 commercial products worldwide. Around US$5 billion are being invested every year in nanotechnology R&D. As the nanomedicine industry continues to grow, it is expected to have a significant impact on the economy.

Medical Uses of Nanomaterials

In Drug Delivery

Nanomedical approaches to drug delivery cover development of nanoscale particles or molecules to improve drug bioavailability. *Bioavailability* refers to the presence of drug molecules where they are needed in the body and where they are most effective. More than US$65 billion are wasted each year due to poor bioavailability. Drug delivery focuses on maximizing bioavailability both at specific places in the body and over a period of time. This can potentially be achieved by molecular targeting using nanoengineered devices. It relates to targeting the molecules and delivering drugs with cell precision.

In vivo imaging is another area where tools and devices are being developed. Using nanoparticle contrast agents, images, such as ultrasound and MRI, have a favourable distribution and improved contrast. The new methods of nanoengineered materials that are being developed might be effective in treating illnesses and diseases, such as cancer. The applications of nanotechnology in developing self-assembled biocompatible nanodevices that shall detect, evaluate, treat and report to the clinical doctor automatically shall be reality in near future.

Drug delivery systems—lipid- or polymer-based nanoparticles—can be designed to improve the pharmacological and therapeutic properties of drugs. The strength of drug delivery systems is their ability to alter the pharmacokinetic and biodistribution of the drug. Nanoparticles have unusual properties that may be used to improve drug delivery. Where larger particles would have been cleared from the body, human cells shall take up these nanoparticles because of their size. The complex drug delivery mechanisms with ability to get drugs through cell membranes and into cell cytoplasm are being developed. Efficiency of drug delivery system is important because many diseases occur due to adverse processes within the cell and can only be countered by drugs that make their way into the cell. The triggered response is one way for drug molecules to be used more efficiently. A drug with poor solubility is replaced by a drug delivery system of nanoparticles where both hydrophilic and hydrophobic environments exist, improving the solubility. If a drug is cleared too quickly from the body, this could force a patient to use high doses. But with nanodrug delivery system, clearance can be reduced by altering the pharmacokinetics of the drug. The particulates from nanodrug delivery systems lower the volume of distribution and reduce the effect on nontarget tissue. Potential nanodrugs work by very specific and well-understood mechanisms. One of the major impacts of nanotechnology and nanoscience will be in leading the development of completely new drugs with more useful behaviour and fewer side effects.

The efficient use of drug delivery of nanoparticles is based upon following three aspects:

1. Efficient encapsulation of the drug
2. Successful delivery of drugs to the targeted region of the body
3. Successful release of the drug

Protein and Peptide Delivery

Proteins and peptides exert multiple biological actions in human body. Their macromolecules are known as *biopharmaceuticals*. They have shown great promise for treatment of various diseases and disorders. Targeted and/or controlled delivery of these biopharmaceuticals using nanomaterials like nanoparticles and dendrimers is an emerging field known as *nanobiopharmaceutics*, and these products are referred to as *nanobiopharmaceuticals*.

Nanoparticle targeting: The nanoparticles are promising tools for the advancement of drug delivery, medical imaging, and as diagnostic sensors. However, the biodistribution of these nanoparticles is mostly unknown due to the difficulty in targeting specific organs in the body. Current research in the excretory system of mice, however, shows the ability of gold composites to selectively target certain organs based on their size and charge. These composites are encapsulated by a dendrimer and assigned a specific charge and size. Positively charged gold nanoparticles were found to enter the kidneys while negatively charged gold nanoparticles remained in the liver and spleen. It is suggested that the positive surface charge of the nanoparticle decreases the rate of osponization of nanoparticles in the liver, thus affecting the excretory pathway. Even at a relatively small size of 5 nm, these particles can become compartmentalized in the peripheral tissues, and therefore accumulate in the body over time.

Neuroelectronic Interfaces

It is a visionary goal dealing with the construction of nanodevices that shall permit computers to be joined and linked to the nervous system. It requires the building of a molecular structure that shall permit control and detection of nerve impulses by an external computer. The computers shall be able to interpret, register and respond to signals the body gives off when it feels sensations. The demand for such structures is huge because many diseases involve the decay of the nervous system (antilymphocyte serum and multiple sclerosis). Also, many injuries and accidents may impair the nervous system resulting in dysfunctional systems and paraplegia. If computers could control the nervous system through neuroelectronic interface, problems that impair the system could be controlled so that effects of diseases and injuries could be overcome.

In Cancer

The small size of nanoparticles endows them with properties that can be very useful in oncology, particularly in imaging. Quantum dots (nanoparticles with quantum confinement properties, such as size-tunable light emission), when used in conjunction with magnetic resonance imaging (MRI), can produce exceptional images of tumour sites. These nanoparticles are much brighter than organic dyes and only need one light source for excitation. This means that the use of fluorescent quantum dots could produce a higher contrast image at a lower cost than today's organic dyes used as contrast media. The negative aspect is that quantum dots are usually made up of quite toxic elements.

High surface area to volume ratio, allows many functional groups to be attached to a nanoparticle that can bind to certain tumour cells. Additionally, the small size of nanoparticles (10–100 nm), allows them to preferentially accumulate at tumour sites, because tumours lack an effective lymphatic drainage system. There is a possibility to manufacture multifunctional nanoparticles that would detect, image, and then proceed to treat a tumour. A promising new cancer treatment that can replace radiation and chemotherapy is edging closer to human trials. Kanzius RF therapy attaches microscopic nanoparticles to cancer cells and then cooks tumours inside the body with radio waves that heat only the nanoparticles and the cancerous cells.

The sensor test chips with thousands of nanowires are able to detect proteins and other biomarkers left behind by cancer cells. They could enable the detection and diagnosis of cancer in the early stages from a few drops of a patient's blood.

The nanoparticles of cadmium selenide (quantum dots) glow when exposed to ultraviolet light. When injected, they seep into cancer tumours. The surgeon can see the glowing tumour, and use it as a guide for more accurate tumour removal.

James Baker from University of Michigan has developed a nanotechnology that can locate and then eliminate cancerous cells. He looked at a molecule known as *dendrimer* that has over 100 hooks on it that allows it to attach to cells in the body for many of purposes. Baker then attached folic acid to a few of the hooks (folic acid, being a vitamin, is received by cells in the body). The cancer cells have more vitamin receptors than normal cells, so Baker's vitamin-laden dendrimer would be absorbed by the cancer cell. To the rest of the hooks on the dendrimer, Baker placed anticancer drugs that would be absorbed with the dendrimer into the cancer cell, thereby delivering the cancer drug to the cancer cell and nowhere else.

Photodynamic therapy has potential for a noninvasive procedure for dealing with diseases, growth and tumours. In photodynamic therapy, a particle is placed within the body and is illuminated with light from outside. The light gets absorbed by the particle and if the particle is metal, energy from the light shall heat the particle and surrounding tissue. Light may also be used to produce high-energy oxygen molecules that shall chemically react with and destroy most organic molecules that are next to them (like tumours). This therapy is liked for many reasons. It does not leave a toxic trail of reactive molecules throughout the body, because it is directed where only the light is shining and the particles exist.

In Surgery

At Rice University, a flesh welder is employed to fuse two pieces of chicken meat into a single piece. The two pieces of chicken are placed together touching. A greenish liquid containing gold-coated nanoshells is dribbled along the seam. An infrared laser is traced along the seam, resulting the two sides to weld together. This could solve the difficulties and blood leaks caused when the surgeon tries to restitch the arteries, he has cut during a kidney or heart transplant. The flesh welder could weld the artery perfectly.

MEDICAL APPLICATIONS OF MOLECULAR NANOTECHNOLOGY

Molecular nanotechnology is a speculative subfield of nanotechnology, and it refers to the possibility of engineering molecular assemblers and machines that could reorder matter at a molecular or atomic scale. Molecular nanotechnology is theoretical, seeking to anticipate what inventions nanotechnology might yield and to propose an agenda for future inquiry.

Nanorobots

The possibility of using nanorobots in medicine would wholly change the world of medicine once it is realized. Nanomedicine would make use of these nanorobots (e.g. computational genes) for introduction into the body to repair or detect damages and infections. A typical blood-borne medical nanorobot could be between 0.5 and 3 μm in size, because that is the maximum size possible due to capillary passage requirement. Carbon can be used to build these nanorobots due to inherent strength and other characteristics of some forms of carbon (diamond/fullerene composites). The nanorobots would be fabricated in desktop nanofactories specialized for this purpose. Nanodevices could be seen at site inside the body using MRI, especially, if their components are manufactured using mostly

^{13}C atoms rather than the natural ^{12}C isotope of carbon, since ^{13}C has a nonzero nuclear magnetic moment. Medical nanodevices after injection into human body shall reach to the site in a specific organ or tissue. The doctor can easily monitor the progress and can scan a section of the body. The doctor can actually see the nanodevices congregated neatly around their target (a tumour mass).

Cell Repair Machines

Molecular machines are capable of entering the cell. With molecular machines, there will be more direct repairs. In the future, it will be possible to build nanomachine based systems that are able to enter cells, sense differences from healthy ones and make modifications to the structure.

The healthcare possibilities of these cell repair machines are impressive. Their compact parts would allow them to be more complex. The machines shall only be able to correct a single molecular disorder like DNA damage or enzyme deficiency. Later, cell repair machines shall be programmed with more abilities, with the help of advanced artificial intelligence systems. Nanocomputers shall be needed to guide these machines. These nanocomputers shall direct machines to examine, take apart and rebuild damaged molecular structures. The repair machines are capable of repairing whole cells by working structure by structure. By working cell by cell and tissue by tissue, whole organ may be repaired. Finally, by working organ by organ, health is restored to the body. The cells damaged to the point of inactivity may be repaired due to the ability of molecular machines to build cells from scratch. Therefore, cell repair machines shall free medicine from reliance on self-repair alone.

Nanonephrology

Nanonephrology is a branch of nanomedicine and nanotechnology that deals with (1) the study of kidney protein structures at the atomic level, (2) nanoimaging approaches to study cellular processes in kidney cells and (3) nanomedical treatments that utilize nanoparticles to treat various kidney diseases. Nanonephrology also covers the science related to the creation and use of material devices at the molecular and atomic levels that may be used for the diagnosis and therapy of renal disease. The possibility of this shall play a role in the management of patients with kidney disease in the future. By understanding the physical and chemical properties of proteins and other macromolecules at the atomic level in various cells in the kidney, novel therapeutic approaches could be designed and developed to combat major renal diseases. The nanoscale artificial kidney is a goal that many physicians dream of. Nanoscale engineering advances will permit programmable and controllable nanoscale robots to perform curative and reconstructive procedures in the human kidney at the cellular and molecular levels. The ultimate goal is designing nanostructures compatible with the kidney cells and that can safely operate in in vivo diseases.

Nanosensor

Nanosensor is any biological, chemical or sensory point used to convey information about nanoparticles to the macroscopic world. Its use mainly includes various medicinal purposes and as gateway to building of other nanoproducts, such as computer chips that work at the nanoscale and nanorobots. There are several procedures for making nanosensors, including top-down lithography, bottom-up assembly and molecular self-assembly.

Predicted Applications

The potential of nanosensors to accurately identify particular cells or places in the body can be medically used. By measuring changes in volume, concentration, velocity, displacement, electrical, gravitational and magnetic forces, temperature or pressure of cells in a body, nanosensors could be helpful in recognizing certain cells and distinguishing between cells, most notably those of cancer at the molecular level, in order to deliver medicine or monitor development to specific places in the body. In addition, nanosensors can detect macroscopic variations from outside the body by communicating the changes to other nanoproducts working within the body.

The fluorescence properties of cadmium selenide quantum dots may be used as sensor to detect tumours within the body. The cadmium selenide dots, however, are highly toxic to the body. So, the researchers are working on developing alternate dots of a different, less toxic material. In particular, the specific benefits of zinc sulphide quantum dots are being investigated even though they are not quite as fluorescent as cadmium selenide. It can be, however, augmented with other metals including manganese and various lanthanide elements. In addition, these newer quantum dots become more fluorescent when they bond to their target cells. The potential predicted functions could also be for detection of specific DNA in order to recognize explicit genetic defects, especially for individuals at high risk. Implanted sensors can automatically detect glucose levels for diabetic subjects more simply than current detectors. Using proteomic patterns and new hybrid materials, nanobiosensors can also be used to enable components configured into a hybrid semiconductor substrate as part of the circuit assembly. The development and miniaturization of nanobiosensors provide interesting new opportunities in detection of diseases. Nanosensors may also be valuable as more accurate monitors of material states for use in systems where size and weight are constrained, as in satellites and other aeronautic machines.

Existing Nanosensors

The most common mass-produced functioning of nanosensors is in the biological world as natural receptors. For instance, sense of smell, particularly in animals such as dogs, functions using receptors that sense nanosized molecules. Certain plant nanosensors detect sunlight. Various fishes use nanosensors to detect minuscule vibrations in the surrounding water and many insects detect sex pheromones using nanosensors.

The first synthetic nanosensor was built by researchers at the Georgia Institute of Technology in 1999 by attaching a single particle onto the end of a carbon nanotube and measuring the vibrational frequency of the nanotube both with and without the particle. The difference between the two frequencies allowed the researchers to measure the mass of the attached particle.

The chemical sensors, also, have been built using nanotubes to detect various properties of gaseous molecules. Carbon nanotubes have been used to sense ionization of gaseous molecules, while nanotubes made out of titanium have been employed to detect atmospheric concentrations of hydrogen at the molecular level.

Production Methods

Today, there are several hypothesized ways to produce nanosensors. Top-down lithography involves starting out with a larger block of some material and carving out the desired form. These carved out devices are used in specific microelectromechanical systems used as microsensors that reach the micro size. However, the most recent of these microsensors have begun to incorporate nanosized components.

The bottom-up method for production of nanosensors involves assembling the sensors out of more minuscule components, individual atoms or molecules. This involves moving atoms of a particular substance, one by one, into particular positions. Though, it has been achieved in laboratory tests using tools, such as atomic force microscopes, it is difficult to do it on large scale both for logistic reasons, as well as economic ones. Most likely, this process would be utilized mainly for building starter molecules for self-assembling sensors.

The third procedure that promises faster results involves self-assembly or growing particular nanostructures to be used as sensors, resulting in two types of assembly. The first involves use of a piece of some previously created or naturally formed nanostructure. It is immersed in free atoms of the same kind. After specific period, the structure with an irregular surface would make it prone to attracting more molecules. It would capture some of the free atoms and continue to form more of itself to make larger components of nanosensors. The second type of self-assembly starts with an already complete set of components that would assemble themselves into a finished product. So far, this has been successful only in assembling computer chips at the micro size, researchers hope to be able to do it at the nanometre size for multiple products, including nanosensors.

Economic Impact

The nanosensor technology is comparatively new field with lots of promises. The global projections for sales of products incorporating nanosensors are expected to be from US$0.6 to 2.7 billion in the next four years. They are likely be included in most modern circuitry used in advanced computing systems. Their potential to provide link between other forms of nanotechnology and the macroscopic world permits developers to fully exploit the potential of nanotechnology to miniaturize computer chips and expand their storage potential.

The present high cost of production of nanosensor has to be overcome in order to make it worthwhile for implementation in consumer products. Additionally, reliability of nanosensor is not yet suitable for widespread use. Because of their scarcity, nanosensors have yet to be marketed and implemented outside research facilities. Consequently, nanosensors have yet to be made compatible with most consumer technologies.

Social Impact

The advancements in detecting and sensing different biological and chemical species with increased capacity and accuracy may transform societal mechanisms that were originally designed on uncertainty and imprecise information. The social issues resulting from the widespread use of nanosensors and surveillance devices may include privacy invasion and security. The ability of nanosensors to measure extremely low contents of air pollutants or toxic materials in water raises questions and dilemmas of risk thresholds, especially if the advancement of such technologies outpaces the ability of the public to respond. The medical sensor shall not only help in diagnosis and treatment, but may also predict the future profile of an individual. This may add to the information used by health insurance companies to grant or deny coverage.

Nanoparticles

A *particle* in nanotechnology is defined as a small object that behaves as a whole unit in terms of its transport and properties. The particles are classified according to their sizes. The fine particles cover a range between 100 and 2500 nm, whereas ultrafine particles are in the range of 1–100 nm. The nanoparticles (Fig. 18.2) are also identical to ultrafine particles in size (between 1 and 100 nm). The nanoparticles may or may not exhibit size-related properties that differ significantly from those observed in fine particles or bulk materials. The nanoparticles have a wide range of applications in medical sciences and are useful in production of liposomes, dendrimers, iron oxide nanoparticles, polymer-drug conjugates, polymeric nanoparticles, etc.

Nanoclusters have at least one dimension between 1 and 10 nm and a narrow size distribution. Nanopowders arc agglomerates of ultrafine particles, nanoparticles or nanoclusters. Nanometre-sized single crystals or single-domain ultrafine particles are often referred to as *nanocrystals*. The research in nanoparticles is currently an area of intense scientific interest due to a wide variety of potential applications in biomedical, optical and electronic fields. The National Nanotechnology Initiative has led to generous public funding for nanoparticle research in the United States.

Background

Although nanoparticles are considered as an invention of modern science, they actually have a very long history. These were used by artisans since 9th century in Mesopotamia for generating a glittering effect on the surface of pots. Today, pottery from the Middle Ages and Renaissance often has a distinct gold or copper coloured metallic glitter. The nanoparticles were created by the artisans by adding copper and silver salts and oxides together with vinegar, ochre and clay on

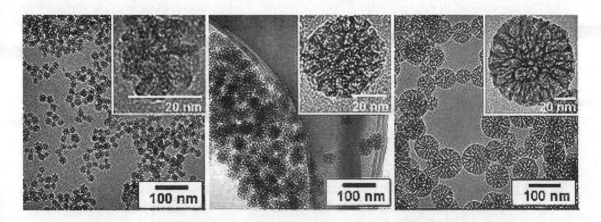

Figure 18.2 Nanoparticles of silica with mean diameter.

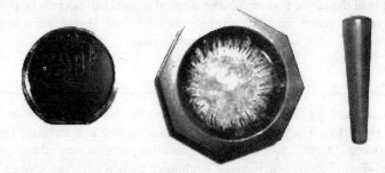

Figure 18.3 Silicon nanopowder.

the surface of previously glazed pottery. The object was then placed into a kiln and heated at about 600°C in a reducing atmosphere. In the heat, the glaze would soften, causing the copper and silver ions to move into the outer layers of the glaze. The ions were reduced back to metals that came together forming the nanoparticles that gave the colour and optical effects.

Michael Faraday in 1857, first described, in scientific terms, the optical properties of nanometre-scale metals in his classic paper.

Properties

1. Nanoparticles (Fig. 18.3) are of great scientific interest as they are a link between bulk materials and atomic or molecular structures. A bulk material should have constant physical properties regardless of its size, but at the nanoscale, size-dependent properties are often observed. The properties of materials change as their size approaches the nanoscale and as the percentage of atoms at the surface of a material becomes significant.
2. The distinct properties of nanoparticles are largely due to the large surface area of the material that dominates the contributions made by the small bulk of the material.
3. The suspensions of nanoparticles are possible since the interaction of the particle surface with the solvent is strong enough to overcome density differences, that otherwise generally results in a material either sinking or floating in a liquid.
4. Nanoparticles possess unexpected optical properties, as they are small enough to limit their electrons and produce quantum effects. For example, gold nanoparticles turn deep red to black in solution.
5. The high surface area to volume ratio of nanoparticles facilitates diffusion, especially at higher temperatures. Sintering can take place at lower temperatures over shorter timescales than for larger particles. This theoretically does not affect the density of the final product, though flow difficulties and the tendency of nanoparticles to agglomerate complicate the matters.
6. The presence of titanium dioxide nanoparticles imparts self-cleaning effect. The size being in nano range, the particles cannot be observed. Zinc oxide nanoparticles have been found to have superior UV blocking properties compared to its bulk substitute. And for this reason, it is often used in the preparation of sunscreen lotions. It is completely photostable.

7. Clay nanoparticles when incorporated into polymer matrices increase reinforcement leading to stronger plastics, verifiable by a higher glass transition temperature and other mechanical property tests.
8. Nanoparticles have also been attached to textile fibres in order to create smart and functional clothing.
9. Nanoparticles with one half hydrophilic and the other half hydrophobic are known as *Janus particles* and are particularly effective for stabilizing emulsions. They can self-assemble at water/oil interfaces and act as solid surfactants.
10. Different types of liposome nanoparticles are now used clinically as delivery systems for anticancer drugs and vaccines.

Synthesis

There are several techniques for creating nanoparticles, including both attrition and pyrolysis. In attrition, macro- or microscale particles are ground in a ball mill, a planetary ball mill, or other size reducing mechanism. The resulting particles are air-classified to recover nanoparticles. In pyrolysis, a vaporous precursor (liquid or gas) is forced through an orifice at high pressure and burnt. The resulting solid (a version of soot) is air-classified to recover oxide particles from by-product gases. Pyrolysis often results in aggregates and agglomerates rather than single primary particles.

Thermal plasma can also release the energy required to cause evaporation of small micrometre-sized particles. The thermal plasma temperatures are in the order of 10,000 K, so that solid powder easily evaporates. Nanoparticles are formed upon cooling while exiting the plasma region. The main types of the thermal-plasma torches used to produce nanoparticles are DC plasma jet, DC arc plasma and radio frequency (RF) induction plasmas.

Inert gas condensation is frequently used to make nanoparticles from metals with low melting points. The metal is vaporized in a vacuum chamber and then supercooled with an inert gas stream. The supercooled metal vapour condenses into nanometre-sized particles that can be entrained in the inert gas stream and deposited on a substrate or studied in situ.

The sol–gel process is a wet-chemical technique (also known as *chemical solution deposition*) widely used recently in the fields of materials science and ceramic engineering. Such methods are used primarily for the fabrication of materials (typically a metal oxide) starting from a chemical solution (*sol*, short for solution) that acts as the precursor for an integrated network (or *gel*) of either discrete particles or network polymers. Sol–gel derived materials have diverse applications in optics, electronics, energy, space, biosensors, medicine (e.g. controlled drug release) and separation (e.g. chromatography) technology.

Colloids

The term *colloid* refers to a broad range of solid–liquid and/or liquid–liquid mixtures that contain distinct solid (and/or liquid) particles that are dispersed to various degrees in a liquid medium. The term is specific to the size of the individual particles that are larger than atomic dimensions, but small enough to exhibit Brownian motion. If the particles are large enough, then their dynamic behaviour in any given period of time in suspension is governed by forces of gravity and sedimentation. But if they are small enough to be colloid, then their irregular motion in suspension is attributed to the collective bombardment of a myriad of thermally agitated molecules in the liquid suspending medium, as described originally by Albert Einstein.

Morphology

The scientists have started naming their particles after the real world shapes that they might represent. Nanospheres, nanoreefs, nanoboxes and more have appeared in the literature. These morphologies sometimes arise spontaneously as an effect of a templating or directing agent present in the synthesis, e.g. miscellar emulsions or anodized alumina pores, or from the innate crystallographic growth patterns of the materials themselves. The study of fine particles is known as *micromeritics*. Amorphous particles usually adopt a spherical shape (due to their microstructural isotropy), whereas the shape of anisotropic microcrystalline whiskers corresponds to their particular crystal habit. At the small end of the size range, nanoparticles are often referred to as *clusters*. Spheres, rods, fibres and cups are just a few of the shapes that have been grown.

Characterization

The characterization of nanoparticle is essential for controlling nanoparticle synthesis and understanding the spectrum of applications. The common techniques used are electron microscopy (TEM, SEM), atomic force microscopy (AFM), X-ray photoelectron spectroscopy (XPS), dynamic light scattering (DLS), powder X-ray diffraction (XRD), Fourier transform infrared spectroscopy (FTIR), marix-assisted laser desorption/ionization time-of-flight (MALDI-TOF) mass spectrometry, ultraviolet–visible (UV–Vis) spectroscopy, dual polarization interferometry and nuclear magnetic resonance (NMR). The technology for nanoparticle tracking analysis (NTA) permits direct tracking of the Brownian motion and this method therefore allow the sizing of individual nanoparticles in solution (Fig. 18.4).

Functionalization

The surface coating of nanoparticles is essential for determining their properties. In particular, the surface coating can regulate solubility, stability and targeting. A coating that is multivalent or polymeric confers high stability. For biological applications, the surface coating should be

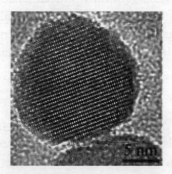

Figure 18.4 Magnetic Fe_3O_4 nanoparticles.

polar to give high aqueous solubility and prevent nanoparticle aggregation. In serum or on the cell surface, highly charged coatings promote nonspecific binding, while polyethylene glycol linked to terminal hydroxyl or methoxy groups repel nonspecific interactions. Nanoparticles when linked to biological molecules can act as address tags to direct the nanoparticles to specific organelles within the cell and specific sites in the body. The common address tags are monoclonal antibodies, aptamers, streptavidin or peptides. These targeting agents should ideally be covalently linked to the nanoparticle and should be present in a controlled number per nanoparticle. Monovalent nanoparticles, bearing a single binding site, avoid clustering and so are preferable for tracking the behaviour of individual proteins. Multivalent nanoparticles, bearing multiple targeting groups, can cluster receptors that can activate cellular signalling pathways and give stronger anchoring.

Safety

Most of the nanoparticles are due to the high surface-to-volume ratio that can make the particles very reactive or catalytic. They are also able to pass through cell membranes in organisms, and their interactions with biological systems are relatively unknown. A recent study revealed the effects of ZnO nanoparticles on human immune cells that were at varying levels of susceptibility to cytotoxicity. Smaller nanoparticles evinced increased cytotoxicity. The lymphocytes (especially naive T cells) were found to be more resistant to nanoparticle cytotoxicity than monocytes. It may be due to the capacity of the latter to produce higher levels of reactive oxygen species in response to internalized nanoparticles. Previously activated memory T cells were more susceptible than naive T cells, implying a relationship between cell cycle and nanoparticle susceptibility. In addition, nanoparticle concentrations below those causing appreciable cell death nonetheless induced the production of proinflammatory cytokines, such as IFN-γ and TNF.

Synthetic Biology

Synthetic biology is a wide area of biological research that combines science and engineering. It encompasses a variety of different approaches, methodologies and discipline, and it describes pinpoints at the design and construction of new biological functions and systems not found in nature.

In 1974, the Polish geneticist Waclaw Szybalski introduced the term *synthetic biology*.

The Nobel Prize in physiology or medicine was awarded to Arber, Nathans and Smith in 1978 for their discovery of restriction enzymes, Waclaw Szybalski then wrote in an editorial in the journal *Gene*, "The work on restriction nucleases not only permits us easily to construct recombinant DNA molecules and to analyse individual genes, but also has led us into the new era of synthetic biology where not only existing genes are described and analysed, but also new gene arrangements can be constructed and evaluated."

Synthetic biology includes the broad redefinition and expansion of biotechnology, with the ultimate goals of being able to design and build engineered biological systems that process information, manipulate chemicals, fabricate materials and structures, produce energy, provide food, and maintain and enhance human health and our environment.

Good examples of engineering in synthetic biology include the pioneering work of Tim Gardner and Jim Collins on an engineered genetic toggle switch, a riboregulator, the Registry of Standard Biological Parts and the International Genetically Engineered Machine (iGEM) competition.

The synthetic biology raises questions for ethics, biosecurity, biosafety, involvement of stakeholders and intellectual property. To date, key stakeholders (especially in the US) have focused primarily on the biosecurity issues, especially the so-called dual-use challenge. For example, while the study of synthetic biology may lead to more efficient ways to produce medical treatments (e.g. against malaria), it may also lead to synthesis or redesign of harmful pathogens (e.g. smallpox) by malicious actors.

There are several key enabling technologies that are critical to the growth of synthetic biology. The key concepts include standardization of biological parts and hierarchical abstraction to permit using those parts in increasingly complex synthetic systems. Achieving this is greatly aided by basic technologies of reading and writing of DNA (sequencing and fabrication) that are improving in price/performance exponentially. The measurements under a variety of conditions are needed for accurate modelling and computer-aided design (CAD).

QUESTIONS

1. Define nanobiotechnology.
2. Define nanolithography and write a note on branches of nanolithography.
3. Write notes on nanomedicine and nanodevices.
4. Write medical uses of nanomaterials.
5. Give an account of challenges in development of nanobiotechnology.

CHAPTER 19
Recent Advances in Biopharmaceuticals

CHAPTER OUTLINE

- Introduction
- Advantages of DNA Technology in the Production of Biopharmaceuticals
- Biomolecules Produced through Recombinant DNA Technology
- Biopharmaceuticals of Animal Origin
- Biopharmaceuticals of Plant Origin
- Pharmaceutical Substances of Microbial Origin
- Recent Advances in Biopharmaceuticals
- Questions

INTRODUCTION

Biopharmaceutical is a pharmaceutical substance used for therapeutic purpose, which is produced by means of direct extraction from native biological sources. The term *biopharmaceutical* is most commonly used to refer to all therapeutic, prophylactics and in vivo diagnostic agents produced using live organisms or their functional components. In general, the biopharmaceuticals are complex macromolecules that are over hundred times larger than small molecule drugs with complex structure activity and maintenance requirements.

The term *biopharmaceutical* was first used in 1980 for the product of pharmaceutical biotechnology or biotechnological medicines. The biopharmaceutical industry can trace its beginning to Edward Jenner who has been credited with the development of small pox vaccine in 1796. The discovery of penicillin by Alexander Fleming in 1928 changed the course of medicine and heralded the start of the age of biopharmaceuticals. Wilhelm Roux cultivated first cells from chick embryos in salt solution in 1885. Ross Harrison successfully undertook the animal cell culture in a frog lymph medium in 1907. It was in 1950 that cell cultures were widely available. During 1960s and 1970s, commercialization of these technologies had further impact on cell culture that continues to this day. The companies started developing and selling disposable plastic and glass cell culture products, improved filtration products and materials, liquid and powdered tissue culture media and laminar flow hoods.

Today, alternatives to large-scale submerged microbial and animal cell cultures are increasing. In addition to vaccines that are grown in animals, embryos and eggs, products grown in transgenic animals and plants, as well as small volume products of developing cell and gene therapies are collectively poised to displace the predominated use of large volume bioreactor.

Biotechnology is profoundly changing the way drugs are produced with the use of chemical synthesis to biomanufacturing. Drugs are now grown and not made. There are also fundamental differences in the average size of these two types of drugs. The chemically synthesized products are known as *small molecule drugs*. Because of this transformative phenomenon in biotechnology, the pharma industry for biotech products is growing more than twice as fast as traditional pharmaceutical companies. The proteomics, genomics, bioinformation and metabolism have led to the identification of hundreds of biological targets and potential biologicals. The potential for improved disease targeting and greater efficacy in many fields is instrumental in biopharmaceuticals accounting for an increasing share of the new drug application as submitted to Health Canada and the US FDA, the latter passing the 50% mark for the first time in 2002. A significant portion of biopharmaceutical drug development is now performed by some experienced big pharma and a small number of established biopharma companies. In addition, there are hundreds of small companies and institutions without sufficient competence to bring novel biopharmaceuticals all the way to market. Today, even these organizations have full access to the required expertise from contract manufacturing and clinical research organizations.

ADVANTAGES OF DNA TECHNOLOGY IN THE PRODUCTION OF BIOPHARMACEUTICALS

The advantages of DNA technological applications to production of biopharmaceuticals are listed below.

1. It provides an alternative to direct extraction from source materials, such as *Escherichia coli*, snake venom, human urine, etc.
2. It overcomes the problem of source availability.
3. Large-scale production of therapeutic proteins, such as insulin, hormones, vaccine and interleukins, using recombinant microorganisms is possible.
4. Production of humanized monoclonal antibodies for therapeutic application is a reality.
5. Its feasibility of generating therapeutic proteins shows advantage over native protein product.
6. Bioremediation by the use of recombinant organisms is possible.
7. Genetic engineering techniques are used in forensic medicine.
8. Production of insect resistant cotton plant by incorporation of insecticidal toxin of *Bacillus thuringiensis* is possible.
9. The golden rice (rice in vitamin A) by incorporating three genes required for its synthesis in rice plant can be produced.

BIOMOLECULES PRODUCED THROUGH RECOMBINANT DNA TECHNOLOGY

The list of important products obtained by recombinant DNA technology is given below.

1. Recombinant hormones: Insulin (and its analogues), follicle-stimulating hormone, growth hormone, salmon calcitonin, etc.
2. Cytokines and growth factors: Interferons, interleukins and colony-stimulating factors (interferons: α, β and γ; erythropoietin; interleukin-2; GM-CSF; G-CSF), etc.
3. Blood products: Albumin, fibrinolytics, thrombolytics and clotting factors (factor VII, factor IX, tissue plasminogen activator, recombinant hirudin), etc.
4. Recombinant enzymes: Dornase alfa (Pulmozyme), α-L-iduronidase (Aldurazyme), acid glucosidase (Myozyme), urate oxidase, etc.
5. Recombinant vaccines: Recombinant protein or peptides, DNA plasmid and anti-idiotype (HBsAg vaccine, HPV vaccine), etc.
6. Monoclonal antibodies and related products: Mouse, chimeric or humanized, whole molecule or fragment, single chain or bispecific and conjugated (rituximab, trastuzmab, infliximab, bevacizumab), etc.
7. Miscellaneous products: Bone morphogenic protein, conjugate antibody, pegylated recombinant proteins, antagonists, etc.

The production of biopharmaceuticals is possible from three different natural origins (Fig. 19.1).

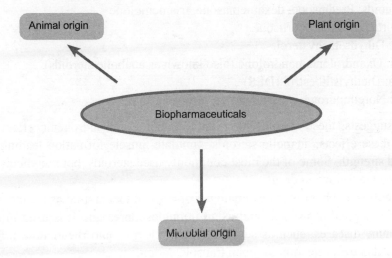

Figure 19.1 Production of biopharmaceuticals.

BIOPHARMACEUTICALS OF ANIMAL ORIGIN

Sex Hormones

Male and female gonads produce a range of steroid hormones, which regulate the development and maintenance of reproductive and related functions. Two main classes of sex steroids are androgens and oestrogens, of which the most important human derivatives are testosterone and estradiol, respectively. Progestogen is a third class of sex steroids, distinct from androgens and oestrogens. It is the most important and naturally occurring. In general, androgens are considered *male sex hormones*, since they exert masculinizing effects, whereas oestrogens and progestogen are known as *female sex hormones*, although all types are present in each gender at different levels.

Androgens

Androgens were first discovered in 1936. The major androgen produced by testes is *testosterone* that is secreted (4–10 mg) daily into blood stream. Androgen, also known as *androgenic hormone* or *testoid*, is the generic term for any natural or synthetic compound (usually a steroid hormone) that stimulates or controls the development and maintenance of male characteristics in vertebrates by binding to androgen receptors. The spectrum of activity covers activities related to accessory male sex organs and development of male secondary sex characteristics. Androgens are also the original anabolic steroids and the precursor of all oestrogens, the female sex hormones. Androgen ablation can be used as an effective therapy in prostate cancer (Table 19.1).

Other examples of medicinally important synthetic steroid hormones are as follows:
1. Glucocorticoids: Prednisone, dexamethasone, triamcinolone
2. Mineralocorticoid: Fludrocortisone
3. Vitamin D: Dihydrotachysterol
4. Androgens: Oxandrolone, nandrolone (also known as anabolic steroids)
5. Oestrogens: Diethylstilbestrol (DES)
6. Progestins: Norethindrone, medroxyprogesterone acetate

As their name suggests, these steroids have both anabolic and androgenic effects. Through a combination of these effects, anabolic steroids stimulate muscle formation leading to increased muscle mass and strength. Some of the most commonly used steroids, both medicinally as well as otherwise, include the following drugs:
1. **Prednisone**: It is a dehydrogenated analogue of cortisol that is used as an anti-inflammatory drug in the treatment of arthritis and as an immunosuppressant. It is used in a wide range of autoimmune diseases, such as severe asthma, allergies and rheumatoid arthritis, and to prevent and also treat rejection in organ transplantation.

Table 19.1 Synthetic medicinal steroids

Hormone	Description	Therapeutic indication
Testosterone	Produced by testes, longer half-life.	Hypogonadism, postmenopausal breast
Methyl testosterone	Synthetic androgen, half-life more than testosterone	Male hypogonadal disorder and female breast cancer

2. **Nandrolone**: It is an androgen (trade names: Durabolin or Kabolin) used to treat testosterone deficiency or breast cancer or osteoporosis. The positive effects of the drug include muscle growth, increased red blood cell production, appetite stimulation and increased bone density.
3. **Anavar**: Being extremely mild and anabolic, it is used for the treatment of alcoholic hepatitis, Turner's syndrome and weight loss caused by HIV. It is used for promotion of strength and duality of muscle mass gains; although its mild nature makes it less than ideal for bulking purposes.
4. **Winstrol**: The anabolic properties of winstrol are mild in comparison to many other steroids. It increases strength without excess weight gain, promotes vasculature and therefore famous amongst bodybuilders.
5. **Steranobol**: Chemically, steranobol has a chloro-group added at the 4-position of testosterone. It is an artificial, synthetic androgenic steroid with anabolic effects similar to testosterone. It is frequently misused to improve physical performance in sports and athletics. It is prohibited by many sports authorities, including the International Olympic Committee.
6. **Clenbuterol**: It is widely used as a bronchodilator in many parts of the world. It is known to directly stimulate the adipose tissue and accelerate the breakdown of triglycerides to form free fatty acids.

Oestrogen

β-Oestradiol is the principal oestrogen produced by the ovary. Oestrogens are synthesized in all vertebrates, as well as some insects. Oestrogenic sex hormones have an ancient evolutionary history. Oestrogens are used in oral contraceptives, oestrogen-replacement therapy for postmenopausal women and in hormone replacement therapy for transwomen. It is 10 times more potent than oestron and 25 times more potent as compared to oestriol. Like all steroid hormones, oestrogens readily diffuse across the cell membrane. Once inside the cell, they bind to and activate oestrogen receptors, which in turn, modulate the expression of many genes.

The medicinally important oestrogens are as follows:
1. Ethinylestradiol is used for oestrogen replacement therapy in deficient states, both pre- and postmenopausal, and breast cancer.
2. Oestradiol is natural oestrogen used in oestrogen-replacement therapy in menopausal and menstrual disorders.
3. Mestranol is a synthetic oestrogen and used in treatment of menopausal conditions.

Progesterones and Progestogens

Progesterone is a hormone produced by corpus luteum. *Progestogens or gestagens* are a group of hormones, including progesterone. The progestogens are one of the five major classes of steroid hormones, in addition to the oestrogens, androgens, glucocorticoids and mineralocorticoids. The progestogens are characterized by their basic 21-carbon skeleton called a *pregnane skeleton* (C21), and they are both naturally occurring and synthetic. They are named for their function in maintaining pregnancy (progestational), although they are also present at other phases of the oestrous and menstrual cycles. The oestrogens possess an estrane skeleton (C18) and androgens, an andrane skeleton (C19).

Table 19.2 Synthetic progesterones

Hormone	Description	Use
Megestrol acetate	Synthetic progesterone	Treatment of endometrial carcinoma and some forms of breast cancer
Medroxyprogesterone acetate	Synthetic progesterone	Treatment of menstrual disorder, endometriosis and contraceptive

Progesterone is mainly used in the treatment of menstrual disorder and breast and endometrical cancer. They are used in hormone replacement in addition to being used as contraceptive agents. The uses of synthetic progesterones are summarized in Table 19.2.

The other steroidal hormones used therapeutically are corticosteroids, catecholamines and prostaglandins. The therapeutic uses of synthetic prostaglandins are in gynaecology. They are used in inhibition of platelet and for induction of vasodilation secretion of gastric acid.

BIOPHARMACEUTICALS OF PLANT ORIGIN

Recombinant Proteins

The plant produces wide varieties of bioactive molecules by secondary metabolic pathways. They are now gaining widespread acceptances as a general platform for the large-scale production of recombinant proteins. The proteins obtained from plants may be used as diagnostic reagents and as vaccine drugs for the production of recombinant proteins on industrial scale. Important advantage with plants in the production of recombinant protein is that vaccine candidates can be expressed in edible plant organs (Table 19.3). The deliveries of recombinant protein in edible plant organs are exceptional because they would be used locally for vaccination campaigns. Potato was the first major system that was used for vaccine and transgenic production.

Plant-derived proteins can be used for the production of insulin, human growth hormones, interferons, monoclonal antibodies, etc.

Plant-Derived Therapeutics

The plants are the sources of many important secondary metabolites, which are used as therapeutic agents in the treatment of wide spectrum of human ailments. Alkaloids, glycosides, flavonoids, steroids, etc., are important groups of secondary metabolites obtained from plants and used systematically in treatment of diseases. Some important drugs of plant origin have been described in Table 19.3.

Alternative Plant-Based Expression System

Most of the plant-derived recombinant pharmaceutical proteins have been produced by biotransformation and from transgenic plant lines. There are some advantages and limitations in use of plant-based production system, which are described in Table 19.4.

Table 19.3 Some important therapeutic agents obtained from plants

Drugs	Biological source	Chemical type	Therapeutic indications
Aspirin	*Salix alba*	Salicylic acid	Anti-inflammatory and analgesic
Atropine	*Atropa belladonna*	Alkaloids	Dilation of pupil
Digitoxin	*Digitalis purpurea, Digitalis lanata*	Steroids	Cardiac stimulants
Morphine	*Papaver somniferum*	Alkaloids	Analgesic
Taxol	*Taxus brevifolia, Taxus baccata*	Terpenoids	Breast cancer, acute lymphoma in children and adult
Vincristine, vinblastine	*Catharanthus roseus*	Alkaloids	Blood cancer, Hodgkin's disease
Sennosides A and B	*Cassia angustifolia, Cassia acutifolia*	Glycosides	Purgative
Podophyllum	*Podophyllum hexandrum*	Resin	Purgative, cholagogue and bitter tonic
Citronellal	*Cymbopogon nardus*	Volatile oils	Insect repellant
Turmeric	*Curcuma longa*	Terpenoids	Anti-inflammatory, condiment and spice
Aloe	*Aloe barbadensis, Aloe vera*	Glycoside	Purgative
Rauwolfia	*Rauwolfia serpentine*	Alkaloids	Antihypertensive and tranquillizer
Scopolamine	*Datura metel*	Alkaloids	CNS depressants and asthma
Xanthine	*Theobroma cacao*	Alkaloids	CNS stimulant and diuretic

Table 19.4 Advantages and limitations of plant-based production system

System	Advantages	Limitations
Transgenic plant, accumulation within plant	Good yield, economy scalability; establishment of permanent lines	Restricted for production, timescale, regulatory compliance
Virus-infected plants	Good yield, manageable timescale	Biosafety, construct size limitation
Cell or tissue culture	Defined timescale, containment, secretion into medium, possibility of bioconversion	High cost of production

PHARMACEUTICAL SUBSTANCES OF MICROBIAL ORIGIN

The microorganisms produce wide spectrum of secondary metabolites of therapeutic significance. Antibiotics are by far the most important group of substances produced by microbial strains, and this group of pharmaceuticals has had the greatest single positive impact on human health care in history. *Antibiotics* are defined as low molecular mass microbial secondary metabolites that at low concentration inhibit the growth of other microorganisms. Around 12,000 antibiotics have been isolated and characterized so far. Several fungal strains are known to produce antibiotics. The first antibiotic to be used medically, penicillin, was extracted from *Penicillium notatum*. Sir Alexander Fleming in 1928 noted the ability of the mould *P. notatum* to produce an antibiotic substance. Approximately, 55% of antibiotics are produced by the genus *Streptomyces* (filamentous bacterial group), 15% from other actinomycetes bacteria and rest from filamentous fungi.

Major Families of Antibiotics

1. β-Lactams
2. Tetracyclines
3. Aminoglycosides
4. Macrolides
5. Ansamycins
6. Peptide and glycopeptides
7. Miscellaneous antibiotic

The major families of antibiotics are described in Figure 19.2.

There are around 124 antibiotics available in the market till August 2010.
1. **β-Lactams**: These are β-lactams that include penicillin and cephalosporins. Penicillin refers to a family of both natural and semisynthetic antibiotics. Naturally produced penicillin includes penicillin G and V. The examples of penicillin are phenethicillin, procillin, oxacillin, etc. The cephalosporins have an antibiotic mechanism of action identical to that of the penicillin. It is the prototypic natural cephalosporins.
2. **Tetracyclines**: They are the family of antibiotics that display a characteristic 4-fused core ring structure. They exert their antimicrobial effect by inhibiting protein synthesis in sensitive microorganisms. Chlortetracycline was the first member of this family to be discovered in 1948. Tetracyclines gained widespread medical use due to their broad spectrum of activity that includes not only Gram-negative bacteria but also mycoplasm, rickettsias, chlamydias and spirochaetes.
3. **Natural tetracyclines**: Oxytetracyclines, chlortetracycline and demeclocyclines.
4. **Semisynthetic**: Methacyclines, doxycyclines and minocyclines.
5. **Aminoglycoside antibiotics**: The aminoglycosides are a closely related family of antibiotics produced almost exclusively by members of the genera *Streptomyces* and *Micromonospora*. They are polycationic compounds of a amino alcohol to which amino sugar is attached.

Therapeutically important aminoglycosides are streptomycin, neomycin, tobramycin, framycetin, paromomycin, gentamicin, sisomicin and aminoglycoside.

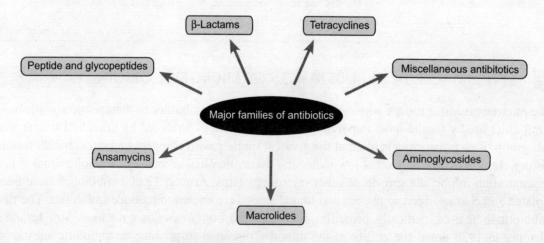

Figure 19.2 Major families of antibiotics.

6. **Macrolides and ansamycins**: They are a large group of antibiotics. They are characterized by a core ring structure containing 12 or more carbon atoms to which one or more sugar is attached.
7. **Ansamycins**: They are synthesized by condensation of a large number of acetate and propionate units. Amongst the best known are the rifampicins that are particularly active against Gram-positive bacteria and mycobacteria.
8. **Peptide and other antibiotics**: They consist of a chain of amino acids that often have cyclized ring-like structure. The first such antibiotics isolated were bacitracin and gramicin.
9. **Vancomycin**: A glycopeptide acts by interfering with bacterial cell wall synthesis and is particularly active against Gram-positive bacteria. Chloramphenicol was first isolated from a culture of *Streptomyces venezuelae* in 1947. But, it is now obtained by direct chemical synthesis and is effective against Gram-negative and Gram-positive bacteria, rickettsias and chlamydias.

Table 19.5 List of recent antibiotics available in market

Generic name	Brand names	Common uses	Mechanism of action
1. Aminoglycosides			
Amikacin	Amikin	In infections caused by Gram-negative bacteria, *E. coli, Pseudomonas aeruginosa*. Effective against aerobic bacteria and tularemia.	Binding to the bacterial 30S ribosomal subunit (some work by binding to the 50S subunit), inhibiting the translocation of the peptidyl-tRNA from the A-site to the P-site and also causing misreading of mRNA, leaving the bacterium unable to synthesize proteins vital to its growth
Gentamicin	Garamycin		
Kanamycin	Kantrex		
Neomycin	Mycifradin		
Netilmicin	Netromycin		
Tobramycin	Nebcin		
Paromomycin	Humatin		
2. Carbacephem			
Loracarbef	Lorabid	A wide variety of bacterial infections	Prevents bacterial cell division by inhibiting cell wall synthesis
3. Carbapenems			
Ertapenem	Invanz	Bactericidal. Empiric broad-spectrum antibacterial coverage.	Inhibition of cell wall synthesis
Doripenem	Doribax		
Imipenem/cilastatin	Primaxin		
Meropenem	Merrem		
4. Cephalosporins (first generation)			
Cefadroxil	Duricef	Gram-positive infections	Same mode of action as other β-lactam antibiotics: disrupt the synthesis of the peptidoglycan layer of bacterial cell walls
Cefazolin	Ancef		
Cefalotin/cefalothin			
	Keflin		
Cefalexin	Keflex		

(Continued)

Generic name	Brand names	Common uses	Mechanism of action
5. Cephalosporins (second generation)			
Cefaclor	Ceclor	Less Gram-positive cover, improved Gram-negative cover	Disrupt the synthesis of the peptidoglycan layer of bacterial cell walls
Cefamandole	Mandol		
Cefoxitin	Mefoxin		
Cefprozil	Cefzil		
Cefuroxime	Ceftin, Zinnat		
6. Cephalosporins (third generation)			
Cefixime	Suprax	Gram-negative organisms, except pseudomonas. Reduced Gram-positive cover	Disrupt the synthesis of the peptidoglycan layer of bacterial cell walls
Cefdinir	Omnicef, Cefdiel		
Cefditoren	Spectracef		
Cefoperazone	Cefobid		
Cefotaxime	Claforan		
Cefpodoxime	Vantin		
Ceftazidime	Fortaz		
Ceftibuten	Cedax		
Ceftizoxime	Cefizox		
Ceftriaxone	Rocephin		
7. Cephalosporins (fourth generation)			
Cefepime	Maxipime	Covers pseudomonal infections	Disrupt the synthesis of the peptidoglycan layer of bacterial cell walls
8. Cephalosporins (fifth generation)			
Ceftobiprole		In methicillin-resistant *Staphylococcus aureus* infection	Disrupt the synthesis of the peptidoglycan layer of bacterial cell walls
9. Glycopeptides			
Teicoplanin	Targocid	Certain kinds of bacterial infection in the bowel	Inhibiting peptidoglycan synthesis
Vancomycin	Vancocin		
Telavancin	Vibativ		
10. Lincosamides			
Clindamycin	Cleocin	Serious streptococcal infections in penicillin-allergic patients. Topically for acne	Binds to 50S subunit of bacterial RNA, thereby inhibiting protein synthesis
Lincomycin	Lincocin		
11. Lipopeptide			
Daptomycin	Cubicin	Gram-positive organisms	Binds to the membrane and cause rapid depolarization, resulting in a loss of membrane potential, leading to inhibition of protein, DNA and RNA synthesis

(Continued)

Generic name	Brand names	Common uses	Mechanism of action
12. Macrolides			
Azithromycin	Zithromax	Streptococcal infections, syphilis. Upper respiratory tract infections. Lower respiratory tract infections, mycoplasmal infections and Lyme disease, pneumonia	Inhibition of bacterial protein biosynthesis by binding reversibly to the subunit 50S of the bacterial ribosome, thereby inhibiting translocation of peptidyl tRNA
Clarithromycin	Biaxin		
Dirithromycin	Dynabac		
Erythromycin	Erythocin, Erythroped		
Roxithromycin	Roxid		
Troleandomycin	TAO		
Telithromycin	Ketek		
13. Monobactams			
Aztreonam	Azactam	In hospital-acquired infections	Same mode of action as other β-lactam antibiotics: disrupt the synthesis of the peptidoglycan layer of bacterial cell walls
14. Nitrofurans			
Furazolidone	Furoxone	Bacterial or protozoal diarrhoea or enteritis	Interferes with several bacterial enzyme systems. It neither significantly alters normal bowel flora nor results in fungal overgrowth
Nitrofurantoin	Macrodantin	Urinary tract infections	Works by killing sensitive bacteria
15. Penicillins			
Amoxicillin	Novamox, Amoxil	In wide range of infections, such as streptococcal infections, syphilis	Common action for all penicillins Disrupt the synthesis of the peptidoglycan layer of bacterial cell walls
Ampicillin	Principen		
Carbenicillin	Geocillin		
Cloxacillin	Tegopen		
Dicloxacillin	Dynapen		
Flucloxacillin	Floxapen		
Mezlocillin	Mezlin		
Methicillin	Staphcillin		
Nafcillin	Unipen		
Oxacillin	Prostaphlin		
Penicillin G	Pentids		
Penicillin V	Pen-Vee-K		
Piperacillin	Pipracil		
Temocillin	Negaban		
Ticarcillin	Ticar		

(Continued)

Generic name	Brand names	Common uses	Mechanism of action
16. Penicillin combinations			
Amoxicillin/ clavulanate	Augmentin	Infections that are proven or strongly suspected to be caused by bacteria	Common action for all penicillin combinations The second component prevents bacterial resistance to the first component
Ampicillin/ sulbactam	Unasyn		
Piperacillin/ tazobactam	Zosyn		
17. Polypeptides			
Bacitracin	Nebasulf	In eye, ear or bladder infections. Applied directly to the eye or inhaled into the lungs; rarely given by injection	Inhibits isoprenyl pyrophosphate, a molecule that carries the building blocks of the peptidoglycan bacterial cell wall outside of the inner membrane Interact with the Gram-negative bacterial outer membrane and cytoplasmic membrane. It displaces bacterial counter ions, which destabilize the outer membrane. They act like a detergent against the cytoplasmic membrane, which alters its permeability
Colistin	Coly-Mycin-S		
Polymyxin B	Neosporine		
18. Quinolones			
Ciprofloxacin	Cipro, Ciproxin, Ciprobay	Urinary tract infections. Bacterial prostatitis, community-acquired pneumonia, bacterial diarrhea, mycoplasmal infections, gonorrhoea	Common action for all quinolones Inhibit the bacterial DNA gyrase or the topoisomerase IV enzyme, thereby inhibiting DNA replication and transcription
Enoxacin	Penetrex		
Gatifloxacin	Tequin		
Levofloxacin	Levaquin		
Lomefloxacin	Maxaquin		
Moxifloxacin	Avelox		
Nalidixic acid	NegGram		
Norfloxacin	Noroxin		
Ofloxacin	Floxin, Ocuflox		
19. Sulphonamides			
Mafenide	Sulfamylon	Urinary tract infections (except sulphacetamide, eye infections, and mafenide and silver sulphadiazine, used topically for burns)	Common action for all sulphonamides Inhibitors of the enzyme DHPS. DHPS catalyses the conversion of PABA to dihydropteroate, a key step in folate synthesis. Folate is necessary for the cell to synthesize nucleic acids (nucleic acids are essential building blocks of DNA and RNA), and in its absence cells will be unable to divide
Sulphadiazine	Micro-sulfon		
Silver sulphadiazine			
	Silvadene		
Sulphamithizole	Thiosulfil forte		
Sulphamethoxazol	Gantanol		
Sulphasalazine	Azulfidine		
Sulphisoxazole	Gantrisin		
Trimethoprime	Proloprim		

(Continued)

Generic name	Brand names	Common uses	Mechanism of action
20. Tetracyclines			
Demeclocycline	Declomycin		Common action for all tetracyclines
Doxycycline	Vibramycin	In syphilis, chlamydial infections, Lyme disease, mycoplasmal infections, acne rickettsial infections, malaria	Inhibit the binding of aminoacyl-tRNA to the mRNA-ribosome complex. They do so mainly by binding to the 30S ribosomal subunit in the mRNA translation complex
Minocycline	Minocin		
Oxytetracycline	Terramycin		
Tetracycline	Sumycin, Achromycin V		
21. Drugs against mycobacteria			
Clofazimine	Lamprene	Antileprotic	Probably interfere with templet function of DNA
Dapsone	Avlosulfon	Antileprotic	Inhibits PABA incorporation into colic acid
Capreomycin	Capastat	Antituberculosis	Inhibits protein synthesis
Cycloserine	Seromycin	Antituberculosis, urinary tract infections	Inhibits bacterial cell wall synthesis
Ethambutol	Myambutol	Antituberculosis	Inhibits arabinogalactan synthesis and to interfere with mycolic acid incorporation in mycobacterial cell wall
Ethionamide	Trecator	Antituberculosis	Inhibits peptide synthesis
Isoniazid	INH	Antituberculosis	Inhibits synthesis of mycolic acid that is unique fatty acid component of mycobacterial cell wall
Pyrazinamide	Aldinamide	Antituberculosis	Inhibits mycolic acid synthesis
Rifampicin (Rifampin in US)	Rifadin, Rimactane	In mostly Gram-positive and mycobacteria	Binds to the β subunit of RNA polymerase to inhibit transcription
Rifabutin	Mycobutin	*Mycobacterium avium* complex	Inhibits DNA-dependent RNA synthesis
Streptomycin		Antituberculosis	As other aminoglycosides
22. Others			
Chloramphenicol	Chloromycetin	In meningitis, MRSA, typhus, cholera. Gram-negative, Gram-positive, anaerobes	Inhibits bacterial protein synthesis by binding to the 50S subunit of the ribosome
Fosfomycin	Monurol	In acute cystitis in women	Inactivates enolpyruvyl transferase, thereby blocking cell wall synthesis
Fusidic acid	Fucidin	In boils, folliculitis sycosis barbae and cutaneous infections	Blocks bacterial protein synthesis
Linezolid	Zyvox	Vancomycin-resistant *Staphylococcus aureus* infection	Inhibits bacterial protein synthesis
Metronidazole	Flagyl	In infections caused by anaerobic bacteria; also amoebiasis, trichomoniasis, giardiasis	Damages DNA in anaerobic microorganisms
Thiamphenicol	Thiocymetin	Used in Gram-negative, Gram-positiver, anaerobes. Widely used in veterinary medicine	Chloramphenicol analogue. May inhibit bacterial protein synthesis by binding to the 50S subunit of the ribosome
Tinidazole	Tindamax, Fasigyn	Protozoan infections	Damages DNA in anaerobic microorganisms

Abbreviations: DHPS, dihydropteroate synthetase; MRSA, methicillin-resistant *Staphylococcus aureus*; PABA, *para*-aminobenzoic acid.

RECENT ADVANCS IN BIOPHARMACEUTICALS

Applications of Carrier Erythrocytes in the Delivery of Biopharmaceuticals

The carrier erythrocytes are resealed erythrocytes loaded by a drug or other therapeutic agents. They have been exploited extensively in recent years for both temporally and spatially controlled delivery of a wide range of drugs and other bioactive agents, because of their remarkable degree of biocompatibility, biodegradability and a number of other potential advantages. The biopharmaceuticals, such as therapeutically significant peptides and proteins, nucleic acid-based biologicals, antigens and vaccines, are being delivered using carrier erythrocytes.

Important Features of Carrier Erythrocytes

1. A remarkable degree of biocompatibility, particularly when the autologous cells are used for drug loading
2. Avoidance of any undesired immune responses against the encapsulated drug
3. Complete biodegradability and the lack of toxic product(s), resulting from the carrier biodegradation
4. Longer lifespan of the carrier erythrocytes in circulation in comparison to the synthetic carriers
5. Considerable protection of the organism against the toxic effects of the encapsulated drug, e.g. antineoplasms
6. Protection of the loaded compound from inactivation by the endogenous factors
7. Desirable size range and the considerably uniform size and shape
8. An easily controllable lifespan within a wide range from minutes to months
9. Relatively inert intracellular environment
10. Possibility of targeted drug delivery to the reticuloendothelial system organs
11. Possibility of an ideal zero-order kinetics of drug release
12. Modification of the pharmacokinetic and pharmacodynamic parameters of the drug
13. Availability of knowledge, techniques and facilities for handling, transfusion and working with erythrocytes
14. Wide variety of compounds with the capability of being entrapped within the erythrocytes
15. Considerable increase in drug dosing intervals with drug concentration in the safe and effective level for a relatively long time
16. Possibility of decreasing drug side effects

Practical Approaches to Dose Selection for First-in-Human Clinical Trials with Novel Biopharmaceuticals

Recent advances in our understanding of disease biology, new therapeutic targets, biomarkers and innovative modalities have fuelled a dramatic expansion in the development of novel human therapeutics. Many biotechnology-derived biologicals with high degree of selectivity and affinity for their intended target are emerging out. As such, they often pose challenges in the development path to approval. One such challenge is the selection of the first-in-human (FIH) dose. This has come under critical scrutiny as a result of a FIH trial with a superagonist monoclonal antibody (TGN1412), which resulted in significant injury to healthy volunteers. Regulatory agencies in such cases have responded with supplemental guidance for the development of novel therapeutics, which are required to be rigidly followed.

Research Advances on Transgenic Plant Vaccines

The advancement of genetics molecular biology and plant biology in recent years has facilitated development of vaccination programs (e.g. genetic engineering subunit vaccine, living vector vaccine, nucleic acid vaccine, etc.). In particular, the technology of the use of transgenic plants to produce human or animal therapeutic vaccines has received considerable attention. Expressing vaccine candidates in vegetables and fruits open up a new approach for producing oral/edible vaccines. The transgenic plant vaccine is a reality today. There are several advantages of transgenic plant vaccines, such as low cost, easiness of storage and convenient immune inoculation. The productions converged in edible tissues can be consumed directly without isolation and purification.

Recent Advances in Analysis of Peptides and Proteins by Mass Spectrometry

The availability of large amounts of purified recombinant proteins together with the possibility of altering their structures promises to revolutionize both the approach for treatment of diseases and our understanding of their molecular bases. Many of these proteins are being developed as potential therapeutic agents, whereas others are being used for diagnostic purposes and for structural studies aimed at elucidation of their mechanisms of actions. It is the cellular machinery of the host organism (e.g. bacteria, yeast or transformed mammalian cells) that defines what modifications of these proteins shall take place. Fast, sensitive and reliable procedures are necessary to bridge the analytical gap between protein chemistry and molecular genetics. The analytical tools of mass spectrometry and tandem mass spectrometry, when used in conjunction with conventional chemical and biochemical techniques, can help to fill this gap, particularly when each approach is used to take advantage of its unique strength.

The engineering of chloroplast genomes is used against herbicides and pathogens. It is used for deveopment of edible crop plants and edible vaccines.

Advances in Chloroplast Engineering

The chloroplast is a pivotal organelle in plant cells and eukaryotic algae to carry out photosynthesis that provides the primary source of glucose. In the last two decades, great progress has been made in chloroplast engineering. The expression of foreign genes in chloroplasts offers several advantages over their expression in the nucleus. The high-level expression, transgene stacking in operons and a lack of epigenetic interference permit stable transgene expression. The transgenic chloroplasts are generally not transmitted through pollen grains because of the cytoplasmic localization.

Chromatography-Free Recovery of Biopharmaceuticals through Aqueous Two-Phase Process

One of the distinct advantages of large-scale aqueous two-phase extraction is its easy scale up together with the possibility of continuous operation using traditional liquid–liquid extractors. This process of extraction would overcome some of the limitations that chromatographic separations are facing. The future of aqueous two-phase process (ATPPs) includes the further development of cost-effective smart polymers, which by responding to a change in the environment (e.g. pH or temperature) are able to change conformation and thereby permitting the recovery of both the biomolecule of interest and the polymer in opposite phases. Another breakthrough could be from the integration of magnetic particles into ATPSs that could speed up phase separation and enhance process throughput.

Recent Advances in Large-Scale Production of Monoclonal Antibodies and Related Products

The biopharmaceuticals and related products commercially produced are listed along with indications and antibody types involved in Table 19.6.

Table 19.6 Monoclonal antibodies and related products in market

Name and year	Target	Indication	Company	Antibody type
Orthoclone OKT3 (1986)	CD3	Acute kidney transplant rejection	Ortho Biotech	Murine
ReoPro (1994)	Platelet GP IIb/IIIa	Prevention of blood clot	Centocor	Murine
Rituxan (1997)	CD20	Non-Hodgkin's lymphoma	Genentech/Biogen-Idec	Chimeric
Panorex (1995)	17A-1	Colorectal cancer	GlaxoSmithKline	Murine
Zenapax (1997)	IL-2Ra (CD25)	Acute kidney transplant rejection	Hoffmann-La Roche	Humanized
Simulect (1998)	IL-2R	Prophylaxis of acute organ rejection	Novartis	Chimeric
Synagis (1998)	RSV	Respiratory syncytial virus	Medimmune	Humanized
Remicade (1998)	TNFα	Rheumatoid arthritis	Centocor	Chimeric
Herceptin (1998)	Her2	Metastatic breast cancer	Genentech	Humanized
Mylotarg (2000)	CD33	Acute myelogenous lymphoma	Wyeth–Ayerst	Humanized

(Continued)

Name and year	Target	Indication	Company	Antibody type
Campath (2001)	CD52	B cell chronic lymphocytic leukaemia	Takeda	Humanized
Zevalin (2002)	CD20	Non-Hodgkin's lymphoma	Biogen-Idec	Murine
Humira (2002)	TNFα	Rheumatoid arthritis	Abbott	Humanized
Bexxar (2003)	CD20	Non-Hodgkin's lymphoma	Corixa/GlaxoSmithKline	Murine
Xolair (2003)	IgE	Allergy	Genentech/Novartis	Humanized
Erbitux (2004)	EGFR/Her1	Colorectal cancer	Bristol-Myers Squibb/Imclone (Eli Lilly)	Humanized
Avastin (2004)	VEGF	Colorectal cancer	Genentech	Humanized
Raptiva (2004)	CD11a	Psoriasis	Genentech/Xoma	Humanized
Tysabri (2004)	A4 integrin	Multiple sclerosis	Biogen-Idec/Elan	Humanized
Vectibix (2006)	EGFR	Colorectal cancer	Amgen	Humanized
Soliris (2007)	C5 complement	PNH	Alexion	Humanized
Stelara (2008)	IL-12 and IL-23	Psoriasis	Centocor	Humanized
Simponi (2008)	TNFα	Rheumatoid arthritis	Centocor	Humanized
Actemra (2009)	IL-6	Rheumatoid arthritis	Roche	Humanized

Abbreviations: IL, interleukin; PNH, paroxysmal nocturnal haemoglobinuria; TNF, tumour necrosis factor.

QUESTIONS

1. Write a note on advantages of DNA technology in the production of biopharmaceuticals.
2. List out biomolecules produced through recombinant DNA technology.
3. Write a note on the recent advances in biopharmaceuticals.
4. Write a note on monoclonal antibodies and related products in market.
5. Give a brief account of plant-derived therapeutics.

Appendix

List of pharma/biotech industries in India and their important products

S. no.	Name of the company	Address	Category of important products/activities
1.	Aurigene Discovery Technologies Ltd.	39-40, KIADB Industrial area, Electronic City, Phase II, Hosur Road, Bangalore 560100, Karnataka www.aurigene.com	Peptide drug discovery, preclinical biology, structural biology, analytical research, formulation and development
2.	Advinus Therapeutics Ltd.	Quantum Towers, Plot No 9, Rajiv Gandhi Infotech Park, Phase-I, Hinjewadi, Pune 411057, Maharashtra www.advinus.com	Preformulation, analytical R&D, drug metabolism, pharmacokinetics, safety pharmacology, clinical pharmacology, bioanalytical
3.	ABL Biotechnologies Ltd.	#1, 2nd Street, Parameshwari Nagar, Adyar, Chennai 600001 Tamil Nadu www.ablbiotechnologies.com	Antibacterials, antivirals, biomolecular products, diagnostics kits, enzymes
4.	Astrazeneca Pharma India Ltd.	Hebbal, Bellary Road, Bangalore 560024 Karnataka www.astrazenecaindia.com	Anti-TB drugs, anti-infectives, anticancer, pain control and anaesthesia, gastrointestinal
5.	Advanced Enzyme Technologies Ltd.	'A' Wing, Sun Magnetica, 5th Floor, Accolade Galaxy, LI Service Road, Louise Wadi, Thane (W) 400604 Maharashtra www.enzymeindia.com	Food enzyme, detergent enzyme, paper enzyme, textile enzyme
7.	Aventis Pharma Ltd.	54/A, Sir Mathuradas Vasanji Road, Andheri (E), Mumbai 400093 Maharashtra www.aventispharmaindia.com	Anti-infectives, antibiotics

S. no.	Name of the company	Address	Category of important products/activities
8.	Bangalore Genei	No. 6, 6th Main, BDA Industrial Suburb, Near SRS Road, Peenya, Bangalore 560058 Karnataka www.bangaloregenei.com	Biotechnological reagents, enzymes, immunoelectrophoresis, teaching kits
9.	Bharat Biotech International Ltd.	Genome Valley, Shameerpet, Hyderabad 500078 Andhra Pradesh www.bharatbiotech.com	Recombinant human epidermal growth factor, vaccines (for hepatitis B, typhoid, poliomyelitis, tetanus, rabies, venom antiserum), recombinant streptokinase
10.	Bharat Serum and Vaccines Ltd.	Hoechst House, 16th Floor, Nariman Point, Mumbai 400021 Maharashtra www.bharatserums.com	Vaccines, plasma products, hormones, diagnostic kits
11.	Biocon India Ltd.	20th KM, Hosur Road, Electronic City, Bangalore 560100 Karnataka www.biocon.com	Biopharmaceuticals, peptides, microbial and proteolytic enzymes, erythropoietin, human insulin, monoclonal antibodies, glycoproteins, probiotics, immunosuppressants
12.	BioGenex India Life Sciences	Plot No. 7, Gunrock Enclave, Secunderabad 500009 Andhra Pradesh www.biogenexindia.com	Molecular pathology, cell and tissue testing, immunohistochemistry detection systems, monoclonal and polyclonal antibodies, cell separation media, immunocolumns
13.	Biological E. Ltd.	18/1 & 3, Azamabad, Hyderabad 500020 Andhra Pradesh www.biologicale.com	Paediatric vaccines (DPT, TT and ATS), anti-snake venom (ASVS), hepatitis B vaccine
14.	Biotech India Pvt. Ltd.	328, Ansal Chamber II, Bhikaji Cama Place, New Delhi 110066 www.biotechindia.com	Gene array, RNA interference, PCR products, biological research reagents, dehydrated culture media for microbiology and molecular biology
15.	Biozeen (Bangalore Biotech Labs Pvt. Ltd.)	Gubbi Cross, Kothanur, Bangalore 560077 Karnataka www.biozeen.com	Consulting, training and research services in biotechnology and pharmaceuticals
16.	Cadila Healthcare	Sarkhej-Bavla N.H. No 8A, Moraiya, Ahmedabad 382210 Gujarat www.zyduscadila.com	Immunodiagnostics and pharmaceuticals
17.	Cipla Ltd.	Mumbai Central, Mumbai 400008 Maharashtra www.cipla.com	Animal products, bulk drugs, flavours and fragrances

S. no.	Name of the company	Address	Category of important products/activities
18.	Dr. Reddy's Laboratories Ltd. (DRL)	Greenlands, Ameerpet Hyderabad 500016 Andhra Pradesh www.drreddys.com	Anticancer, autoimmune diseases, monoclonal antibodies, CNS stimulants, haemopoietic growth factors
19.	Emcure Pharmaceuticals	184, M.I.D.C., Bhosari, Pune 411026 Maharashtra www.emcure.co.in	Antivirals, erythropoietin, recombinant human granulocyte macrophage, colony stimulating factors, vaccines
20.	Genotypic Technologies Pvt. Ltd.	80 Ft Road, R.M.V. 2nd Stage, Bangalore 560094 Karnataka www.genotypictech.com	Microarray and sequencing data analysis, microarray and sequencing services, gene expression, miRNA, CGH, CNV and chip-on-chip profiling, transcriptome, methylome sequencing services, whole genome and low coverage de novo sequencing, biointerpreters
21.	Gland Pharma Ltd.	6-3-865/1/2, Flat No 201, Green Land Apts, Ameerpet, Hyderabad 500016 Andhra Pradesh www.glandpharma.com	Antibiotics, antifungals, antiarthritic agents, viscosupplement low molecular weight heparin
22.	GVK Biosciences Pvt. Ltd.	Plot No. 28A, IDA Nacharam Hyderabad 500076 Andhra Pradesh www.gvkbio.com	Advanced bioinformatics, biostatistics, protein modelling and rational drug design, resins and chemicals
23.	GlaxoSmithKline Pharmaceuticals Ltd. (GSK India)	Dr. Annie Besant Road Worli, Mumbai 400030 www.gsk-india.com	Anti-infective, antidiabetic vaccines, vitamins
24.	Intas Pharmaceuticals Ltd.	Chinubhai Center, Off Nehru Bridge, Ashram Road, Ahmedabad 380009 Gujarat www.intaspharma.com	Antibacterials, anticoagulants, antiplatelets, anti-infectives
26.	Ipca Labs Ltd.	142 AB, Kandivli Industrial Estate, Kandivli (W), Mumbai 400067 Maharashtra www.ipcalabs.com	Anti-infectives, anticoagulant, antifungal, antibiotics
27.	Jubilant Biosys Ltd. (JBL)	#96, Industrial Suburb, 2nd Stage, Yeshwantpur, Bangalore 560022 Karnataka www.jubilantbiosys.com	Collaborative research with major biotechnological and pharmaceutical industries in discovery biology, medicinal chemistry, structural biology, ADME, toxicology, pharmacology, molecular modelling, pharma information technology

S. no.	Name of the company	Address	Category of important products/activities
28.	Jupiter Biosciences Ltd.	Cherlapally, Hyderabad 500061 Andhra Pradesh www.jupiterbioscience.com	Amino acids, peptide synthesis, fine chemicals, drug intermediates
29.	Karnataka Antibiotics & Pharmaceuticals Ltd. (KAPL)	Nirman Bhawan, 80 Feet Road, 1st Block, Rajaji Nagar, Bangalore 560010 Karnataka www.kaplindia.com	Antimicrobials, antibiotics, diagnostic kits, growth promoters
30.	Kopran Limited	Parijat House, 1076, Dr. E. Moses Road, Worli, Mumbai 400018 Maharashtra www.kopran.com	Antibiotics, antivirals, life-saving medical equipments, diagnostics reagents
31.	Khandelwal Laboratories Pvt. Ltd.	79/87 D. Ladpath, Mumbai 400033 Maharashtra www.khandelwallab.com	Antibacterials, application of nanoparticles in cancer hyperthermia, targeted drug delivery, diagnostics
32.	Lupin Ltd.	Laxmi Towers, "C" Wing, Bandra Kurla Complex, Mumbai 400051 Maharashtra www.lupinworld.com	Antibiotics, anti-infectives, fibrinolytics, hormones, pregnancy and diagnostic kits
33.	Life Cell Ltd.	No. 26. Vandalur Kelambakkam, Main Road, Keelakottaiyur Chennai 600048 Tamil Nadu www.lifecellinternational.com	Stem cell therapy and stem cell banking, haematopoietic cell therapy, bone marrow stem cell therapy, umbilical cord blood stem cell therapy, peripheral blood stem cell therapy
34.	Magene Life Sciences	D. No. 14-59/3, Pudur Post, Raval Kole X Roads, Medchal Mandal, Hyderabad 501401 Andhra Pradesh www.magenelifesciences.com	Biotechnology research, drug discovery, pharmaceutical and clinical research
35.	Mediclone Biotech Pvt. Ltd.	Block-B, 36/37, No. 144 "Millennium House", M.K. Srinivasan Nagar, Old Mahabalipuram Road, Chennai 600096 Tamil Nadu www.mediclonebiotech.org	Immunobiotechnological products, antibody-based kits
36.	Novartis India Ltd.	Sandoz House, Shivsagar Estate, Dr. Annie Besant Rd., Mumbai 400018, Maharashtra www.novartis.com	Vaccines, diagnostics, antibiotics

S. no.	Name of the company	Address	Category of important products/ activities
37.	Nicholas Piramal India Ltd.	10th Floor, Nicholas Piramal Towers, G.K. Marg, Lower Parel, Mumbai 400013 Maharashtra www.piramalpharmasolutions.com	Preformulations, formulation development, clinical trial services, hormonal formulations
38.	Nycomed Pharma Pvt. Ltd.	Plot No. 29-31, Suren Road, Andheri (E), Mumbai 400093 Maharashtra www.nycomed.in	Tissue management, gastroenterology, pain management, drug discovery research and clinical trials
39.	Omega Biotech Ltd.	#56, 3rd Cross, Kumar Layout, Talaghattapura, Kanakapura Road, Bangalore 560062 Karnataka www.omegabiotec.com	Plasmid isolation kits, genomic DNA isolation kits, RNA isolation kits, DNA/RNA cleanup kits, electrophoresis kits
40.	Ocimum Biosolutions Ltd.	6th Floor, Reliance Classic, Road No. 1, Banjara Hills, Hyderabad 500034 Andhra Pradesh www.ocimumbio.com	Gene and microRNA expression profiling, SNP genotyping, copy number analysis, gene sequencing, support for clinical trials and biomarker discovery, gene expression databases, toxic genomics database, predictive toxic genomics, models/ tools for genomic data analysis
41.	Orchid Pharmaceuticals Ltd.	Plot No. 476/14, Old Mahabalipuram Road, Sholinganallur, Chennai 600119 Tamil Nadu www.orchidpharma.com	Active pharmaceuticals, ingredients such as oral cephalosporins, sterile veterinary cephalosporins, anti-infectives, noncephalosporins, beta-lactams, coenzymes
42.	Panacea Biotech Ltd.	B-1 Extn. A-27, Mohan Co-opp. Industrial Estate, Mathura Road, New Delhi 110044 www.panacea-biotec.com	Vaccines, cephalosporins
43.	Pragati Biocare Pvt. Ltd.	No. 81/3, Govindappa Road, Basvangudi, Bangalore 560004 Karnataka www.pragatibiocare.com	Biotechnology-based animal nutrition, plant nutrition, biopharmaceuticals
44.	Ranbaxy Laboratories Ltd.	Plot 90, Sector 32, Gurgaon 122001 Haryana www.ranbaxy.com	Antibiotics, anti-infectives, cephalosporins, statins
45.	RPG Life Sciences Ltd.	# 463, Ceat Mahal, Dr. Annie Besant Road, Worli, Mumbai 400030 Maharashtra www.rpglifesciences.com	Biopharmaceuticals in cancer therapy, cardiology and immunosuppressants

S. no.	Name of the company	Address	Category of important products/activities
46.	Reliance Life Sciences	Dhirubhai Ambani Life Sciences Centre, MIDC, Rabale, Navi Mumbai 400701 Maharashtra www.relbio.com	Biopharmaceuticals, hormones, peptides, steroids, clinical services, embryonic stem cells, ocular stem cells, haematopoietic stem cells, skin and tissue engineering, assisted reproduction
47.	Serum Institute of India Ltd.	212/2, Hadapsar, Off Soli Poonawalla Road, Pune 411028 Maharashtra www.seruminstitute.com	Antisera, plasma and hormonal products, anticancer, vaccines, proteins and recombinant DNA products
48.	Sami Labs Ltd.	I Main, II Phase, Peenya Industrial Area, Bangalore 560058 Karnataka www.samilabs.com	Enzymes, probiotics and synbiotics, functional nutrition supplements, specialty fine chemicals, cosmeceuticals
49.	Shantha Biotechniques Ltd.	H.No. 5-10-173, 3rd & 4th Floor, Vasantha Chambers, Fateh Maidan Road Basheerbagh, Hyderabad 500004 Andhra Pradesh www.shanthabiotech.com	Vaccines, recombinant human erythropoietin, recombinant interferons, pentavalent vaccines
50.	Sudershan Biotech Ltd.	4-100, C/9 & C/10, Doctor's Colony, L. B. Nagar, Hyderabad 500074 Andhra Pradesh www.sudershanbio.com	Industrial enzymes, diagnostic proteins, therapeutic proteins, human serum albumin (HSA), interferon beta, hepatitis C vaccine, erythropoietin
51.	Syngene International Ltd.	Biocon Park, Bommasandra Industrial Estate-Phase-IV, Bangalore 560099 Karnataka www.biocon.com/syngene	Collaborative research with leading biotech and pharmaceutical industries for biopharmaceuticals preparations
52.	Titan Biotech Ltd.	A-2/3, 3rd Floor, Lusa Tower, Azadpur Commercial Complex, Delhi 110033 www.titanbiotechltd.com	Laboratory chemicals, molecular biology chemicals, biotechnological reagents
53.	Torrent Pharmaceuticals Ltd.	Nr. Kanoria Hospital, Village Bhat, Gandhinagar 382428 Gujarat www.torrentpharma.com	Antimicrobial, anti-infectives, antifungal, antihaemophilic, pain management
54.	Themis Laboratories Pvt. Ltd.	Khira Industrial Estate, B.M. Bhargava Road Santacruz (W), Mumbai 400054 Maharashtra www.themislabs.com	Active pharmaceutical ingredients (API), anti-infectives, antibiotics

S. no.	Name of the company	Address	Category of important products/ activities
55.	USV Ltd.	B.S. Devshi Marg, Govandi, Mumbai 400088 Maharashtra www.usvindia.com	Biotherapeutics, biosimilars, peptides, growth hormone inhibitors, small molecules
56.	Vimta Labs	142, IDA, Phase II, Cherlapally, Hyderabad 500051 Andhra Pradesh www.vimta.com	CRO in genomics, proteomics, chemistry, molecular biology, microbiology and informatics, bioanalyticals
57.	Wockhardt Healthcare Ltd.	Wockhardt Towers, Bandra-Kurla Complex, Bandra (East), Mumbai 400051 Maharashtra www.wockhardt.com	Biopharmaceuticals, vaccines, hepatitis B vaccine, erythropoietin, recombinant insulin
58.	Zydus Cadila Ltd.	Sarkhej-Bavla, N.H. No 8A, Moraiya, Ahmedabad 382213 Gujarat www.zyduscadila.com	Vaccines, biologics, biosimilar therapeutic proteins, biosimilar monoclonal antibodies, biobetters and novel biologics, immunodiagnostics agents, pharmaceuticals

Glossary

Adjunct: It refers to an auxiliary treatment that is secondary to the main treatment.

Adjuvant: Insoluble substance capable of increasing the formation and persistence of antibodies when injected with an antigen. It is a substance or drug that aids another substance in its action.

Aerobe: The microorganism that can grow and multiply in the presence of oxygen.

Agglutinate: To cause to unite and adhere as if glued; to gather into clump or mass as a protoplast and bacteria in the presence of specific antibodies.

AIDS: Acquired immune deficiency syndrome.

Allopolyploid: Hybrids between two (or more) species or genera with the sum of the complete diploid chromosome complements of each parent.

Anaerobes: The microorganisms that can grow and multiply in the absence of oxygen.

Aneuploid: The nucleus of a cell when does not contain an exact haploid number of chromosomes, one or more chromosomes being represented more or less times than the rest is referred to as *aneuploid*. The chromosomes may or may not show rearrangements.

Antibiotic: Natural organic substance or its synthetic analogue produced by plants or microorganism that is toxic to other species and is capable of retarding or preventing its growth and presumably functioning as defence mechanism (e.g. bacitracin, gentamycin, mycostatin, nystatin, penicillin, phosphomycin, etc.).

Antibody: A protein (immunoglobulin) synthesized by a B lymphocytes that recognizes a specific site on an antigen. The basic immunoglobulin molecule consists of two identical heavy and two identical light chains.

Antigen: A compound that elicits immune response and induces the production of antibodies.

Antioxidant: A substance such as ascorbic acid, citric acid, etc. that is sometimes added to a sterilizing solution or to an isolation medium to inhibit or prevent oxidative browning of the culture medium.

Autoclave: A closed chamber in which substances to be sterilized is heated under pressure to above their boiling point.

Autoimmunity: An abnormal immune response against human body's own components.

Autologous cells: Cells taken from an individual, cultured, stored and sometimes genetically manipulated.

Autonomous replicating sequence (ARS): Any cloned DNA sequence that initiates and supports extrachromosomal replication of DNA molecule in a host cell (often used in yeast cells).

Autoradiography: Refers to a technique that captures the image formed in a photographic emulsion as a result of the emission of light or radioactivity from a labelled component that is placed next to unexposed film. The process of detection of radioactively labelled molecules by exposure of an X-ray-sensitive film.

Autosomes: The chromosomes that are not involved in sex determination.

Autotroph: An organism capable of synthesizing its own food (e.g. plants through photosynthesis).

Auxins: A category of plant growth regulators or phytohormones that are involved in cell enlargement and elongation, root initiation, etc. (e.g. indole acetic acid, indole butyric acid, etc.).

Axenic culture: It refers to a culture of an organism that is entirely free of all other contaminating organism

B and T cell lymphomas: The cancers caused by proliferation of the two principal types of white blood cells: B and T lymphocytes.

Bactericide: A substance or agent that rapidly kills bacteria.

Bacteriophage: A virus that infects a bacterium. It is also called *phage*.

Bacteriostat: An agent that prevents bacterial growth and its multiplication without killing it.

Baker's yeast: The living cells of aerobically grown yeast, *Saccharomyces cerevisiae*, used in preparation of bread.

Batch culture (batch fermentation): A cell or microbial suspension culture of a set volume grown in liquid medium for a limited time. The inoculum of successive subculture is of similar size, and the culture contains about the same cell mass at the end of each passage.

Bioassay: It is performed on living cells or on a living organism, used to detect and quantify minute amounts of substances that influence or are essential to growth.

Biomass: All organic matter that is produced from the photosynthetic conversion of solar energy.

Bioprocess: A process of biological significance in which living cells or components thereof are used to produce a desired end product.

Bioreactor: It is a fermenter or containment system for fermentation processes.

Biosynthesis: It refers to biological synthesis capable of the building or forming of a biochemical compound in a living organism.

Biotechnology: It refers to application of technology to biosciences and covers industrial use of biological processes; such as yeast fermentation for alcohol production or plant cell culture for extraction of secondary products.

Callus: Unorganized plant cell mass capable of in vitro repeated cell division and growth.

Callus culture: The systematic in vitro cultivation of callus on solidified nutrient medium initiated by inoculation of small explants or sections from established organ or other cultures. These cultures can be maintained indefinitely by regular subdivision and subculture using fresh nutrient medium and may be used as the basis for organogenetic cultures, cell cultures or proliferation of embryoids.

Cell culture: The culture of single or groups of cells grown on solid nutrient or dispersed in liquid nutrient medium (cell suspension culture) under defined laboratory conditions.

Cell hybridization: The fusion of two or more dissimilar cells, leading to the formation of a synkaryon.

Cell line: It refers to cells that acquire ability to multiply indefinitely in vitro.

Cell number: The absolute number or approximation of the number of cells per unit area of a culture or medium volume.

Cell strain: It is derived either from a primary culture or from a cell line by the selection or cloning of cells having specific properties or markers. The properties or markers usually persist during subsequent cultivation.

Chromosomes: The threads of DNA in the nucleus that carry genetic inheritance.

Clone: It is a group of cells that descended from a common ancestor. In genetic engineering, it usually refers to a cell carrying a foreign gene.

Cloning vector: It is a plasmid or virus into which a foreign gene is placed to ensure its replication in a new host cell.

Complementary DNA (cDNA): It is a DNA strand formed from messenger RNA using the enzyme reverse transcriptase.

Conjugation: The transfer of genetic material from one cell to another cell.

Contaminant: It refers to an undesirable bacterial, fungal or algal microorganism accidentally introduced into a culture or culture medium that may overgrow the plant cell. It can inhibit the growth of desired cells through release of toxic metabolites.

Culture medium: A chemically well-defined and prepared nutrient solution for in vitro growth of plant tissues or other organisms.

Dehumidifier: An equipment that removes moisture from the air.

Deionized water: The water that is passed through an ion-exchange device to remove soluble minerals and some organic salts. Such a process is known as *deionization*.

Detergent: It is a synthetic cleaning agent chemically different from soap. It is commonly used to prepare work areas for aseptic plant tissue culture manipulations and to remove dirt and microorganisms from plant material prior to explantation.

Distilled water: The water that has been converted to steam and recovered as condensed vapours. This process removes dissolved materials, particulates and microorganisms from water. The process is usually repeated for added purity, often in a glass apparatus, to minimize metal recontamination. This purified distilled water is commonly used to make nutrient media for plant tissue culture.

Dual culture: It is a culture system that includes plant tissue and one organism (nematode species) or microorganism (fungus). Dual cultures are usually used for studying host–parasite interactions or in the production of axenic cultures for a variety of purposes.

Endogenous: Arising from within the body.

Enzyme: A class of specialized proteins produced in living cells that controls biological processes. The rate of specific chemical reaction, even at very low concentration (organic catalyst) is monitored by enzyme, but it is not used in the reaction.

Equimolar: The same amount of solute per litre of solution (the same molar concentration).

Erlenmeyer flask: It is named after Erlenmeyer (1825–1909). It is a wide-necked variety of conical, flat-bottomed glass flask, commonly used for preparation of medium. The smaller versions of these flasks are used as culture containers.

Established culture: It a suspension culture subjected to several passages and adjusted so well that the cell number per unit time is constant from subculture to subculture.

Excise: It refers to cutting, severing or otherwise removing or extracting an organ or a segment of tissue from a plant or plant part; as the surgical removal of shoot tips, the process is known as *excision*.

Explant: It is the excised plant portion used to initiate a tissue culture. The process of dissection and removal to culture of these small organs or tissue sections is known as *explantation*.

Explant donor: The source plant or mother plant from which explant is used to initiate a culture.

Ex vitro (Latin *from glass*): The organisms removed from culture and transplanted into soil or potting mixture.

Fermentation: The process of biological significance by which microorganisms turn raw materials such as glucose into products such as alcohol.

Gene: It is a unit of heredity and a segment of DNA coding for a specific protein.

Gene therapy: It refers to therapy at the intracellular level to replace or inactivate the effects of disease-causing genes or to augment normal gene functions to overcome illness.

Gene transfer: The process involving use of genetic or physical manipulations to introduce foreign genes into host cells for achieving characteristics in progeny.

Genetic engineering: A branch of biotechnology (including DNA technologies) used to isolate genes from an organism, manipulate them in laboratory and insert them into another cell system.

Genetic engineering or recombinant DNA technology: It refers to technology involving man-made changes in the genetic constitutions of cells (apart from selective breeding). The technology usually employs a vector (e.g. the Ti plasmid of *Agrobacterium tumefaciens*) for transferring useful genetic information from a donor organism into a cell or organism that does not posses it.

Genetic selection: The process of selection of genes, cell sections, clones, etc. by man within population or between populations or species. The purpose is usually to alter a specific phenotypic character. Such selection usually results in differential success rates of the various genotypes.

Genetic transformation: It refers to transfer of extracellular DNA (genetic information) amongst and between species with the use of bacterial or viral vectors.

Genome: It refers haploid set of chromosomes with its associated genes.

Germplasm: The reproductive body tissues distinct from nonproductive tissues (somatic). The genetic material, basis of heredity of an organism, passed on through previous generations.

Gibberellic acid: Gibberellin A3 or GA_3 ($C_{19}H_{22}O_6$, mw 346.37) is one of the gibberellins; a group of growth hormones promoting cell division and elongation. The first of the group to be isolated and most widely used gibberellins in plant tissue culture isolated from the fungal pathogens *Gibberella fujikuroi*. It dissolves in base (ca. 1M KOH or NaOH).

Growth inhibitor: Any substance by its own or from another organism that inhibits the growth of an organism. The inhibitory effect can range from mild inhibition (growth retardation) to severe inhibition or death (toxic reaction). Two hormones that may act as inhibitors are ethylene and abscisic acid (ABA).

HEPA filter: An acronym for high-efficiency particulate air filter. A filter capable of screening out particles larger than $0.3\ \mu$ (designed for fine filtration up to $0.3\ \mu$ at efficiency of 99.99%). They are used in laminar air flow cabinets (hoods) for sterile transfer work.

IAA: Indole-3-acetic acid.

IBA: Indole-3-butyric acid.

Immobilized enzyme: An enzyme that is physically defined or localized in a defined region, enabling it to be reused in a continuous process.

Immunomodulator: An agent capable of causing general stimulation of the immune system to counter disease.

In situ: In the natural, original place or position.

In vitro (Latin *in glass*): Experimentation on organisms or their portions in glassware or culture growing under artificial conditions. The antonym of in vivo.

In vivo (Latin *in life*): Experimentation on organisms under natural conditions within intact living organisms. The antonym of in vitro.

Incubator: An equipment providing controlled environmental conditions, (light, photoperiod, temperature, humidity, etc.) suitable for incubating plant cells and cultures.

Inoculum: Inoculated material introduced onto or into a host or nutrient medium.

Interferon: A glycoprotein naturally produced by cells that interfere with the ability of a virus to reproduce after it invades the body. Interferon can curtail the spread of certain types of cancer.

Interleukin: It is a type of lymphokine whose role in the immune system is being extensively studied. Two types of interleukins have been identified: (1) interleukin-1 (IL-1), derived from macrophages, amplifies the production of other lymphokines and (2) interleukin-2 (IL-2). It regulates the maturation and replication of T lymphocytes.

Kinetin (K, Kn): (2-furanylmethyl)-1H-purin-6-amine.

Kinin: The original class name for substances (growth hormones) promoting cell division to which the prefix *cyto-* has been added (cytokinins) to distinguish them from kinins in animal systems.

Laminar air flow cabinet or hood: A structure capable of providing a uniform flow of filtered air in the work area. The air is filtered through prefilter (furnace-filter quality), then a high-efficiency particulate air (HEPA) filter that strains out particles greater than 0.3 μ. These hoods are employed in plant tissue culture for aseptic manipulations

Lipoproteins: A class of serum proteins capable of transporting lipids and cholesterol in the blood streams. Any abnormalities in the lipoprotein metabolism have been implicated in certain heart diseases.

Liposomes: These are thermodynamically stable vesicular structures, consisting of one or more concentric spheres of lipid bilayers separated by water or aqueous compartments.

Liquid culture: The culture of plant cells in liquid medium, in suspension or on supports.

Lysis: It refers to breaking a part of cells.

Macrophage: It is a type of white blood cell produced in blood vessels and loose connective tissues capable of ingesting dead tissues and cells. It is involved in production of interleukin-1 (IL-1). When exposed to lymphokine macrophage activating factor. Macrophage also kills tumour cells.

Meristem culture: This generally implies to the culture of meristemoidal regions of plants or meristematic growth (associated with or sharing the characteristics of meristem) in culture.

Messenger RNA (mRNA): Nucleic acid segment that carries instructions to a ribosome for synthesis of particular protein.

Micropropagation: It refers to a technique involving use of small pieces of tissue, such as meristem, grown in nutrient culture to produce large number of plants.

Monoclonal antibodies: These are antibodies that are obtained from a single source or clone of cells capable of recognizing only one kind of antigen.

Mutagen: A substance capable of inducing mutations.

Mutant: It refers to a cell that manifests new characteristics due to change in DNA.

Mutation: It refers to the phenomenon of stable changes of a gene inherited on reproduction.

Natural active immunity: Immunity that is acquired after the occurrence of a disease.

Organ culture: The maintenance or growth of organ primordial or whole or part of an organ in vitro in such a way that it allows differentiation and preservation of the architecture and the cellular function.

Ovulary culture: It is a culture (in vitro) in which the explant is an ovary containing the ovules. This technique is primarily employed when the ovary is essential for proper embryo development.

Ovule culture: It is a culture (in vitro) derived from an explanted ovule. This technique is useful in study or development of zygotes and young embryos.

Packed cell volume (PCV): It is a quantitative method of estimating cell growth based on the total cell volume in an aliquot of suspension culture.

Papain: It is a water-soluble proteolytic enzyme (protease) obtained from papaya fruit (*Carica papaya*) and used especially as a meat tenderizer.

Parathormone: A substance with hormone-like properties that is not a secretary product (e.g. ethylene or carbon dioxide).

Passive immunity: Immunity acquired after receiving performed antibodies.

Peptide: A compound consisting of two or more amino acids covalently linked. Peptides with three or more amino acids are known as *polypeptides*.

pH: It is a measurement of the degree of acidity or alkalinity of a solution. The negative log of the hydrogen ion concentration, in moles per litre (mol/L).

Phenotype: It refers to a phenomenon of physical characteristics of an organism, affected by both genotype and environment.

Phytokinin: An absolute term for cytokinin.

Phytotron: It is a controlled environment chamber or building for studying plant cells, tissues, organs and whole plantlets.

Plantlet: A shoot, sometimes exclusively a rooted shoot, growing in culture or derived from culture.

Polygene: A system of genes associated with quantitative character variation in which each gene individually affects the many labware products.

Polypeptide: These are long chain of amino acids joined to peptide bond.

Primary culture: A culture initiated from cells, tissues or organs taken directly from an organism. Until then, it is referred to as *primary culture*. A primary culture is taken up for subculturing.

Probe DNA: A radioactively labelled (usually $32p$) DNA molecule used to detect complementary sequence nucleic acid molecules.

Promeristem or protomeristem: It is the embryonic meristem containing organ initials or foundation cells.

Proteins: These are large biomolecules consisting of amino acids, the products of genes.

Protoplast: It is plant or microbial cell whose wall has been removed so that the cell assumes a spherical shape.

Protoplast culture: It refers to the isolation and culture of plant protoplast by mechanical means or by enzymatic digestion of plant tissues or organs, or cultures derived therefrom. Protoplasts are usually utilized for selection or hybridization at the cellular level.

Protoplast fusion: The coalescence of the plasmalemma and cytoplasm of two or more protoplasts in contact with one another. Initial adhesion is a random process, but coalescence may be promoted in various ways (including fusion). In the absence of a fusion agent, when adhesion occurs between adjacent protoplasts during enzymatic wall degradation or between freshly isolated protoplasts, it is termed *spontaneous fusion*.

Pure culture: Axenic culture.

Radiation: It refers to rays of heat, light or particles in wave form.

Radioimmunoassay (RIA): It is a technique for quantification of a substance by measuring the reactivity of radioactively labelled forms of the substance under investigation with antibodies. Moreover, it includes a variety of techniques that use radiolabel reagents to detect antigen or antibody.

Recombinant DNA: It is genetic material with novel gene sequences produced by crossovers, chromosome reassortment, by other natural means or through genetic engineering. It is hybrid DNA produced by joining pieces of DNA from different organism.

Restriction enzyme: It is an enzyme that breaks DNA in highly specific locations, creating gaps into which new genes can be inserted.

Root culture: Isolated root tips of apical produce in vitro root system with indeterminate growth habits. These were among the first kinds of plant tissue cultures and remain an important research tool in the study of development phenomena.

Somatic cell embryogenesis: It is process of the production of embryos from somatic cells of explants (direct embryogenesis) or by induction on callus formed by explants (indirect embryogenesis).

Somatic cell variant or embryoid: It is an organized embryonic structure morphology similar to a zygotic embryo but, initiated from somatic (nonzygotic) cells. These structures ultimately develop into plantlets (in vitro) through developmental processes that are similar to those of zygotic embryos.

Somatic hybrid: A cell or plant product formed as the result of cell or protoplast fusion and implying genomic integration. The process is known as *somatic hybridization*.

Suspension culture: A type of culture in which the cells multiply while suspended in the medium without becoming attached to an inert surface.

T lymphocytes (T cells): White blood cells are produced in the marrow but mature in the thymus. They are important in the body's defence against certain bacteria and fungi. They help B lymphocyte make antibodies and help in the recognition and rejection of foreign tissue. T lymphocytes may also be important in the body's defence against cancers.

Tissue culture: Process where individual cells or clumps of plant or animal tissue are grown artificially using a nutrient medium.

Toxin: It is an antigenic substance produced by microorganisms.

Toxoid: It refers to inactivated antigen.

Transfer RNA (tRNA): RNA molecule that carries amino acids to sites on ribosomes, where protein is synthesized.

Transformation: It refers to change in genetic structure of an organism by incorporation of foreign DNA.

Transgenic organism: It is an organism formed by the insertion of foreign genetic material into the germ cell lines of organisms. Recombinant DNA technique is commonly employed to produce transgenic organism.

Vaccine: A preparation that contains an antigen consisting of whole disease-causing organisms (killed or weakened), or parts of such organisms, used to confer immunity against the disease that the organism cause. Vaccine preparation can be natural, synthetic or derived by recombinant DNA technology.

Vector: It is the agent (e.g. plasmid or virus) used to carry new DNA into a cell.

Bibliography

1. Abhilash M. Applications of proteomics. *The Internet Journal of Genomics and Proteomics*. 2009;4(1)1–4.
2. Albert B. *Molecular Biology of the Cell*. 4th ed. New York: Garland Publication Inc.; 2002, pp. 1065–1125.
3. Ana MA, Paula AJ, Rjosa I, et al. Chromatography-free recovery of biopharmaceuticals through aqueous two-phase processing. *Trends Biotechnol*. 2009;27(4).
4. Anson BD, Ma J, He JQ. Identifying cardiotoxic compounds. *Genet Eng Biotechnol News*. (Mary Ann Liebert). 2009;29(9):34–35.
5. Arakawa T, Chong DKX, Langridge WHR. Efficacy of a food plant-based oral cholera toxin B subunit vaccine. *Nat Biotechnol*. 1998;16.
6. Arora DR. Quality assurance in microbiology, Indian. *J Med Microbiol*. 2004;22(2).
7. Asian Development Bank (ADB). *Water for All: The Water Policy of the Asian Development Bank*. Manila: ADB; 2001.
8. Astbury WT. Molecular Biology or Ultrastructural Biology. *Nature*. 1961;190:1124.
9. Barz W, Reinhard E, Zenk MH. *Plant Tissue Culture and Its Biotechnological Applications*. Berlin: Springer; 1977.
10. Bernard RG, Pasternak JJ, Patten CL. *Molecular Biotechnology, Principles and Applications of Recombinant DNA*. 4th ed. Washington, DC: ASM Press; 2009.
11. Bernard SG, Dan ST. Catalytic monoclonal antibodies: Tailor-made, enzyme-like catalysts for chemical reactions. *Trends Biotechnol*. 1989;7(11):304–310.
12. Bradford CB, Fernando AG, Bi-Xing C, et al. X-ray crystal structure of an anti-Buckminsterfullerene antibody, Fab fragment: Biomolecular recognition of C60. *Proc Natl Acad Sci*. 2000;97:12193–12197.
13. Bridson EY. Culture Media. In: *The Oxoid Manual*. 8th ed. England: Oxoid Limited; 1998.
14. Cass AEG. *Biosensors, A Practical Approach*. Oxford: IRL Press; 1990.
15. Chaerkady R. Applications of Proteomics to Lab Diagnosis, Annual review of pathology. *Mech Dis*. 2008;3:485–498.
16. Christina R. Immunomodulators in the treatment of cutaneous lymphoma. *J Euro Acad Dermatol Venereol*. 1999;13(2):83–90.
17. Cournoyer D, Thomas Caskey MD. Gene Transfer into Humans: A first step. *N Engl J Med*. 1990;323:601–603.
18. Cristina B, Ivan P, Kevin R. Nanomaterials and nanoparticles, Sources and toxicity. *Biointerphases*. 2007;2(4):MR17–MR71.
19. Bruce D, Bruce A. *Engineering Genesis, The Ethics of Genetic Engineering*. London: Earthscan Publications; 1999.
20. Daan JAC, Robert DS. *Pharmaceutical Biotechnology*. 2nd ed. London: Routledge; 2002.
21. Daniel CL. *Introduction to Proteomics, Tools for the New Biology*. New York: Humana Press; 2001.
22. De Aizpura HJ, Russell-Jones GJ. Oral vaccination, identification of classes of proteins that provoke an immune response upon oral feeding. *J Exp Med*. 1998;167:440–451.
23. Cournoyer D, Caskey CT. Gene transfer into humans: A first step. *N Engl J Med*. 1990;323:601–603.

24. Drury H. Topical immunomodulators progress towards treating inflammation, infection, and cancer. *Lancet Infect Dis.* 2002;2(4):259.
25. Dubey RC. *A Text Book of Biotechnology.* New Delhi: S. Chand and Company Ltd.; 2001.
26. Duguid JP, Collee JG, Fraser AG, et al. Organization of the clinical bacteriology laboratory; Quality assurance. In: *Mackie & McCartney Practical Medical Microbiology.* 14th ed. London: Churchill Livingstone;1996.
27. European Commission. Wonders of Life. Stories from Life Sciences Research (from the Fourth and Fifth Framework Programmes). Luxembourg: Office of Official Publications of the European Communities; 2002;27.
28. Fiechter A. *Advances in Biochemical Engineering, Plant Cell Cultures.* Berlin: Springer-Verlag; 1980;16.
29. *First Human Trial* shows that an edible vaccine is feasible, National Institute of Allergy and Infectious Diseases (NIAID); 1998.
30. Gamborg OL, Wetter LR. Plant Tissue Culture Methods. Saskatoon, Sask. National Research Council of Canada, Prairie Regional Laboratory; 1975.
31. Walsh G. *Pharmaceuticals Biotechnology: Concepts and Applications.* England: John Wiley & Sons Ltd.; 2007.
32. Zylstraa GJ, Kukor JJ. What is environmental biotechnology? *Curr Opin Biotechnol.* 2005;16(3):243–45.
33. Gursel I, Gursel M, McCormac B. Liposomes as immunological adjuvants and vaccine carriers. *J Control Release.* 1996;41(1–2): 49–56.
34. Daniell H, Chase C. Molecular Biology and Biotechnology of Plant Organelles, Chloroplasts and Mitochondria. Springer; 2004.
35. Daniell H, Khan MS, Allison L. Milestones in chloroplast genetic engineering: An environmentally friendly era in biotechnology. *Trends in Plant Science.* 2002;7(2):84–91.
36. Hongbao Ma, Guozhong Chen. Gene transfer technique. *Nat Sci.* 2005;3(1).
37. Hopkinson J. Hollow fiber cell culture systems for economical cell-product manufacturing. *Biotechnology.* 1985.
38. Denyer SP, Hodges NA, Gorman SP. *Hugo and Russell's Pharmaceutical Microbiology.* 7th ed. USA: Wiley-Blackwell; 2004.
39. Hwee-Ling K. Current trends in modern pharmaceutical analysis for drug discovery. *Drug Discov Today.* 2003;8(19):889–897.
40. Ishihara K. Production of anti-insulin monoclonal antibodies and its clinical application. *Diabetes Research and Clinical Practice.* 1989;7(1):S73–S76.
41. Sinisterra JV. Application of ultrasound to biotechnology. *Ultrasonic.* 1992;30(3):180–185.
42. Jalalpure SS, Salahuddin MD Antidiabetic activity of aqueous fruit extracts of *Cucumis trigonus. Roxb.* In: *Streptozotocin-Induced-Diabetic Rats. Journal of Ethnopharmacology.* 2010;127(2):565–567.
43. Jalalpure SS, Salahuddin MD, Shaikh MI, et al. Anticonvulsant effects of *Calotropis procera* root in rats. *Pharmaceutical Biology.* 2009;47(2):162–167.
44. Jalalpure SS, Alagawadi KR, Mahajanshetty CS, et al. *In vitro anthelmintic property of various seed oils. Iranian Journal of Pharmaceutical Research.* 2006;4:281–284.
45. James RS. *Clinical use of immunosuppressants in autoimmune diseases. Perspectives in Drug Discovery and Design.* 1992;2(1).
46. Tibbitts J, Cavagnaro JA. Practical approaches to dose selection for first-in-human clinical trials with novel biopharmaceuticals. *Regulatory Toxicology and Pharmacology.* 2010;58(2).
47. Jeremy MB, Tymoczko JL, Stryer L. *Biochemistry.* San Francisco: W. H. Freeman; 2007.
48. Jogdand SN. *Gene Biotechnology.* 2nd ed. Bangalore: Himalaya; 2007.
49. Karp G. *Cell and Molecular Biology.* New York: John Wiley and Sons;1996.
50. Kaznessis YN. Models for synthetic biology. *BMC Systems Biology.* 2007; 1: 47.
51. Kimmelman J. Recent developments in gene transfer, risk and ethics. *BMJ.* 2005;330.
52. Kohli DV. *Methods in Biotechnology and Bioengineering.* 2ed ed. New Delhi: CBS Publication Ltd; 2005.
53. Kokate CK, Purohit AP, Gokhale SB. *Textbook of Pharmacy.* 4th ed. Pune: Nirali Prakashan; 1996.

54. Kokate CK, Purohit AP, Gokhale SB. *Pharmacognosy*. Vol. I and II, 5th ed. Pune: Nirali Prakashan; 2009.
55. Lambert G, Fattal E, Couvreur P. Nanoparticulate systems for the delivery of antisense oligonucleotides. *Advanced Drug Delivery Reviews*. 2001;47:99–112.
56. Landridge W. Edible vaccines. *Scientific AM*. 2000; 283:66–71.
57. Loo C, Lin A, Hirsch L, et al. Nanoshell-enabled photonics-based imaging and therapy of cancer. *Technol Cancer Res Treat*. 2004:3 (1):33–40.
58. Martinez-Perez IM, Zhang G, Ignatova Z, et al. Computational genes: A tool for molecular diagnosis and therapy of aberrant mutational phenotype. *BMC Bioinformatics*. 2007;8:365.
59. Mason HS, Ball JM, Shi JJ, et al. Expression of Norwalk virus capsid protein in transgenic tobacco and potato and its oral immunogenicity in mice. *Proc Natl Acad Sci USA*. 1996;93:5335–5340.
60. McGarvey PB, Hammond J, Dienelt MM, et al. Expression of the rabies virus glycoprotein in transgenic tomatoes. *BioTechnol*. 1995;13:1484–1487.
61. Meyers RA. *Molecular Biology and Biotechnology*. New York: VCH Publishers; 1995.
62. Groves M. *Pharmaceutical Biotechnology*. CRC Publishers; 2005.
63. Mortazavi A, Williams BA, McCue K, et al. Mapping and quantifying mammalian mranscriptases by RNA-Seq. *Nat Methods*. 2008;5 (7):621–28.
64. Murray K. Application of recombinant DNA, techniques, the development of viral vaccines. *Vaccine*. 1998;6(2):164–174.
65. Moo-Young M. *Comprehensive Biotechnology: The Principle, Application and Regulation of Biotechnology in Industry, Agriculture and Medicine*. Vol. 1. New York: Pergamon Press; 1985.
66. Nickoloff JA. *Plant Cell Electroporation and Electrofusion Protocols*. Humana Press; 1995.
67. Noble D. Genes and causation. Philosophical transactions, series A. Mathematical, Physical, and Engineering Sciences. 2008; 366(1878):3001–3015.
68. Nolting B, Nalting B. Biophysical anotechnology: In: *Methods in Modern Biophysics*. 2nd ed. Heidelberg: Springer-Berlin; 2005.
69. North JR. Immunosensors antibody based biosensors. *Trends in Biotechnology*. 1985;3(7):180–86.
70. Ochiai K, Yamanaka T, Kimura K. Inheritance of drug resistance (and its transfer) between Shigella strains and Between Shigella and *E. coli* strains (in Japanese). *Hihon Iji Shimpor*. 1959;1861:34.
71. Pais A. *Subtle is the Lord, the Science and the Life of Albert Einstein*. Oxford University Press; 2005.
72. Pascual DW. Vaccines are for dinner. *Proc Natl Acad Sci USA*. 2007;104.
73. Patwardhan B. *Drug Discovery and Development: Traditional Medicine and Ethnopharmacology*. New Delhi: New India Publishing Agency; 2009.
74. Pearson H. Genetics: What is a gene? *Nature*. 2008;441(7092):398–401.
75. Perkins JJ. Sterilizer control, sterilization indicators and culture tests. In: *Principles and methods of sterilization in health sciences*. Illinois: Charles C. Thomas Publishers;1969;483–500.
76. Pisarchik ML, Thompson NL. Binding of a monoclonal antibody and its Fab fragment to supported phospholipid monolayers measured by total internal reflection fluorescence microscopy. *Biophys J*. 1990;58(5):1235–1249.
77. Prescott LM. Harley JP, Klien DA. *Microbiology*. 2nd ed. Dubque, Iowa: William C. Brown Publishers;1993.
78. Prove P, Faust U, Sittig W, et al. *Fundamentals of Biotechnology*. Weinheim: VCH Verlag; 1987.
79. Panchagnula R, Dey CS. Monoclonal antibodies in drug targeting. *J Clin Pharm Ther*. 1999;22(1):7–19.
80. Fischer R, Stoger E, Schillberg S, et al. Plant based production of biopharmaceuticals. *Curr Opin Plant Biol*. 2004;7(2):152–158.
81. Ratner MA, Ratner D. *Nanotechnology: A Gentle Introduction to the Next Big Idea*. Prentice Hall Publisher; 2002.

82. Reinetrt J, Bajaj YPS. *Applied and Fundamental Aspects of Plant Cell, Tissue and Organ Culture*. Berlin: Springer-Verlag; 1977.
83. Aaron OR, Jeffrey PD. Horizontal gene transfer in plants. *J Exp Bot*. 2007;58(1):1–9.
84. Freitas Jr, RA. *Nanomedicine Basic Capabilities*. Vol. 1. Landes Bioscience; 1999.
85. Robert B. *For Common Uses and Possible Side Effects Reference is, the Merck Manual of Medical Information*. Home Edition, Pocket; 1999.
86. Satyanarayana U. *Biotechnology*. Kolkata: Books and Allied Pvt. Ltd.; 2008.
87. Scheller F, Schubert F. *Biosensors*. New York: Elsevier; 1992.
88. Schmidt M. Diffusion of synthetic biology: A challenge to biosafety, *Systems and Synthetic Biology*. 2008;2.
89. Shiny KJ, Remani KN, Nirmala E. Bio-treatment of wastewater using aquatic invertebrates: daphnia magna and paramecium caudatum. *Bioresour Technol*. 2005;96(1):55–8.
90. Siest G. These products can also be used as reference materials in clinical chemistry laboratories. *Clin Chem*. 1991;39(8):1573–1589.
91. Rogers S, Girolami M, Kolch W, et al. Investigating the correspondence between transcriptomic and proteomic expression profiles using coupled cluster models. *Bioinformatics*. 2008;24(24):2894–2900.
92. Smith VH, Tilman GD, Nekola JC. Eutrophication: Impacts of excess nutrient inputs on freshwater, marine, and terrestrial ecosystems. *Environmental Pollution*. 1999;100(1–3):179–96.
93. Spier R, Hennessen W. Developments in biological standardization. In: *Advances in Animal Cell Technology and Cell Engineering Evaluation and Exploitation*. Vol. 66. San Diego, California: Academic Press; 1996: 273–78.
94. Staba EJ. *Plant Tissue Culture as a Source of Biomedicinals*. Florida: CRC Press;1980.
95. Stackebrandt E. Diversification and focusing: Strategies of microbial culture collections. *Trends in Microbiology*. 2010;18(7).
96. Stelwagen K, Ann MVG, McBride BW. Applications of recombinant DNA technology to improve milk production. *Livestock Production Science*. 1992;31(3–4):153–178.
97. Steven AC. Recent advances in the analysis of peptides and proteins by mass Spectrometry. *Advanced Drug Delivery Reviews*. 1989; 4(2):113–147.
98. Storb R, Yu C, Wagner JL. Stable mixed hematopoietic chimerism in DLA-identical littermate dogs given sub lethal total body irradiation before and pharmacological immunosuppression after marrow transplantation. *Blood*. 1997;89(8):3048–3054.
99. Street HE. *Tissue Culture & Plant Science*. London: Academic Press; 1974.
100. Strohl WR. *Industrial Antibiotic Today and the Future in Biotechnology of Antibiotic*. 2nd ed. New York: Marcel Decker; 1997.
101. Suh WH, Suslick KS, Stucky GD, et al. Nanotechnology, nanotoxicology, and neuroscience. *Progress in Neurobiology*. 1999;87(3): 133–170.
102. Suhara H. Gene transfer of human prostacyclin synthase into the liver is effective for the treatment of pulmonary hypertension in rats. *J Thorac Cardiovasc Surg*. 2002;123:855–861.
103. Teiger E, Deprez I, Fataccioli V, et al. Gene therapy in heart disease. *Biomed Pharmacother*. 199; 55(3):148–154.
104. Terando A, Chang AE. Applications of gene transfer to cellular immunotherapy. *Surg Oncol Clin N Am*. 2002;11(3):621–643.
105. Thanavala Y, Yang Y-F, Lyons P, et al. Immunogenicity of transgenic plant-derived hepatitis B surface antigen. *Proc Natl Acad Sci USA*. 1995;(92):3358–3361.
106. Thieman WJ, Palladino MA. *Introduction to Biotechnology*. San Fransisco: Pearson, Benjamin Cummings; 2008.
107. Tsuchihara T, Ogata S, Nemoto K, et al. Nonviral retrograde gene transfer of human hepatocyte growth factor improves neuropathic pain-related phenomena in rats. *Molecular Therapy*. 2008;17(1):42–50.
108. Tyagi AK, Mohanty A, Bajaj S, Choudhary A, Maheshwari SC. Transgenic rice: A valuable monocot system for crop improvement and gene research. *Crit Rev Biotechnol*. 1999;19:41–79.

109. Vyas SP, Dixit VK. *Pharmaceutical Biotechnology*. New Delhi: CBS Publishers and Distributers; 2006.
110. Wee YJ, Kim IN, Ryu HW, et al. Biotechnological production of lactic acid and its recent application. *Food Technology Biotechnology*. 2006; 44:163–172.
111. Winter PC, Hickey GI, Fletcher HL. *Genetics*. 2nd ed. Academic Book Review; 2009.
112. Wira CR, Crane-Godreau M, Grant K. Endocrine regulation of the mucosal immune system in the female reproductive tract. In: Ogra PL, Mestecky J, Lamm ME, Strober W, McGhee JR, Bienenstock J, eds. *Mucosal Immunology*. San Francisco: Elsevier; 2004.
113. Wolfe JC, Craver BP. Neutral particle lithography: A simple solution to charge-related artefacts in ion beam proximity printing. *J Phys D Appl Phys*. 2008; 41.
114. Wurm FM. Production of recombinant protein therapeutics from mammalian cells. *Nat Biotechnol*. 2008; 22:1393.

Index

A

Aberrant gene, mutation, 271
Absolute growth index, 108
Acceleration phase, 101, 102
Acetobacter sp., 106, 254
Acinetobacter sp., 146
Acylating agents, 147
Adaptive immunity, 20
Adhesion phenomenon, 59
ADME–Tox software, 235
Adrenomedullin, 101
Adult organs, 71
Aerobic organisms, 106
Aerobic treatment, 37
Agrobacterium radiobacter, 147
Agrobacterium rhizogenes, 95, 190, 191
Agrobacterium tumefaciens, 84
Agrochemicals, 231
AIDS vaccine, 227
Alcoholysis, 148
Allele, 117
Amalgamation, 55
Aminolysis, 147
2-Aminopurine, 77, 79
Amphibian embryonic cells, 59
Analytical applications of mAbs, 160
Androgens, 290
Aneuploid chromosome, 56, 57, 59
Aneuploidy, 59
Animal cell culture, 55, 288
Antibiotics, 9, 141, 167–187
Antibodies, 24, 302
Antibody–nanoparticle computational modelling, 270
Anticodons, 121
Antifoaming agent, 101
Antifoams, 101
Antigens, 15, 221, 229
Antigen–antibody reactions, 24
Antilymphocyteglobulin, 28
Antisense oligonucleotides, 271
Antithyroglobulin, 28
Antony van Leeuwenhoek, 4
Apoenzyme, 242
Applied research, 55
Artemisinin, 237
Arthritis, 162
Artificial nanomaterial, 270
Asexual hybridization, 82
Asymmetric biocatalytic hydrolysis, 147
Atherosclerosis, 161
Atropa belladonna, 79, 293
Augustino Bassi, 7
Autotrophic host cells, 133
Auxotrophic mutants, 181
Avian cells, 59
Axons, 56

B

Bacteria, 57, 100, 173, 200, 205, 246
Bacterial antigens, 23, 161
Baeyer–Villiger oxidation, 146, 152
Baker's yeast, 146
Batch culture, 101
B-endorphins, 82
Binary fission, 115
Biocatalyst, 40, 96, 239
Biochemical constraints, 271
Biochemical reactions, 87, 239
Biocompatibility, 270, 300
Bioconversions, 145, 153, 255
Biodegradability, 270, 300
Biohydrolysis, 147
Biological diversity, 99

Biomanufacturing, 288
Biopharmaceuticals, 6, 287
Biopulping, 178
Bioreactor, 40, 145, 176
Bioremediation, 36, 143
Biosensors, 262
Biotransformation, 143, 150
Biotrickling filters, 40
Blending inheritance, 112
Blood plasma, 61
Blood products, 229, 289
Blood serum, 61
B lymphocytes (B cells), 17
Bone morphogenetic proteins, 101
Bottom-up method, 273
Bovine somatotropin, 140
Breeding, 75, 78, 79
Broad-spectrum antibiotics, 169, 171
5-Bromouracil, 77, 79

C

Caffeine, 79
Callus culture, 13, 89, 92, 97
Candida antarctica, 147, 148
Capsicum, 79
Capsicum annuum, 79
Carbon nanotubes, 279
Carbon source, 90, 100
Carrel flask cultures, 69
Carrier erythrocytes, 300
Catabolic diversity, 143
Catalytic mAbs, 162
Cell alteration, 57
Cell cultures, 59
Cell cycle, 115, 285
Cell division, 58, 295
Cell lines, 57
Cellomics, 4
Characterization of nanoparticles, 284
Chemical solution deposition, 283
Chemodemes, 79, 81
Chemoprotective agents, 237
Chemostat, 104
Chemostat bioreactor, 104
Chick embryo, 63, 71, 198
Chick embryo extract, 63
Chicken plasma, 61
Chimeric DNA, 130

Chiral esters, 147
Chloroplast engineering, 301
Cholesterol, 59
Chromatin, 113
Chromosomal aberration, 77
Chromosomal mutations, 77
Cistron, 77
Citronellyl, 147
Classical growth, 57
Clay nanoparticles, 283
Coding strand, 120
Codons, 121
Colchicine, 80
Colchicum, 80
Cold trypsinization, 66
Collagenase, 60
Colloid, 283
Colony-stimulating factors, 227, 289
Competence factor, 145
Computational gene, 270
Computational methodology, 233
Conjugation, 125
Connective tissue cells, 56
Contact inhibition, 58
Continuous culture, 104
Contraceptive vaccines, 229
Copolymerization, 250
Corkscrew, 75
Corn steep liquor, 149
Corticosteroids, 82, 292
Cortisone, 151
Covalently closed circles, 131
Crosslinking, 253
Crown gall disease, 84
Culture, 2, 13, 60, 287, 293 295
Culture medium, 55, 60, 90
Culture pieces, 70
Cystic fibrosis gene, 140
Cytofluorometry, 150
Cytokines, 18, 135, 289
Cytotoxic T cells, 18

D

Death phase, 102, 103
Deceleration phase, 102
Dendrimer, 275, 276, 277, 281
Diagnostic applications, 160
Dihybrids, 81

Dioscorea bulbifera, 79
Diploid karyotype, 59
Disaggregation, 66
DMEM, 62
DNA, 111, 189
DNA libraries, 130
DNA ligase, 132
DNA polymerase, 13, 270
DNA replication, 12, 115
DNA synthesis, 13, 58
DNA technological applications, 288
DNA technology, 82, 130, 289, 303
DNA uptake, 85
DNA vaccines, 229
Donor cell, 202
Down's syndrome, 72
Downstream processing, 3
Drug discovery, 233
Drug targeting, 163

E

Eastern blotting, 127
E. coli, 131
Economic environment, 232
Electron tunnelling, 266
Electroporation, 191, 193
Electropores, 193
ELISA, 49, 264
Embryo culture, 13
Embryo extract, 63, 71
Embryonic organ, 70
Embryos, 13, 287, 288
Enantiomers, 148
Enantiotopic group, 148
Encapsulation, 229, 250, 254, 275
Endogenous factors, 300
Endonuclease enzyme, 130
Environmental biotechnology, 35
Enzymatic digestion, 66
Enzymatic disaggregation, 66
Enzyme, 131
Enzyme engineering, 259
Epoxide hydrolase, 147
Erlenmeyer flask, 257
Erythropoietin, 45, 82, 289
Escherichia coli, 40, 288
Established cell lines, 57
E-test, 173

Ethnobotany, 236
Ethnopharmacology, 236
Ethyl ethane, 77
Euploidy, 80
Eutrophication, 36
Excised root, 89
Exons, 270
Explant, 68
Ex-situ bioremediation, 36
Extracapillary spaces, 180
Extrinsic factor, 185

F

Fast neutron, 79
Fed-batch culture, 103
Feeder layers, 65
Fermentation, 176
Ferments, 240
Fibre entrapment, 253
Fibroblasts, 56
Flow cytometry, 159
F-minus (F-negative), 203
Formaldehyde, 40
F-plasmid, 203
F-plus (F-positive), 203
Functionalization, 284
Fusion and culture of hybridomas, 159
Fusion proteins, 134

G

GALT, 220
Gamma irradiation, 64
Gas phase, 65
Gel entrapment, 253
Gel strength, 109
Gene, 116
Gene cloning, 10, 130, 207
Gene expression, 119
Gene knockin, 207
Gene knockout, 207
Gene product, 50, 112
Generalized transduction, 202
Genetic code, 121
Genetic engineering, 2, 141, 260, 288, 301
Genetics, 59
Gene transfer, 5

Genome, 4, 10, 301
Genomics, proteomics, 4, 43
Geranyl acetate, 147
Gerhardt Domagk, 9
Germination, 91
Germplasm storage, 93
Glucuronidase, 150
Glycolytic, 59
Growth factors, 101
Growth hormones, 134

H

Haberlandt's hypothesis, 13, 88
Ham's F-12, 62
Hank's balanced salt, 63
Hank's BSS, 63
Haptens, 26
Hapten inhibition, 26
HAT, 157, 160, 209
Headful packaging, 202
Helper T cells, 18
Hemicellulase, 93
Hepatitis B vaccines, 82, 228
Heterokaryon research, 157
Heterokaryons, 67
Hexaploidy, 80
High-throughput screening (HTS), 236
Histones, 118
Histotypic culture, 55
Holoenzyme, 242
Homocystinuria, 72
Homogeneity, 55
Horizontal shaker, 92
Human and animal tumours, 161
Human artificial chromosome (HAC), 119
Human fibroblast, 57
Human follicle-stimulating hormone (FSH), 139
Human growth hormone, 82
Human immunodeficiency virus, 161
Human insulin, 82
Human mAbs, 158
Human therapeutics, 229
Human viruses, 138
Humatrope, 212
Humicola lanuginosa, 148
Hunter's syndrome, 72
Huntington's disease, 140
Hurler's syndrome, 72
Hyaluronidase, 60, 258

Hybridization, 75, 81
Hybridoma technique, 155
Hydrolysis, 147, 148, 261
Hydroperoxyl, 79
Hydrophobic, 48, 270, 275, 283
Hypersensitivity, 15, 23
Hypersensitivity reaction, 23

I

I-cell disease, 72
Identification of clones, 133
IgA, 220
IgE, 220
IgG, 220
Immobilization, 249
Immobilized lipase, 147
Immune, 15
Immune globulin intravenous, 28
Immune response, 15
Immune system, 15
Immunity, 15
Immunochemistry, 156
Immunofluoroscence, 159
Immunogenicity, 31, 129, 271
Immunoglobulins, 19
Immunomodulators, 15
Immunostimulators, 28
Immunosuppressants, 16
Immunotoxins, 28
Impetigo, 168
Incubation phase, 181
Induced fusion, 205
Induced mutation, 77
Inducer T cells, 18
Innate immunity, 19
Inoculum, 40, 249
In-situ bioremediation, 36
Insulin, 6, 229, 289
Insulin-like growth factor, 101
Interferon, 28, 29, 227, 229, 289, 292
Interleukin-2, 28, 213, 229, 289
Interleukins, 137
Intones, 118
Introns, 120, 270
In vivo imaging, 275
Ionizing radiation, 219
Isodiametric, 150
Isolation of gene sequence, 131
Isolation of tissue, 66

J

Janus particles, 283
Japanese mint, 79
John Needham, 7

K

Kanzius RF therapy, 276
Karyotype, 59
Killed vaccines, 138
Kinetic resolution, 147
Klinefelter's syndrome, 72
Krebs pathways, 59

L

Lactalbumin hydrolysate, 63
Lactic acid, 59, 176, 263
Lag phase, 58, 101
Laminar flow cabinet, 65
Lanatosides A and B, 80, 81
Latent viruses, 57
Late-phase clinical trials, 237
Lepraemurium, 72
Limitations, 69
Limiting dilution, 158
Lipopolysaccharide, 20
Liposomal vaccines, 229
Liposomes, 196
Liquid medium, 68, 283
Locus, 72
Log phase, 58, 102
Lord Joseph, 8
Louis Pasteur, 2, 240
Lymphocytes differentiation, 162
Lymphocytes phenotyping, 162
Lymphoid system, 17
Lymphotoxin, 138
Lysogenic function, 133

M

Macrophages, 56
Magnetic resonance imaging, 276
Malaria vaccines, 229
Malvoside, 168
Mammalian plasma, 61
Mannose receptors, 20
Mass spectrometry, 49, 301
Matrix, 40
Matrix-assisted laser desorption/ionization (MALDI), 49
McFarland, 173
Mechanical disaggregation, 67
Medicinal steroids, 290
Medium no. 612, 63
Medium no. 635, 64
Medium no. 858, 64
Medium no. 866, 64
Meiosis, 76, 77
Mentha, 79
Meristem culture, 89
Messenger RNA, 11, 120
Method of production of interferon, 136
Michaelis constant, 244
Microbial contamination, 65, 77
Microcapsule entrapment, 253
Microelements, 101
Microfold cells, 220
Micromeritics, 284
Miles and Misra technique, 109
MIM catalysed, 147
Minor gene, 181
Mitosis, 76, 77
Modern biotechnology, 231
Molecular biotechnology, 4
Molecular machines, 278
Molecular nanotechnology, 277
Monoclonal antibodies (mAbs), 155, 277
Morphine, 79
Mouse fibroblasts, 57
Mucor miehei, 147
Mucor polymophosporus, 152
Mucor spinosus, 152
Multifaceted approach, 237
Muscle cells, 56
Mutagens, 77
Mutant, 10, 181
Mutation, 77
Mutation frequency, 182
Mutation rate, 182
Mycelium, 147
Mycobacteria, 72, 295, 299
Myeloma, 28
Myoblasts, 67
Myocardial infarction, 139
Myo-inositol, 90
Mystery, 75

N

Nanobiopharmaceutics, 275
Nanobiotechnology, 269
Nanoclusters, 281
Nanocomputers, 278
Nanoimaging, 278
Nanolithography, 269, 272, 273
Nanolithography techniques, 273
Nanomedical approaches, 274
Nanomedicine, 269, 274, 277
Nanonephrology, 278
Nanoparticle targeting, 275
Nanorobots, 277
Nanoscale engineering, 278
Nanosensor, 278
Nanosphere lithography, 273
Nanotechnology, 269, 274, 281
Narrow spectrum, 171
Neuroelectronic interfaces, 276
Neurons, 56
Newton–Raphson method, 270
Next-generation lithography, 272
Nicotiana, 13
Niemann-Pick disease, 72
Nitrogen mustard, 77
Nitrogen source, 100
Nitrous acid, 77
Nonsegregation, 80
Northern blotting, 126
Novel biopharmaceuticals, 300
Nucleoproteins, 63
Nucleosides, 148
Nucleotide sequence, 120
Nutritive substrate, 61

O

Octaploidy, 80
Oestrogen, 291
Oligonucleotides, 134
Operons, 118, 301
Optical lithography, 272
Organ, 13, 89, 290, 302
Organ culture, 70
Organogenesis, 88
Organotypic culture, 55
Orotic aciduria, 72
Orthogonal matrix, 149
Oscar Low, 169

P

Palladium, 64
Papaver somniferum, 79, 230, 293
Parasitic protozoans, 72
Passaged, 56, 57, 197
Pattern recognition receptors, 20
Paul Ehrlich, 8
Pectinesterase, 261
Pectin *trans*-eliminase, 261
Penicillin, 169
Penicillium notatum, 228
Pentaploidy, 80
Perfusion bioreactor, 180
Pharmacokinetic advances, 235
Pharmacokinetics, 235
Phenol, 79
Phenotypes, 271
Phenotypic variations, 77
Phospholipids, 148
Photodynamic therapy, 277
Physical barriers, 19
Physical disruption, 66
Physiochemical, 56
Phytohormone treatment, 150
Piperascens, 79
Plant-derived therapeutics, 292
Plant gene vector, 192
Plant tissue culture, 86
Plasma coagulum, 61
Plasmids, 5
Plasmodium falciparum, 161
Point mutations, 77
Pollutants, 28
Polyaromatic hydrocarbons, 143
Polychlorinated biphenyls, 28
Polyclonal antibodies, 127
Poly-D-lysine, 64
Polygalacturonase, 239, 261
Polyhybrids, 81
Polymerase chain reaction (PCR), 5
Polymethyl galacturonase, 239
Polyploidy, 79
Polystyrene, 64
Polyvinyl chloride, 64
Precipitation, 39
Precursors, 87
Pregnane skeleton, 291
Preproinsulin, 134
Preservation of antibodies, 160

Presterilized culture media, 65
Pretreated, attenuated vaccine, 22
Primary cell lines, 56
Primary techniques, 67
Primary transcript, 120
Production of insulin, 5
Production of mAbs, 155
Productivity ratio, 108
Progesterones and progestogens, 291
Prokaryotes, 99
Proliferation, 19
Prostate cancer, 28, 290
Prosthetic group, 45
Protein and peptide delivery, 275
Proteins and peptides, 275
Protein Data Bank, 270
Protein synthesis, 58, 294, 296, 299
Proteomics, 50
Protoplast, 1
Protoplast fusion, 13
Protozoa, 7
Purification, 41, 301
Purification of proteins, 162
Purified fibronectin, 64
Pyrolysis, 283
Pyruvic acid, 59

Q

Quantum dots, 276, 279
Quartz sand, 62
Quinolones, 298

R

Racemic amines, 147
Radiation mutagens, 79
Radiobacter, 147, 192
Rational approach, 235
Rat tail collagen, 64
Recalcitrant plants, 150
Recent antibiotics, 295
Recipient cell, 143
Recombinant, 82
Recombinant DNA, 36
Recombinant DNA technology, 130, 289
Recombinant enzymes, 229, 289
Recombinant live vaccines, 138
Recombinant proteins, 99, 292, 301

Recombinant vaccines, 229, 289
Refsums disease, 72
Regioisomers, 146
Regioselective reactions, 152
Relative growth index, 108
Relaxosome, 204
Restriction endonuclease, 127
Restriction enzyme, 125, 132
Reticuloendothelial cells, 56
Reverse transcriptase, 132
Rhizopus delemar, 147
Rhizopus niveus, 147
Ribonucleic acid (RNA), 111
Richard Petri, 8
Rickettsiae, 72
Ri-DNA, 95
RNA, 11, 296, 298, 299
Robert Koch, 8
Roller test tube cultures, 68
RPMI 1640, 62
Rudolf Emmerich, 169

S

Saccharomyces cerevisiae, 134
Safety, 138, 173, 285
Sahachiro Hata, 8
Scaffolding, 241
Secondary screening, 181
Semisolid culture, 101
Sendai virus, 67
Sensititre, 173
Serum, 24
Seven-transmembrane spanning receptors, 20
Sex hormones, 290
Sex pheromones, 279
Shikonin, 94
Silicon carbide, 86
Simms' ox serum, 63
Single-celled algae, 37
Single-cell proteins, 176
Slide or coverslip cultures, 68
Sodium dodecyl sulphate, 49, 126
Soft agar method, 158
Solanum khasianum, 79
Solasodine, 79, 82
Sol–gel process, 283
Solid culture medium, 92, 101
Solid-state fermentation, 109
Somatic cell fusion, 67

Somatic mutations, 139
Somatostatin, 82
Southern blotting, 126
Spatial orientation, 57
Specificity, 20
Spirochete borrelia, 131
Spontaneous mutation, 77
Stationary phase, 58
Steady state, 104
Sterilization, 7
Steroid hormone, 82
Strategy in cloning, 137
Streptomyces, 9, 293, 294, 295
Structural elucidation, 237
Structure of proteins, 44
Sulphonamides, 298
Suppressor T cells, 18
Suspension culture, 90
SV40 genome, 137
Synthetic biology, 285
Synthetic steroid hormones, 290

T

Tagetes erecta, 13, 88
Telomeres, 119
Template strand, 13, 115, 120
Termination (stop or nonsense) codons, 122
Test tube culture, 70
Testing of pharmaceuticals, 232
Tetraploid cells, 59, 108
Tetraploidy, 80
Thermal neurons, 79, 232, 283
Thick clot, 69
Thin clot, 69
Thrombin, 61
Thrust pump, 181
Ti plasmid, 85
Tissue culture, 2, 230, 287, 293
Tissue explant, 91
Tissue extracts, 62
Tissue plasminogen activator, 212
T lymphocytes (T cells), 18
Toll-like receptors, 20
Totipotency, 13
Transcription, 30
Transduction, 200, 202
Transfection, 65
Transfer RNA (tRNA), 121

Transformation, 200
Transgene, 301
Transgenic plants, 83, 229
Transgenic plant vaccines, 301
Transient transfection, 125
Transition, 57
Translation of codons, 122
Translocations, 59
Traumatic damage, 61
Triazine, 79
Trichothecium roseum, 152
Triggered response, 275
Triolein, 148
Triploidy, 80
Trypsin, 66
Tryptone soy agar (TSA), 108
Tumefaciens, 84, 215
Tumour-inducing plasmid, 84
Tumour necrosis factor, 138
Tumours, 66
Turbidostat bioreactor, 105
Turner's syndrome, 72, 291

U

Ultrasonic waves, 232
Ultrasound, 232, 275
Upstream, 118, 145
Urethane, 79

V

Vaccines, 30
Vaccine storage, 32
Vaccinia virus of smallpox, 138
Vector, 85
Vicinal diols, 147
Vinyl esters, 147
Virologists, 72
Virtual screening techniques, 236
Virulent, 145, 259
Virus-free plants, 93
Vulnerability of crops, 231

W

Western blotting, 126, 127
Whiskers, 86, 284
Whole-embryo culture, 71

Whole-organ culture, 70, 278
Wild-type virus, 138, 183, 192, 214, 271
Woodchips, 178, 254

X

Xenobiotics, 60, 144, 203
Xenopus laevis, 130
Xeroderma pigmentosa, 72

Y

Yeast, 57

Z

Zinc sulphide quantum dots, 279